WITHDRAWN

Paradox of Plenty

Paradox of Plenty

A Social History of Eating in Modern America

HARVEY LEVENSTEIN

New York Oxford
OXFORD UNIVERSITY PRESS
1993

Oxford University Press

Oxford New York Toronto
Delhi Bombay Calcutta Madras Karachi
Kuala Lumpur Singapore Hong Kong Tokyo
Nairobi Dar es Salaam Cape Town
Melbourne Auckland Madrid

and associated companies in
Berlin Ibadan

Published by Oxford University Press, Inc.,
200 Madison Avenue, New York, New York 10016

Oxford is a registered trademark of Oxford University Press

Library of Congress Cataloging-in-Publication Data
Levenstein, Harvey A., 1938–
Paradox of plenty: a social history of
eating in modern America /
Harvey Levenstein.
p. cm. Includes bibliographical references and index.
ISBN 0-19-505543-8
1. Food habits—United States—History—20th century.
2. Diet—United States—History—20th century.
3. United States—Social life and customs.
I. Title. GT2853.U5L47 1993
394.1'2'0973—dc20 91-48170

2 4 6 8 9 7 5 3 1

Printed in the United States of America
on acid-free paper

To the Memory of David Shannon

Preface

The sympathetic reader may well imagine the trepidation with which, having completed this manuscript, I open the daily newspaper each morning. Will a new study in the *New England Journal of Medicine* undermine that generalization I made in chapter 13? Has the surgeon general changed her mind about cholesterol? Has the new head of the Food and Drug Administration, about whom I was cautiously optimistic, quit in disgust over his inability to rein in misleading corporate nutritional claims? The normal perils of writing contemporary history seem to be magnified manyfold by writing about food and nutrition. Yet it is precisely the changeability of ideas about food and health that is a major theme of this book.

On one level this work deals with how broad changes in nutritional ideas have affected what and how Americans eat. In *Revolution at the Table: The Transformation of the American Diet*, I dealt with the years from 1880 to 1930, when the idea that the American people must be taught to eat not what they like but what was good for them gained the ascendancy. It told of the emergence of the New Nutrition—the result of the discovery that foods could be broken down into proteins, carbohydrates, and fats—and the Newer Nutrition, based on vitamins. As we shall see, the Newer Nutrition, with its emphasis on eating enough nutrients, continued to reign supreme for the first three decades covered by this volume, until it was challenged by the Negative Nutrition, a fluctuating set of admonitions to eat less of things.

Other themes in *Revolution at the Table* also resonate again, including the alacrity with which experts, often backed by the flimsiest of evidence, have recommended fundamental changes in the national diet. Like the previous volume, the book highlights the important role commercial and professional interest groups play in setting the national dietary agenda: food producers and processors, retailers, physicians, research scientists, faddists, home economists, and political pressure groups. Much of the book also deals with the broad social, economic, and political changes that helped determine how Americans responded to these ideas: the Great Depression, for example, which reinforced family-oriented values and led to a renewed reverence for "home cooking"; the challenge World War II food shortages presented to the idea that the cornucopia was a proper symbol for America; the way the food of the 1950s and early 1960s reflected the prevailing mood of pride in the achievements of corporate America and the "American Way of Life"; the connection between the challenges to this perception in the later 1960s and 1970s and doubts about the adequacy and healthfulness of the food supply; the 1980s tug-of-war between the forces of ostentatious excess and those of equally ostentatious restraint; and finally the chastened, gloomier mood of the past few years, and what may be a growing reluctance to try to put experts' dietary advice into practice.

Revolution at the Table also dealt with a period during which tastes in food became progressively less important as a sign of social distinction. As we shall see, this continued to be the case during the thirty-odd years following the onset of the Great Depression. But this tide was then reversed, as food tastes once again became significant symbols of status. As social pendulums such as these have swung back and forth and the dictums of nutritional science have oscillated, metamorphosed, burst onto the horizon, and disappeared, one factor has remained constant: their intimate connection with the changing role of women in America. The rise of industry, cities, and towns in nineteenth-century America meant the triumph of a culture in which the food roles of each gender were quite clearly demarcated: With few exceptions, the ideal was that men would ensure that the home was well supplied with food, while women were in charge of preparing and serving it. As I showed in *Revolution at the Table*, by the turn of the century a number of factors were beginning to eat away at this, particularly the servant shortage, which made it difficult for middle-class women to fulfill unrealistic expectations of them. By the 1920s changing ideas of what made for a happy marriage—abetted by the steady march of food production to outside the home—were diminishing the relative importance of food preparation in assessments of women's success or failure as wives and mothers. Although, as we shall see, the Depression helped revive some of the older expectations, the steady growth in the number of married women working outside the home came to have an enormous impact on American food habits. It not only helped break down older conceptions of which foods were proper for men and women and stimulate attempts to redefine gender roles in the kitchen, it was also reflected in where, how, and what Americans ate in public, and a goodly portion of this book is devoted to eating out.

The connection of women's changing role to fluctuations in ideal body type is more problematic, but no history of modern American eating can ignore the question of why manias for slimness have periodically arisen, even, as we shall soon see, during the Depression. But first, the book begins with another repeated theme, the paradox that has never failed to disturb the citizens of a nation who, as David Potter pointed out some years ago, have always conceived of themselves as the "People of Plenty": that hunger, breadlines, and food banks can exist in a land of such agricultural abundance.

Again, I would like to thank my three faithful readers: Joseph Conlin, my wife Mona, and Claude Fischler. Joe, whose breadth of knowledge about American history and society never ceases to amaze me, read the entire first draft and made many excellent suggestions for revision, excision, addition, and correction. Mona also read it all. She pointed out passages that did not make sense to her, argued a number of points with me, urged me to make my paragraphs shorter, and (perhaps most important) persuaded me to excise an unfortunate reference to Donald Trump. Claude was facing an inexorable deadline for completion of his own book for much of the time I was working on this one, but he still managed to read the last seven chapters, making useful suggestions and saving me from some potentially embarrassing errors—even, to my chagrin, in my English usage. The reader will also notice that I am indebted to him for many of my ideas about contemporary food habits. India Cooper, my editor at Oxford, read the manuscript with a knowledgeable eye and made many intelligent suggestions for improving it.

As usual, the staff of the Interlibrary Loan Department at McMaster University's Mills Library were immensely helpful. Much of my field research was done at the Schlesinger Library of Women's History at Radcliffe College, where the librarians were unfailingly cooperative. So too were the librarians at the Library of Congress, the National Archives, the Boston Public Library, the Hamilton Public Library, the Metropolitan Toronto Library, the University of Guelph Library, the University of Toronto's Robarts Library, the New York City Public Library, Columbia University Libraries, and Cornell University Libraries. Nancy Jenkins encouraged me to write on French food in America and arranged interviews with Julia Child and Craig Claiborne. Not only were both as gracious as could be, they also prepared lunch for me—something that elevated my status around here immeasurably.

Much of the research travel for this book was supported by grants from the Social Sciences and Humanities Research Council of Canada and the McMaster University Arts Research Board. Another SSHRCC grant freed me from undergraduate teaching duties for a year to complete the manuscript. The fact that George Shephard, the younger scholar who was employed to teach my courses, won that year's award as the best teacher in the university helped dispel any compunctions I might have had regarding whether my undergraduate students were being short-changed.

Hamilton, Ontario, Canada H. L.
February 1992

Contents

Paradox of Plenty

Prologue: Depression Paradoxes

August 1930. The Great Depression is almost one year old. An eighteen-year-old unemployed shoe worker from Massachusetts who has slept for six nights in New York's Central Park and not eaten for two days purloins some clothing from Macy's. Pursued for only one block by a female store detective, the famished boy collapses in a heap on the street. When he is arraigned in the Jefferson Market courthouse, sympathetic court employees take up a collection to buy him sandwiches and coffee.[1] Within a few months there are some eighty-two breadlines operating in New York City—two more than in Philadelphia, the nation's third-largest city.[2] Halfway across the continent, food riots break out in small towns called Henryetta, Oklahoma, and England, Arkansas, as hungry crowds shouting "We want food" and "We will not let our children starve" threaten local relief agencies and merchants.[3] Severe drought has dried up crops over much of the South and brought hardship to farmers as far north as Pennsylvania and as far west as Montana. In Alabama sharecroppers scrape by on their historic diet of the three M's: meat (fat salt pork), corn meal, and molasses. Shriveled gardens stop producing green vegetables, and fruit is but a memory. When "rations" run out before the Saturday payday, people simply go without eating.[4]

Meanwhile, the members' dining room of the House of Representatives in Washington is serving a very popular "dieter's menu"—a slimming regime heavy on fresh fruit and vegetables originated by the House physician for "brain workers."[5] Henry Ford announces that by eating the right combinations of foods he expects to live to see one hundred.[6] Indeed, there is so much talk of dieting in the United States that a number of speakers at that fall's American Bakers' Association convention blame the agricultural depression on the weight-loss mania. A practical program of farm relief, they all agree, would be one that tells Americans to "eat more and fast less."[7]

Well, no one has ever said that the United States is not a land of stark contrasts and paradox.

And then things got worse. By 1933 average family income had dropped by 40 percent from its 1929 level. During 1932 some 28 percent of the nation's households did not have a single employed worker.[8] By early 1933 well over fourteen million workers, close to 30 percent of the labor force, were unemployed.[9] In early 1931 Americans were shocked to read of people digging for rotten food in garbage dumps in St. Louis. In Harlem men and women could be seen competing with dogs and cats for the contents of garbage cans. Orderly lines would form at some municipal garbage dumps as people waited patiently throughout the day for their turn at the scraps disgorged by the dump trucks. At others, the hungry would simply rush pell-mell at the new piles, digging frantically with sticks or their hands.[10] Kentucky hill people were reported to be living on dandelions and blackberries. Conditions in the Appalachian coal fields were said to be particularly grim. The miner's diet often consisted only of beans ("miner's strawberries") and "bulldog gravy" (flour, water, and grease), with a "water sandwich" (stale bread soaked in lard and water) for his lunch pail. Miners' families were even worse off. One food-conserving strategy was to have siblings eat only on alternate days.[11] In Oklahoma investigators found Indians so weak from hunger they could not arise from bed.[12]

Government, particularly the one in Washington, seemed helpless in the face of the crisis. There is particular irony in this, for the president, Herbert Hoover, had gained a reputation as a brilliant administrator during previous food crises: his success in coordinating relief supplies to Belgium in 1915 and 1916, heading the U.S. Food Administration in 1917 and 1918, and directing the postwar food relief effort in Central and Eastern Europe had been the springboard for his political career. However, he resolutely opposed any suggestion that the federal government distribute food to the needy during the Depression. This was the very kind of activity, the Quaker president thought, at which America's voluntary agencies and local communities traditionally shone. Neighbor helping neighbor through hard times was the American way, not handouts from the central government. Indeed, so adamant was he about the dangerous precedent that federal feeding of people would set that when, after much cajoling, he finally approved a congressional appropriation of twenty million dollars to feed livestock starving in the southwestern drought, he stood firm against demands for funds to feed the farmers as well. "The administration would feed jackasses, but it wouldn't feed starving babies," said a disgusted congressman from Arkansas.[13]

Yet even at the outset the local and private institutions to whom Hoover assigned responsibility could barely cope with the emergency. The Salvation Army expanded its network of "soup kitchens," joined by other gospel missions trying to save souls by filling stomachs.[14] Other organizations gave out uncooked food. The gangster Al Capone opened one such breadline in Chicago, as did the local Hearst newspaper. (Both evoked suspicions of ulterior motives.)[15] St. Louis society women dished out restaurants' unsold food to the poor; the vegetarian physical culturist and publisher Bernarr McFadden served one-cent meals at his restaurants in Boston and New York City. (When Eleanor Roosevelt dined in one of them during the 1932 campaign she ate in the prescribed healthy way, standing up. However, she seems to have passed up his latest "naturopathic" cure for constipation—eating sand and dirt.)[16] The private charities of Muncie, Indiana, provided the unemployed with free seed and vacant lots to raise vegetables "so that they [could] profit from self-help."[17] So impressed was Hoover by this kind of activity that in January 1932 he suggested that five or ten million dollars donated "by substantial men" would see the country through its relief need for the year "without a breakdown in our fundamental ideas of government."[18]

Men like Hoover assumed that another private agency—the American Red Cross—would take the lead in distributing this private largesse, were it forthcoming (which it was not). But its previous record should have called this into question, for the Red Cross saw its mandate as dealing with acts of God and war, not the business cycle. People fleeing floods, earthquakes, and pestilence were its cup of tea, not the casualties of capitalism. It steadfastly refused all pleas to use its funds to help the merely unemployed. The most it would do was hire dietitians to advise the poor on how to eat economically.[19] Indeed, it would not even give out *government* funds to feed the needy. When Arkansas Senator Joseph T. Robinson pushed for Congress to grant the Red Cross twenty-five million dollars to feed the hungry, both President Hoover and Red Cross chairman John Barton Payne opposed the measure—the latter on the grounds that the Red Cross must "continue its historic voluntary role and refuse to be drawn into politics."[20] A disgusted John L. Lewis, the fiery head of the United Mine Workers union, said that "the only thing that apparently inspires the Red Cross to extend assistance is a conflagration, flood, pestilence, and war. It doesn't make any difference to them how many people die of starvation, how many children suffer from malnutrition or how many women are weakened."[21] However, the Red Cross did help fund the National Smiles Award, which gave a five-hundred-dollar prize to a New Jersey woman whose upbeat demeanor and suggestions for "spreading the psychology of cheer and good will" bested those of sixty thousand other contestants.[22]

The Red Cross shared Hoover's reverence for the tradition of self-help. Thus it sent vegetable seeds to victims of the southwestern drought of 1930 and 1931, but only in those counties whose local Red Cross chapters could demonstrate that "local people showed some disposition to help themselves" by footing part of the bill—something that was well-nigh impossible in the hardest-hit places. When local chapters did ignore national policy—as they often did—and gave food or money to help feed the hungry, the local elites who dominated the chapters were not always equitable in doling out the assistance. In drought-struck Lonoke Country, Arkansas, the white plantation owners who ran the chapter demanded that black sharecroppers clean the streets or clear their plantations in return for aid. A black woman who refused to clean a white woman's yard was denied assistance. One black 'cropper was told to work the planter's farm for three days before aid would be proffered. When his starving mules could not plow because of lack of feed, he too was denied help.[23] Even those chapters that tried to distribute aid fairly—and there were many—had to rely on unpaid, part-time staff, who had more pressing demands on their time, to put their good intentions into action.[24] In March 1932, however, Congress practically forced the national organization to act on the food front. It called on the federal Farm Board to donate wheat it had bought to maintain wheat prices to the Red Cross for distribution among the needy. Reluctantly, the Red Cross agreed. In the next eleven months it distributed over 8.5 million barrels of flour to 5,140,855 families in almost every county in the nation.[25]

But this provided little more than a brief respite, as the economy spiraled downward into the disastrous winter of 1932–1933. The underfunded municipal welfare authorities that were expected to carry the burden of relief tottered and collapsed under the new load. Private charity organizations, which normally played a major role in dealing with the very poor in the cities, were even harder hit. Not only were their finances inadequate, they were also burdened by a new professionalized philosophy of social work, which regarded poverty as an individual aberration whose origins were often psychological. The individual casework method this necessitated was hardly relevant to the new wave of "Depression poor," whose problem was simply that although they were willing and able to work they were unable to find jobs. Accustomed to doling out small amounts of cash to the poor and supervising their budgets, only slowly did the social workers begin to set up commissaries to distribute food.[26]

Other professionals were hardly better equipped to deal with the crisis. By the 1920s the idea that poor peoples' inadequate diets were more the result of their ignorance of nutritional science than of low income prevailed among dietitians and home economists. Malnutrition was regarded as a clinical disorder—caused by ignorance or poor childhood training, not poverty—which could strike individuals of any status.[27] A prominent home economist from Iowa assured the New York Federation of Women's Clubs in 1930 that New York City's municipal relief payments were more than adequate to feed a family. Entire families in the city suffered "malnutrition and semi-starvation," she said, not because of "undereating" but because of "ignorance."[28] Home economists and social workers therefore plied the poor with recipes, menus, and budgets.[29] In Detroit, where twenty-seven thousand families were said to be living on less than fifteen cents per person per day, home economists at the Merrill-Palmer School put on demonstrations for six hundred relief workers to teach them how to instruct welfare recipients to cook economically and healthfully.[30] In 1931, while the U.S. Department of Agriculture joined in blocking distribution of free surplus food to the poor, its home economists began churning out guides to healthy eating at minimal cost.[31] The home economist–dietitian of the New York chapter of the Red Cross organized a ceremony to honor nine young women—the wives of disabled or unemployed veterans—who, having attended her lectures on the subject, had been able to feed their families a healthy diet on one dollar a day. First prize went to a young woman with a disabled husband and two small children who herself had just returned from a tuberculosis sanitarium. The prize was a certificate.[32]

By 1933 producing food budgets for the poor had become one of the nation's few growth industries. Demonstrating how to eat well on next to nothing even became rather chic. In December 1932 one hundred prominent New Yorkers gathered in the posh Waldorf-Astoria Hotel for what—at eight cents a person—was billed as the cheapest meal ever served in that temple of extravagance. Assembling over luncheon to inaugurate International Golden Rule Week, they dined on one of the twenty-one menus devised by home economists to provide a well-balanced diet for a family of five for $8.80 a week. According to the *New York Times*, the menu was "prepared by the Waldorf chefs with the same artistry that in bygone days has gone into the luncheons of princes, Presidents, and prima donnas." It consisted of tomato juice, stuffed flank steak with gravy, browned potatoes, mashed turnips, bread and butter, and butterscotch pudding with milk. Similar dinners took place simultaneously in Washington, Chicago, and other large cities, and the accompanying speeches were broadcast nationally by NBC.[33] Relief recipients lucky enough to have radios could also tune in to "Betty Crocker," who devoted two of her weekly shows entirely to recipes and menus for families on relief.[34] But by then most of the recipes, menus, and budgets looked like pipe dreams, for although food prices had dropped, relief payments had dropped even further. In May 1932 the normal weekly relief payment for a family of five in New York City was $6.00.[35] That month the average weekly grant in Philadelphia was reduced to $4.39 per week. A month later the city's relief money ran out completely, and fifty-seven thousand families were abandoned to scrape along on their own.[36] Yet curiously, while the nation's heart may have gone out to people such as these, their plight seems hardly to have affected its stomach.

CHAPTER 1

Depression Dieting and the Vitamin Gold Rush

The signs of deprivation all around them—the breadlines, the people rummaging through garbage cans and selling apples on the streets, the hobos at the back door asking politely for a bite to eat, the heartrending stories from Appalachia—hardly altered most Americans' deep-rooted attitudes toward food. These continued to reflect the economy of agricultural abundance that had shaped them. Yes, there had been depression before, both agricultural and industrial, and there had been hardship and even hunger, particularly on the agricultural frontiers and in the swollen slums of the expanding cities. But working-class America was still very much a land of immigrants and their offspring—people to whom America still represented an unparalleled abundance and variety of food. Few could think of it as a land of hunger and want, particularly in the light of family memories of life in the "old country." As for the middle classes, for almost fifty years they had been bombarded with warnings against the perils of overindulging in this abundance. Since the 1880s the scientists, home economists, cookery writers, advertisers, and faddists to whom they turned for dietary wisdom had been propagating the ideas of the New Nutrition. These taught that all foods could be broken down into proteins, carbohydrates, and fats, and that one should eat only as much of each of them as the body required. The idea that the body's energy needs could be measured in calories took hold, along with the notion that one would gain weight if one ingested more of these than the body burned. Ideals of feminine beauty changed markedly, as the heavily corseted matronly ideal of the late nineteenth century gave way first to the more lithe and athletic prewar Gibson Girl and then, in the early 1920s, to the positively skinny "flapper." It was not just females who were affected. Excess male girth came to denote sloth, immobility, and ill health rather than substantive achievement. "A whole new anti-avoirdupois philosophy has grown up," wrote one observer in 1931, "until the stoutish individual who used to be considered a peculiarly good-natured fellow has come to be looked upon merely as lacking self-control."[1] "Even the middle-aged can remember the time when a slight excess of avoirdupois was regarded as a sign of an amiable, easy-going nature, rather than an error of judgement," she said elsewhere.[2]

Yet it is often said that corpulence tends to be regarded as attractive in cultures of economic scarcity.[3] One social scientist even correlated the ups and downs of ideal female body shapes in early modern European painting with times of feast and famine.[4] One might therefore expect the Depression to have brought a shift toward significantly heftier ideals than those of the 1920s. Indeed, some women's fashion designers bet on this and in 1931 proclaimed a return to the nineteenth-century ideal of plumpness and plumes. Yet despite (or because of?) support from the Bureau of Home Economics of the U.S. Department of Agriculture, which hoped it might persuade women to start chomping their way through the wheat surplus, the new style flopped.[5] While the Depression never saw the reemergence of the pencil-thin extremes of the flapper style, for the most part ideal body types did not diverge much from those of 1929, by which time dress lengths had been lowered and some curves were being admired. During the next decade, shoulders (male and female) assumed more padding and curves became somewhat more pronounced, but slimness remained the ideal, particularly among women. College girls "determined to become fashionably thin overnight" went from one crash diet to another, wrote a concerned dietitian.[6] Males might want to be thinner but seemed to do little about it. "Perhaps the most dangerous fad just now is limited almost entirely to women—that of dieting to get or keep thin," said a physician in 1933. "The practice of medicine would be made much easier if I could persuade my male patients to diet more and my female patients to diet less."[7] Two years later, in a book called *Diet and Die*, Carl Malmberg warned that the "craze for slimness" was leading many people who were not "made to be slim" to follow dangerous diets. "It is better to be fat than dead," he wrote, but it was clear that many women would disagree.[8]

Why the continuing struggle for slimness? It was not simply that the upper and middle classes (that is, the classes who could afford to be fashionable) were relatively unaffected by Depression deprivation. It was also that the slim ideal was becoming rooted in more than mere fashion—it was based on health concerns as well. Nutrition experts, reinforced by studies of mortality by insurance companies, were now warning that excess weight, particularly among the middle-aged, led to early death.[9] Other scientists buttressed the pressure to lose weight by shifting responsibility for obesity directly onto the eater. Whereas many of the previous decade's experts had been willing to ascribe excess weight to malfunctioning thyroid glands or lazy metabolisms, in the 1930s there was a decisive shift toward blaming it all on consuming too many calories. "The weight is made up of tangible material," wrote Dr. Frank Evans, one of the era's main proponents of reducing, "and there is no gland with an aperture through which it can be introduced." The only opening capable of doing this was the mouth, and it was around what went into it that scientific and public attention now centered.[10]

The result was paradoxical: In the midst of the greatest economic crisis the nation had ever seen, its middle and upper classes—particularly the female members—continued to regard eating less and losing weight as an elusive goal, rather than a tragedy. Indeed, no sooner did the Depression strike then a wave of reducing diets swept the middle class. One of the most popular was that touted by Dr. William Hay, who had created dishes such as "Fountain of Youth Salad" for patrons of his sanitarium, Pocono Hay-Ven, in Pennsylvania and elaborated his philosophy in his book *How to Always Be Well.* The Hay Diet's central feature was one that would go in and out of fashion over the next half-century: It prohibited eating proteins and carbohydrates at the same time. A third type of food, "alkalines" (mainly fruits and vegetables), was also to be consumed separately. As an added fillip, it called for taking a large enema or strong cathartic every day.[11] A major competitor, the Hollywood Eighteen-day Diet, was promoted by California citrus fruit interests. Its followers could live on fewer than six hundred calories a day by limiting each meal to half a grapefruit, melba toast, coffee without cream or sugar, and, at lunch and dinner, some raw vegetables. Then there were the two-food diets: the pineapple and lamb chop diet, the baked potato and buttermilk diet, the "Mayo diet" of raw tomatoes and hard-boiled eggs. There was even a coffee and doughnuts diet. The United Fruit Company helped popularize a reducing diet built around bananas and skim milk developed by Dr. George Harrop of Johns Hopkins University.[12] Although the good doctor was reluctant to exploit it for commercial gain, at the end of 1934 it was declared to be by far the most popular diet of the year, easily outdistancing its nearest rival, the grapefruit juice diet.[13] Even processors of relatively high-calorie foods managed to scamper aboard the reducing bandwagon. Advertisements for Wonder Bread featured professional models in bathing suits telling how, because it was "slo-baked," it gave them the quick infusions of energy that allowed them to "diet with a smile."[14] "Betty Crocker," General Mills' invented spokesperson, took a similar tack. She called bread "the outstanding energy food" and used her popular radio show to denounce the canard that it was "fattening."[15] Manufacturers of fruit juices and even candies claimed that their products yielded "quick energy" yet were "never fattening."[16] Welch's grape juice was even more effective, so it seemed: Its "predigested grape sugar" actually "burned up ugly fat."[17]

Radio supplemented the printed word. One program on reducing elicited thirty-five thousand letters. Victor Lindlahr, whose dieting book sold over half a million copies, held radio "reducing parties."[18] Local radio was a particular favorite of the hucksters promoting thyroid products, laxative salts, cathartic drugs, the drug dinitrophenol—which increased the metabolic rate—and other supposed aids to quick weight loss. Because local shows were hardly ever recorded, they were practically impervious to censorship for false claims.[19] Fortunately for consumers, the promoters of the phenomenally successful Helen's Liquid Reducer Compound promised to "Gargle Your Fat Away" in the print media. The company could therefore be forced out of business when Food and Drug Administration investigators discovered that the gargle contained hydrogen peroxide, a disinfectant, and bleach. The handsome and charming lecturer Gayelord Hauser—who capitalized on screen star Greta Garbo's devotion to his diet of mushroom burgers, broiled grapefruit, and his own brands of cathartic salts—was more difficult to nab. He continued to dine out, as it were, on his laxative diet until the 1960s, becoming a favorite of the Duchess of Windsor and other high-society figures.[20]

But food companies and hucksters were by no means alone in promoting reducing for health and beauty. So did some of the most respected medical and public health authorities. In a bizarre episode in April 1934, while breadlines still snaked around street corners not far from the Loop and people scoured the city's garbage dumps for food scraps, Chicago's municipal health commissioner, Dr. Herman Bundesen, announced that three lucky girls had been selected to participate in a scientifically supervised weight-losing "derby." For one month they were put on a particularly grim version of the bananas and skim milk diet. At the end of the month, when they had collectively lost a total of thirty-two pounds, the proud doctor told the public to "remember that every pound lost is health gained, beauty added. Dieting to reduce is dieting for health."[21]

In emphasizing the importance of dieting for health, the commissioner was reflecting one aspect of the tidal change that had swept middle-class American attitudes toward food in the previous decades: the conviction that you should eat what is good for you, not what you like. "Taste and habit, long the sole arbiters of the dining table, seem overthrown," wrote one observer in 1930. "Man, and perhaps more particularly woman, of the 1930 genus no longer eats what he likes in nonchalant abandon, fancy free. He eats what he thinks is good for him, on some scientific or pseudo-scientific hypothesis."[22] "The willingness to eat not for pleasure but for health is doubtless due to a fundamental U.S. trait: the fear of being sickly," said *Fortune* magazine in 1936. "Perhaps in England, but certainly not in France or Spain or Germany or Russia will you find people so anxious to believe that by eating in a certain way they can achieve the life buoyant and vigorous. Here it is the gourmet who is a curiosity, the dietitian who is a prophet."[23]

One of the more bizarre manifestations of this was that one of the nation's main concerns as it entered the Depression decade seemed to be something called "acidosis." It was supposedly caused by eating improper combinations of foods—mainly unbalanced proportions of proteins and carbohydrates—which caused the acids to overwhelm the alkalines in the stomach. This sapped its victims' vitality and made them susceptible to a number of awful afflictions, including excess weight.[24] Much of the credit for originating the acidosis scare seems to lie with Alfred W. McCann, a New York pure food crusader and unabashed quack. Since the mid-1920s he had been warning—in newspaper columns, books, and long-winded exhortations on his popular radio show—that it led to "kidneycide" and heart failure. By 1931, when the fifty-two-year-old McCann collapsed and died from a heart attack after an hour-long radio harangue, fear of acidosis had spread into the mainstream.[25] "Where once we prated about calories and vitamins, we are now concerned with an alkaline balance," said a bemused correspondent for the *New York Times* in May 1930.[26] When Kenneth Roberts surveyed the raft of diet books swamping the Library of Congress in 1932, he found that the most recent ones, mainly written by doctors, believed "nearly every disease in the world to be not only the result of eating improper foods, but also the result of eating proper foods in improper combinations. . . . If a person is so ignorant as to permit fermentable foods to pass his lips he is doomed. . . . They give him acidosis and what acidosis will do to him, in a quiet way, almost passes belief."[27]

Among the experts sounding the alarms about acidosis was America's best-known nutrition researcher, Professor Elmer McCollum of Johns Hopkins, whose famous experiments with vitamin-deprived rats and popular column in *McCall's* magazine had helped make him one of the culture heroes of the twenties. Food producers, with whom McCollum worked closely, quickly picked up the acidosis beat. The California Fruit Growers' Exchange's claim that its Sunkist brand citrus fruits, although apparently acidy, had a beneficial alkaline effect in the stomach received the endorsement of the federal Bureau of Home Economics. Welch's grape juice promised to "correct" acidosis while also fighting fat.[28] However, laboratory experiments soon began to undermine the acidosis scaremongers.[29] In May 1933 nutrition experts advised that "true acidosis is much less frequent than is commonly believed."[30] By 1935 McCollum was distancing himself from the acidosis scare, criticizing the Hays-type "compatible eating" diets, which claimed to combat acidosis, as baseless.[31]

By then, however, new discoveries in vitamin research were thrusting those tantalizing little things back into the limelight. Apparently crucial in maintaining vision, vitality, and even life itself, these tasteless and invisible items had gradually come to world attention from 1911 to 1921, and they proved to be a boon for food advertisers in the 1920s. Because so little was known about what they did and how much of them was needed for good health (there were no standardized methods for measuring them), they provided immense scope for exaggerated health claims. Thanks to food advertising and home economics in the schools, vitamin-consciousness was widespread by the end of the 1920s. In the early 1920s there had been much concern over deficiencies in calcium and vitamin A. Experts had therefore recommended drinking enormous quantities of milk and stuffing oneself with green vegetables. By the late 1920s the importance of vitamin C had been discovered and duly exaggerated. In the early 1930s vitamin G underwent the same process.[32]

New claims were also made for the longer-known vitamins. Although the only major affliction known to result from a deficiency in vitamin B was beri-beri—a polished-rice eaters' ailment practically unknown east of Pago-Pago—Standard Brands spent enormous sums trying to get consumers to rely on its Fleischmann's brand of compressed yeast cakes as a source of this vitamin. In the mid-1930s it was claiming that eating three of the soft, slimy cakes a day would clear up pimples, boils, and acne, increase energy levels, and cure poor digestion, "fallen stomach," "underfed blood," and constipation.[33] (Only the last claim contained more than a shred of truth, for it did have a pronounced laxative effect.) Quaker Oats ploughed the same furrow, claiming to provide "the precious yeast-vitamin (B)" much more economically than yeast cakes.[34]

That advertisers would trumpet their foods' vitamin B content so loudly is no surprise, for scientists were making some truly extraordinary claims on its behalf. One group reported that it increased "brain power." In April 1937 two Boston physicians reported that it had effected "a rapid and spectacular cure" of heart disease in 120 of their patients.[35] Two days later these claims were rivaled by a renewed push for vitamin G (later recognized as part of the vitamin B complex and renamed vitamin B_2, riboflavin.) Dr. Agnes Fay Morgan of the University of California announced that the black hair of rats deprived of this vitamin turned gray. Then, when they were fed it again, their hair turned black![36] This was exactly the kind of discovery the chemist Henry C. Sherman, one of the nation's leading nutritionists, was awaiting. For some years he had been arguing that a "well-rounded diet" could lengthen a person's life by at least seven years. The Fountain of Youth, he had said, lay not in Ponce de León's mythical land of eternal spring but "in every man's kitchen." Milk, eggs, fruits, and vegetables were more than "protective foods"; they would extend the normal life span.[37] He now rushed to his lab and deprived his rats of vitamin G. Lo and behold, he soon announced, early in their lives they took on the characteristics of little old people. "In fact," he said, "they look older than any man I ever saw." Yet when vitamin G was restored to their diet "they became regular Beau Brummels." Now, he announced, if liver and kidneys, high in vitamin G, were added to the list of foods he had earlier recommended, "early onset of senility" could definitely be headed off.[38]

But advocates of other vitamins battled for center stage. Vitamin C's effects were not clear for some years after its discovery, providing ample scope for claims of all kinds. In 1934 the California Fruit Growers' Exchange, citing a University of Chicago doctor who said that Sunkist orange and lemon juice had drastically reduced children's tooth and gum problems, advertised it as fighting "gum troubles" and tooth decay.[39] At the American Dental Association convention at the end of that year, fifteen hundred dentists, doctors, and nutritionists debated whether it was more effective than the toothbrush in combating tooth decay.[40] Some researchers said it cured stomach ulcers; others suspected that a deficiency of it turned people into alcoholics.[41] As for vitamin D, it was hailed as an anticoagulant that could save "bleeders" from certain death and was also credited with combating lead poisoning.[42]

As vitamin-mania increased in intensity, it became apparent that it might be a mixed blessing for large food processors. Yes, it provided tempting opportunities for outrageous health claims, but the main thrust of vitamin research was on deprivation: Rats (and presumably people) deprived of certain vitamins went blind, lost their vitality, teeth, and hair, developed scurvy, pellagra, beri-beri, and so on. Processors might encourage people to eat their products to head off these horrific consequences, but there were still disturbing indications that modern food processing, particularly milling and canning, itself robbed foods of vitamins. The large processors tried to reassure the public in a number of ways. The millers enlisted organized medicine, securing an official endorsement of white bread from the American Medical Association, which declared it "a wholesome, nutritious food [with] a rightful place in the normal diet of the normal individual" and called aspersions on it "without scientific foundation." Then, in 1930, they elicited support from the U.S. Public Health Service for a U.S. Department of Agriculture statement praising both white and whole wheat bread as "economical sources of energy and protein in any diet."[43] A battery of nutrition experts were brought on board. McCollum, who in 1928 had warned that white flour had been deprived of most of its vitamin content, was hired by General Mills in 1930 to encourage its consumption. So was Lafayette Mendel, the renowned vitamin expert from Yale. In 1934 the two famous scientists appeared with a galaxy of Hollywood stars on a Betty Crocker radio special to assure the public that white bread was a healthful diet food.[44] The next year McCollum wrote a well-publicized letter to Congress denouncing "the pernicious teachings of food faddists who have sought to make people afraid of white-flour bread." He also provided the canning industry with a useful statement assuring the public that nutritional research supported "the high favor of canned goods among consumers."[45] In 1938 the grateful Grocery Manufacturers Association presented him with an award for his contributions to knowledge of food.[46]

The professional home economists, who controlled nutrition education in schools and colleges, were also more or less co-opted by the food processors. Ruth Atwater, the daughter of Wilbur O. Atwater, the revered founder of human nutrition research in America, had taught home economics at Pratt, Skidmore, and the University of Chicago before she was hired in 1927 by the National Association of Canners to promote their products. Her sister, Helen, was editor of the *Journal of Home Economics* and saw nothing untoward about publishing Ruth's assurances that "research has shown conclusively that commercially canned foods have the same food value as similar foods prepared in home kitchens, with the possibility of added energy value due to the presence of sugar syrups in many canned fruits and a few canned vegetables."[47] But then, no one seems to have thought twice about the food processors having become an indispensable source of funding for the American Home Economics Association, publisher of the journal.[48] Since they also provided an increasing number of jobs for home economists—who developed recipes and instructional materials using their products for home economists in the schools—there was never a shortage of professional "dietitians" (a rubric used increasingly by home economists who specialized in food) to provide similar testimonials to canning and other forms of processing.[49] Dr. Walter P. Eddy, a nutrition expert at Columbia Teachers College, a leading center of home economics research, reported that people could rely on all of the forty-nine kinds of canned foods he tested to provide more than adequate supplies of vitamins A, B, C, and G. After all, his rats had "thrived" on them for a whole year.[50]

The mass media, especially the women's magazines, which profited mightily from food ads, also helped still public concern.[51] Dr. Eddy wrote a regular column for *Good Housekeeping* magazine, which assured readers that advertised items such as Jell-O were excellent and inexpensive sources of nutrients.[52] In 1934 the *Ladies' Home Journal* ran an article on canning that told readers, "Here's food so meticulously prepared that no suspicion of loss of nutriment or purity or wholesomeness can be laid to it." Only the exceptionally beady-eyed would have noticed that it was actually a paid advertisement of the National Association of Canners set up by the *Journal* exactly like one of its own articles.[53] Earlier, an article that amounted to an ode to large food manufacturers had informed readers that any of their advertising making claims for vitamins was submitted to "cool scientific men" who told the copy writers, "Thus far may you go and no farther."[54] Continental Can's cool scientific men seem to have allowed their copy writers considerable latitude, however. They assured consumers that canned fruits and vegetables were cooked in sealed cans "to retain the vitamins."[55]

All of this helps explain one of the great mysteries of the 1930s. To wit: Here was a nation swept by anti–Big Business sentiment. Giant corporations and banks were commonly accused of having brought on the economic collapse, and the reputation of American [illegible]n sank to an all-time low. It was certainly a propitious time t[illegible] these villains seemed to be doing to that most precious part of the [illegible]an heritage—its supply of good, wholesome food. Yet those who warned of the pernicious effects of Big Business on American food achieved remarkably little.

It was not that voices were not raised. To decry the processing of foods and proclaim the superiority of the "natural" represents an age-old current in America. The 1930s were no exception. Indeed, two such assaults—Arthur Kallett and Frederick J. Schlink's 1933 book *100,000,000 Guinea Pigs* and Schlink's *Eat, Drink, and Be Wary* in 1935—were by far the most effective denunciations of processors since Upton Sinclair's 1905 exposé of meat packers, *The Jungle*. The two engineers had helped found Consumers Research, Inc., which Kallett directed, and had close connections with Stuart Chase, Robert Lynd, and other leaders in the consumer movement. Their books, articles, and new product reports helped stimulate a miniwave of "guinea pig journalism."[56] But their emphasis was on dangerous additives to food, not nutrient deprivation. "Poison for Profit," the title of one of their articles in the *Nation* magazine, neatly summarizes their thrust.[57] They saw themselves as continuing the crusade against dangerous drugs and food additives, begun by the revered, recently deceased chemist Dr. Harvey Wiley, which resulted in the Pure Food and Drugs Act in 1906. Convinced that new developments in science and business had left that act behind, they denounced the spraying of fruits with arsenic trioxide and lead arsenate, the "sulphuring" of dried fruits, and the surreptitious use of sodium sulphite to keep meats looking fresh, as well as Wiley's bête noire, the use of sodium benzoate as a preservative. They warned that white bread was made with yeast fed with a chemical—potassium bromate—whose effects on humans were unknown, and they condemned white flour, not because it had been deprived of the nutrients of the whole wheat germ but because "poisonous" chemicals were used to bleach it.[58] (They did not attack millers for removing the roughage in the bran of the whole grain, for one of Schlink's more peculiar hobbyhorses was that bran was bad for you.) Yet by mid-decade their warnings seemed only to have spurred sales of additive-free and pesticide-free foods in health food stores.[59]

The large canners themselves had inadvertently contributed to the beginning of the campaign. In 1930, concerned by competition from small, low-cost canners who packed cheaper fruits and vegetables, they had their friends in Congress amend the Food and Drug Act to require canned foods that did not meet certain standards to be labeled "Below U.S. Standard, Low Quality but Not Illegal"—a tag hardly calculated to set shoppers reaching for the product.[60] But the big canners turned out to be a little too cute for their own good, for this so-called Canners' Bill drew attention to the fact that nothing on their own labels gave the consumer even a hint of the quality of what lay inside the opaque tin. Calls for labeling canned goods as A, B, and C in quality—or some other system—inevitably arose. When Assistant Secretary of Agriculture Rexford Tugwell, a Columbia professor who had served on Preside[illegible] D. Roosevelt's "Brain Trust," turned his attentio[illegible] Drug Act, this was among his proposals. But the [illegible]w realized that grade labeling would undermine their main a[illegible]vant[illegible] over the small fry: their large advertising budgets, and the confidence these created in their brand names. If the government guaranteed that the quality of the contents of two different cans was equal, why should consumers pay more for the brand-name product?[61]

Tugwell, a brilliant man who took few pains to mask his disdain for his intellectual inferiors, was already one of the most unpopular figures in the administration. He was now duly reviled as "Rex the Red" by the food industry and charged with aiming to communize food processing. He soon left the bill in the hands of its less abrasive sponsor in the Senate, Royal Copeland, a mild-mannered homeopathic physician from New York. Like the consumer leaders, Copeland was particularly concerned with protecting the public from "poisons" masquerading as healthful additives, and he rewrote the bill to place greater emphasis on this. Yet Schlink, Kallett, and their left-wing associates at Consumers Research opposed the bill as wishy-washy and ineffective. On the other hand, despite President Roosevelt's personal assurances that the bill was aimed only at "a small minority of chiselers and evaders," the large food companies joined the United Medicine Manufacturers of America in organizing a powerful lobby against it.[62] Among the provisions they found most objectionable was a proposal that false and misleading advertising be penalized (the 1906 act banned only false labeling), something that would jeopardize their extravagant health and nutrition claims.[63]

In the political context of the time, business opposition was not, in itself, enough to doom the legislation. But the bill's supporters could never point to much organized concern about the quality of the food supply. True, a number of middle-class women's organizations supported it, as did the American Home Economics Association, the American Dietetic Association, and the American Nurses Association—all somewhat radicalized by the Depression.[64] But conspicuously absent from the list was the powerful American Medical Association, which probably swung more weight in Congress than all the others combined. Indeed, the AMA helped undermine the case for government regulation of food claims with a system whereby advertisers voluntarily submitted their copy to an AMA committee for "acceptance." This merely meant certification that the product contained what it said it contained, provided on the basis of chemical analyses submitted by the companies themselves. For most products the hurdle was hardly a high one. A typical favorable report was "*Product:* Heinz Pure (Virgin) Olive Oil (Imported). *Description:* Imported first cold pressed (Virgin) Spanish olive oil." It was this system that allowed General Mills to claim AMA endorsement for Gold Medal white flour.[65]

But most important in damaging the bill's prospects were the media—particularly the mass circulation magazines, which bombarded it with calumny. Their outrage was directly linked to their balance sheets, for by the mid-1930s the food industries had become their largest advertisers.[66] The editors of Hearst-owned *Good Housekeeping* magazine—for whom Dr. Wiley had served as a columnist—discovered what was afoot in late 1933, after having been carried away by Wiley's widow's emotional appeal that they support the bill in his memory. Within weeks they were ordered to reverse themselves and publish a condemnation of the bill written by a New York City advertising man.[67] The *Ladies' Home Journal* was forced into an equally embarrassing comedown. Its editors—recalling the *Journal*'s leading role in agitating for the 1906 law—had also rashly supported the Tugwell bill, only to be forced by the publisher into a humiliating reversal. As if to atone for the initial faux pas, the head of the *Journal*'s parent company, Curtis Publications, testified to Congress on the bill's evils.[68]

The struggle over the bill lasted for more than five years, exhausting Copeland, who died shortly after its passage, and giving consumer advocates a chastening view of the difficulties involved in taking on entrenched interests. Thanks to their attacks on the Hearst magazines, they had even drawn the ire of the newly created House Un-American Activities Committee, which set about trying to prove that they were a Communist "front."[69] By the time the bill passed, many doubted that it had been worth the effort, for by then most of its teeth had been pulled.[70] The bowdlerized version finally became law after a wave of public outrage greeted the punishment—the only one available under the old rules—meted out to a Tennessee druggist who produced a patent medicine that killed seventy-three people in seven states: a two-hundred-dollar fine for mislabeling the product.[71]

The legislative battle, with its emphasis on poisonous products and additives, had attracted most of the consumer movement's attention. Nevertheless, an undercurrent of popular suspicion that processing made foods less nutritious had persisted, spurring a search for ways to reinsert vitamins and minerals into foods. A major breakthrough had come in 1928, when chemists at the University of Wisconsin were able to irradiate canned and pasteurized milk with vitamin D—ironically, a nutrient it had never contained. Pet Milk led the way in irradiating its canned milk and was soon followed by much of the rest of the industry, providing a bonanza in royalties for the university's research foundation. With financing from the state's powerful dairy interests, the university chemists soon simplified the method and developed ways of irradiating cheese and other food products.[72] Other food companies subsidized further research, hoping to discover new nutritious qualities in their products or how to vitalize them.[73]

The process was echoed throughout the nation's research establishments, as Depression-battered university scientists and threadbare government agricultural experiment stations turned to food processors for funding. In 1933, when the famed New York state agricultural experiment station at Geneva was reduced to desperate straits—with a budget that allowed a mere three hundred dollars a year for equipment and chemicals—the Birds Eye subsidiary of General Foods stepped in with three thousand dollars a year to subsidize research into how to preserve the vitamin content of frozen foods. Then the fruit juice processors, corn syrup manufacturers, and even wine producers joined in, funding studies of how vitamins fared in the processing of their products. The experiment station used some of the new money to fund graduate research by its scientists at nearby Cornell University, helping to establish a mutually beneficial relationship between Cornell scientists and the packers and processors.[74] By the end of the decade the industry had come up with the funding necessary to create a School of Nutrition.[75] By then the powerful Grocery Manufacturers Association had pledged $250,000 a year to underwrite a well-endowed industry-supported nutrition research foundation.[76]

For most processors, the payoff from vitamin research would be some time in coming. In 1935 scientists began to come up with commercial methods for synthesizing vitamins, but food processors hesitated to use them. Even health food producers were slow on the uptake, for their businesses revolved around their own special foods. The industry leader, the Battle Creek Food Company, subsisted mainly on ersatz foods such as "psylla seed" and "Feroclyst," an "iron preparation with copper and chlorophyll."[77] One of the most popular lines of health foods in California (even then a mecca for food faddists) was a range of dehydrated foods produced in capsule form by Anabolic Food Products of Glendale. Available only through doctors, the capsules of brown powder were said to contain all of the nutrients in lettuce, endive, Cape Cod cranberries, Irish kelp (a popular cure-all of the 1920s), or whichever other of the forty-three available vegetables the doctor prescribed.[78]

But once a number of vitamins were available in pill or liquid form, their attraction was soon manifest. Trend-setting southern California led the way in what one observer called "the quick change-over from counting calories to supplementing the diet with vitamins and minerals in capsules," behavior he thought "came from the same wellspring as the cults of its religion."[79] Chemical producers scrambled to churn out vitamins for the retail trade. Drugstore trade associations stole a march on other retailers by having a number of state legislatures declare vitamins to be drugs, thereby restricting their sale to pharmacies and keeping prices high. In 1938 over a hundred million dollars' worth of the rather expensive new pills was sold by druggists—making them second only to laxatives in drugstore sales—more than a quarter of them on prescription.[80] (Common sense showed that "those individuals who can afford a dollar a week for vitamin pills don't need them," said a New York hospital director, "and those who might be benefited can't afford them.")[81] In 1939 the large grocery chains, led by Kroger's and IGA, counterattacked, challenging the druggists and the state laws by stocking thousands of their stores with vitamin pills and potions.[82] The new Food and Drug Act backed them up by declaring that, if they were not prescribed for illness or sold with health claims attached, vitamins should be considered foods, not drugs. When giant Lever Brothers began to manufacture vitamins and wholesale them through its grocery distribution network, the *Journal of the American Medical Association* commented that the "vitamin gold rush of 1941" made that of 1849 pale by comparison.[83]

But Lever was selling vitaminized pills, not foods. Indeed, as an industry observer noted, food processors had "failed to cash in adequately on the [vitamin] trend."[84] Their most visible response was a major campaign by Kellogg urging shoppers, "Get your vitamins in food—it's the thriftier way."[85] Only slowly did it dawn on the processors that the best defense might be a kind of co-option: putting nutrients into their foods. This tardiness in awakening to the possibilities of nutrification is quite understandable. After all, it meant acknowledging that their critics had been right and their advertising wrong, that processing often did deprive foods of nutrients. But scientific advances were making it impossible to maintain the old stance: Methods had been developed for measuring most vitamins in standard "units." The amounts of vitamins lost in processing could now be calculated, as could the nutritional content of liquids or powders added to restore what had been lost. Consumers could no longer be fooled, the vitamin manufacturer Hoffmann–La Roche warned food processors in 1939. The American housewife now knew that vitamins B_1 and C, for example, were essential for growth, "nerve stability," teeth, and gums. She would "insist on specific declarations" in "units," and she would "know herself whether the amounts named are a meaningless gesture or worth while." She knew that these vitamins "may be destroyed or lost in modern processing and cooking, that they can now be restored."[86] The Scott and Browne Vitamin Corporation also sounded the jig-is-up theme. Consumers now knew, it told the processors, that while, with few exceptions, Mother Nature "included all the necessary vitamins in basic *raw* food materials, modern food processing and preserving methods impaired the potency of these vitamins." People now "*want* and *buy* food products fortified with vitamins."[87] A top vitamin researcher warned millers and sugar refiners that to continue to ask consumers "to disregard Nature's laws" or to make up for the deficiencies in their staples "by judicious use of other foods" was bad business. They must restore the nutrients to their products, for "to blink at the scientific facts, which will presently become common knowledge, will be suicidal for the commercial enterprises concerned."[88]

As if to reinforce that warning, in August 1939 the AMA recommended "restoring" processed foods with enough nutrients to bring them back up to their "high natural levels."[89] But the move was intended as a spur, not a rebuke, to the food industries. They must be enlisted, thought the doctors, to battle the real enemy—the vitamin pill vendors, particularly those who peddled vitamins as cures for illness. The organized doctors also looked askance at "fortification," adding more nutrients to processed foods than they originally contained or adding ones they never had. (Some manufacturers had begun to add vitamin D to frankfurters and chewing gum.)[90] But in 1940 and 1941 they were forced to reexamine this position, as one vitamin—B_1, or thiamin—came to be regarded as absolutely essential for national defense.

In retrospect it is amazing that so much could have been concluded from the experience of so few. In mid-1939 three doctors at the Mayo Clinic in Rochester, Minnesota, put four teenagers on a diet low in thiamin and found that they became sluggish, moody, "fearful," and "mentally fatigued." Whereas some parents of teenagers might not have seen anything extraordinary in this, the doctors thought they were onto something important. They repeated the experiment with six female housekeepers at the clinic, aged twenty-one to forty-six. When they were deprived of thiamin their ability to work—measured by having them do chest presses—declined markedly. At the early stage of thiamin deprivation, the doctors reported, their symptoms resembled those of neurasthenia, while the later stage resembled anorexia nervosa.[91] When, after eleven days, two of the six were put on a diet much higher in thiamin than normal, their chest-pressing ability rose.[92]

By October 1940, when the results of the second experiment were made public, much of Europe had fallen to the Nazis, the Japanese were on the move in Asia, and America was feverishly rearming. An editorial in the *Journal of the American Medical Association* was quick to see a connection. Carefully avoiding mention of the paltry number of people tested, it warned that the "moodiness, sluggishness, indifference, fear, and mental and physical fatigue" induced by cutting the thiamin intake "in a group of healthy subjects" were "states of mind and body . . . such as would be least desirable in a population facing invasion, when maintenance of stamina, determination and hope may mean defeat or successful resistance."[93] One of the Mayo experimenters, Dr. Russell Wilder, declared that a deficiency in thiamin was "a principal cause for" the majority of cases "commonly spoken of as loss of morale." It was Hitler's secret weapon in occupied Europe, rumor had it, where the Nazis were "making deliberate use of thiamin starvation to reduce the populations . . . to a state of depression and mental weakness and despair which will make them easier to hold in subjection." Canadian colleagues had told him that some Canadian soldiers recruited directly from the relief rolls who had initially been "defiant" or "depressed" had "after satisfactory attention to their nutritional deficiencies" become "perfectly manageable and effective."[94] This confirmed his own experiment, in which adding thiamin to young peoples' diets increased mental alertness and almost doubled their capacity for physical work. (It seems not to have affected their moodiness.)[95]

In the North American diet, this "morale vitamin," as it came to be called, was found in beans, legumes, and, most commonly, whole wheat flour. Yet in 1940 only 2 percent of the bread Americans bought was of a whole wheat variety.[96] Modern milling processes removed from 70 to 80 percent of wheat's thiamin to produce the white wheat flour that was the American staple. Wilder warned that thiamin consumption had been steadily declining for over one hundred years and had now dropped to critical levels.[97] Nutrition experts were duly alarmed, but they thought it impossible to convince Americans to switch back to whole wheat bread, especially since the mere mention of whole grain breads evoked memories of the heavy, grim-tasting loaves of World War I. The best solution, it seemed, was to put thiamin back into white wheat flour. In 1940 the average American ate two hundred pounds of it,and it constituted about one-quarter of his or her caloric intake.[98] It did not take much persuasion for flour millers, who had resisted all previous efforts to modify white flour, to come around. They did not relish being accused of leaving the country defenseless in the face of foreign invasion.[99] So in early February 1941 they began turning out flour "enriched" not only with vitamin B_1 but also with iron and pellagra-preventing nicotinic acid.[100]

By May 1941 Vice-President Henry Wallace thought the benefits of thiamin-awareness were becoming apparent. Addressing a national conference on nutrition, he extolled a radio commentator who said, "What puts the sparkle in your eye, the spring in your step, the zip in your soul? It is the oomph vitamin!" It did seem, said Wallace, that to many Americans the addition of the B vitamins to the diet "makes life seem enormously worth living."[101] That November, when Gallup asked Americans to name a vitamin they had heard a lot about in recent months, the overwhelming majority named vitamins B_1 and B_2. But the poll also indicated that 84 percent of housewives could not explain the difference between calories and vitamins.[102] To them, vitamins seemed to provide "pep" and "energy."

Alas, pumping thiamin back into the national diet had no discernible effect on the morale of the nation. Within a few years, no one—not even Wilder—was linking it with morale. In any event, Wilder's estimates of how much was required were later shown to be quite inflated; the normal American diet provided it in more than adequate quantities. Nevertheless, enrichment did give official blessing to an important idea: that one could look well fed and actually be starving. In 1941 the AMA warned that "hidden hunger" struck those who "satiate[d] themselves with vast quantities of food" but did not eat enough essential nutrients.[103] Washington's announcement of the flour enrichment program was hailed as "designed to rescue some 45,000,000 Americans from hungerless vitamin famine."[104] It also represented official acknowledgment of the idea that processing deprives essential foods of important nutrients. The two ideas, joined together, would become the basis for every future revival of concerns over food and health. "The discovery that tables may groan with food and that we may nevertheless face a form of starvation has driven home the fact that we have applied science and technology none too wisely in the preparation of food," said a *New York Times* editorial in December 1941.[105]

The Depression had begun with Americans concerned over people who were *feeling* hunger and might be experiencing starvation—that is, the deterioration of health that resulted from the wasting away of the body. It ended with the diffusion of completely different concepts of hunger and starvation: hunger that could not be felt, starvation that could not be seen. It had begun with media paeans to the wonders of modern American food processing. It ended with official warnings that science and technology had deprived American food of its healthful properties. The war and the conditions of the postwar era would push these concerns to the back burner, but they would ultimately reemerge to play important roles in shaping the modern American diet.

CHAPTER 2

The Great Regression: The New Woman Goes Home

In 1935, after an absence of almost ten years, Robert and Helen Lynd returned to Muncie, Indiana, the middle-sized midwestern city they had made famous in their acclaimed sociological study, *Middletown*. They were immediately struck by how little domestic life seemed to have been affected by the Depression. "As one walked Middletown's residential streets in 1935," they wrote, "one felt overpoweringly the continuities with 1925 these homes represented. Whatever changes may have occured elsewhere in the city's life . . . here in these big and little, clean and cluttered houses in their green yards one gained that sense of life's having gone on unaltered in one's absence."[1] Of course, as they soon discovered, the life beyond the lawns was not altogether unchanged, but in terms of the essentials—what people ate, where they lived, and how family members related to each other—the remarkable thing was how little, rather than how much, had changed. Had they returned again five or six years later, their impression would likely have been much the same, particularly with regard to the middle class. A study of the impact of the Depression on a group of Minneapolis middle-class families confirmed the Lynds' view. While they had cut back on "non-essentials" such as insurance, spending on food and clothing had remained more than adequate. Indeed, if anything, diets had improved.[2]

While the United States was not Muncie writ large, much less Minneapolis, the Lynds' observation could have been applied to its eating habits. As indicated in the previous chapter, by the late 1920s the key ideas of the twentieth-century revolution in eating habits were already well entrenched, particularly among the middle classes. During the Depression years the great food-producing and -processing organizations, which had become major forces in disseminating these new attitudes, continued to expand and reinforce them, while also encouraging the kind of cooking that would make the maximum use of their products.

Initially the Depression struck these firms hard, forcing vicious price-cutting. Swift and Co. reported a five-million-dollar loss in 1932, mainly the result of price-cutting.[3] California canners stuck with mountains of canned fruits and vegetables dumped them in such volume that by early 1933 it cost them more to ship them to the East than they would fetch from eastern wholesalers. The Sun-Maid Raisin cooperative was forced to sell raisins as animal fodder for a mere six cents a pound.[4] But Sun-Maid survived, and so did the most of the large processors, mainly by paying producers much less for their crops. Campbell's slashed the prices it paid its New Jersey tomato growers by almost one-third, to well below their cost of production. Midwestern canners cut them even more drastically.[5] The cuts were then passed down to pickers in the fields, mainly migrants, whose desperate protests were crushed. Even with these reductions, though, many smaller canners did not have the resources to ride out the storm, and the giant canners such as the California Packing Corporation (then Cal Pak, soon to become Del Monte) increased their dominance of the industry, particularly in the West.[6] Dairy farmers dumped their milk on highways in desperate attempts to raise its price, but the same low prices helped Pet Milk show a profit during each year of the Depression.[7]

While price-cutting sent farmers and small processors to the wall, it did help cushion consumers' diets from the effects of the Depression, for it meant that despite falling income most could still buy the processors' foods in more or less the same volume as before. The president of Armour Packing reported that, while his company's sales had dropped from $668 million in 1931 to $468 million in 1932, the actual quantity of food shipped had hardly changed. The Rath Packing Company reported a 30 percent decline in sales and an *increase* in tonnage sold.[8] Indeed, per capita consumption of meat increased by 10 percent from 1929 to 1934. In 1929 beef consumption, which had been declining for many years, began a steady march upward that lasted through the Depression and continued for over forty more years.[9] Americans did cut back quite sharply on canned vegetables and soups in 1931 and 1932, but by 1934 sales of these relatively expensive items were again on the upswing. By the end of the decade Americans were eating 50 percent more processed fruit and vegetables than at the beginning—almost as much as the fresh kind.[10]

The 1920s had been vintage years for large food-processing corporations. Even at the beginning of that decade a few large enterprises had dominated the meat-packing, sugar, and flour-milling industries. Then the same thing came to pass in dairy, baking, tropical fruits, and breakfast cereals. It all climaxed in a veritable corporate feeding frenzy, which saw the emergence of two massive conglomerates: General Foods (which alone gobbled up twenty other food companies in the space of about four years) and Standard Brands, each of which produced a host of brand-name products. Although the Crash of 1929 and the Depression brought an end to the leveraged buyouts financed by "watered" stocks, it only slowed and did not halt the long-term trend toward domination by fewer and fewer giants.

Abraham Hoffman, a government economist who studied the process in the late 1930s, concluded that it was inevitable that large processors would dominate the food industries because their size allowed them to manufacture on a very large scale, cut prices, and drive smaller competitors out of business. Moreover, conglomerates like Standard Brands could distribute a number of foods through a nationwide network of warehouses and salespeople more cheaply than those selling and storing just one or two foods.[11] But he underestimated a crucial factor: Oligopolies and conglomerates dominated certain sectors of the foods industries not because of their "economies of scale" but because they could afford to spend vast amounts on advertising to promote their brand names.[12] The giants tended to arise in industries such as flour-milling, sugar- and salt-refining, bread-baking, canning, and milk-processing in which, because of mass production techniques, there was little to differentiate products from each other.[13] Significantly, when Standard Brands and General Foods set about snapping up other firms it was the targets' brand names—household symbols such as Jell-O and Chase and Sanborn—and not their production facilities that commanded most of the price.[14]

A good example of the importance of size and advertising arose in the mid-1930s in the rather unappetizingly named edible oils industry. Procter & Gamble had created Crisco, a tasteless shortening that could substitute for lard and butter in frying and baking. The market it tapped was enormous. The frying pan still reigned supreme in most American kitchens, and—because techniques for large-scale commercial pastry-making had not caught up with those for bread-making—80 percent of cakes and pies were still baked at home.[15] The process for making the stuff was not particularly arcane, and the nation was virtually awash in edible oil, so a number of smaller companies were able to compete successfully with P&G in the restaurant and other bulk markets, where price was the determining factor. But none could afford the advertising campaign necessary to challenge Crisco's hold on the country's housewives.

Only Lever Brothers, the other giant soap maker, could hope to crack Crisco's hold on the mass market. Like P&G, it already produced enormous amounts of oils for soap-making and had a national sales and distribution network in place in the grocery trade. Most important, though, it too was accustomed to spending enormous amounts on advertising and promotion. Surveying housewives to find Crisco's weaknesses, it discovered that Crisco was not as white as they would have liked, was unevenly packed in the can, and hardened inordinately in the refrigerator. Lever then came up with Spry, which was lily white, evenly packed, and—to indicate that it was softer—had a little curl on top. It then mounted a massive advertising campaign and blanketed the nation with salespeople giving out free samples and discount coupons. Crisco reacted by whitening and softening its product and becoming "double creamed," but Spry then became "triple creamed." When Crisco reached "super creamed," a standoff ensued, both companies panting. Both products were now almost exactly the same and, thanks to enormous advertising budgets, shared the huge new market almost equally.[16] It would have been foolhardy for any other company, except another giant with an equally formidable promotion budget, to challenge their dominance of the retail market.[17]

The impact of the rise of giant food companies such as these on what people actually ate harbors an apparent contradiction: On the one hand, their success has often been built on providing consumers with new foods or older foods prepared in new ways, which has added variety to the diet.[18] On the other hand, their mass production and distribution techniques have contributed mightily toward standardizing the national diet. By the 1930s improved transportation and food preservation processes were well on the way toward eradicating perhaps the oldest distinction in human diet: seasonality. Although the fresh produce on store shelves still reflected the changing seasons to a certain degree, the expansion of fruit- and vegetable-growing in California, the Southwest, and Florida had made many fruits and vegetables available for much longer periods than before. Fresh corn grown in Texas arrived in New York City in mid-May.[19] The American agricultural empire in the Caribbean and South America made tropical fruits affordable almost year round. Much of what was not available fresh could be bought in cans. Housewives marveled at the choice of canned "fresh picked" foods facing them on the grocer's shelves all year—spinach, tomatoes, peas, corn, grapefruit, cherries, and pineapple—an endless lineup that never varied with the seasons.[20] The traditional women's magazine features on "seasonable" recipes appeared less and less frequently during the decade. In April 1931, for example, *American Cookery* ran articles on "Seasonable Menus for a Week in April," "Vegetables Now in Season," and "Eat More Fish in April." By mid-decade seasonality had virtually disappeared except in its holiday recipes or in terms of the weather, as in hot- or cold-weather cooking. In 1941 the title of its monthly feature "Seasonable and Tested Recipes" was changed to "Tested Recipes of the Month."[21]

The same forces—improvements in transportation, preservation, and distribution—liberating Americans from seasonality also continued to free them from the dictates of regional geography. Milk, cheese, and green vegetables poured into the South from the Mid-Atlantic states and Midwest. Practically the entire nation was blanketed with immature citrus fruits and indestructible iceberg lettuce from southern California, canned fish and vegetables from central and northern California, canned tomatoes and peas from New Jersey, Wisconsin cheese, western beef, midwestern ham and sausage, Florida oranges, Hawaiian pineapple, Central American bananas, and Cuban sugar. The shelves of an A&P in Louisville, Kentucky, were hardly distinguishable from the shelves of one in Utica, New York, or Sacramento, California.

The gap between city and farm diets continued to narrow as farming became more specialized and farmers continued to find a bowl of corn flakes and bananas as enticing at breakfast as did city folk. Thanks in part to the influence of the media, processed "urban" foods such as canned salmon had become high-status foods in rural areas, and even those who could little afford them sacrificed to purchase them.[22] Poor Appalachian farmers shunned tasty country hams in favor of water-logged canned ones; they sold home-grown vegetables to buy the brand-name canned variety.[23] Better-off farm women seemed even more taken with the convenience of processed foods than city ones, abandoning much of traditional home food production with apparent alacrity. As early as 1932 the gap in the amount of time spent on preparing food in city and farm homes had been reduced to virtually nil. On average, farm women did spend slightly more time in the kitchen, but mainly, it seemed, because most of them had to bake their own bread.[24]

Class differences also continued to blur. There had been a kind of flattening at the top during the 1920s, as most of the upper class abandoned their prewar attachment to ostentatiously elegant French dining. By the 1930s an appetite for sophisticated food was no longer a mark of their social distinction; their food tastes had became practically indistinguishable from those of the well-off upper middle class. Indeed, anthropologists studying the social life of a New England city in 1933 discovered that the upper class spent considerably less per person on food, both inside and outside the home, than the upper middle class.[25] Distinctions were maintained, of course, but they came in *where* one ate, not what one ate. As an observer later wrote: "For those born to wealth and position, the hurly-burly of an unmannerly world can be mitigated by a stately progress from one social refuge to another—from the men's club, university club, or luncheon club to the country club, yacht club, or beach club."[26] What kind of food one ate while sheltered in these places hardly mattered.

On the other end of the class ladder, working-class food tastes also continued to lose their distinctiveness, particularly as the food habits of the immigrants who had been the backbone of the prewar urban working class became less distinctive. By 1930 immigration from Europe had been cut to a trickle for almost fifteen years, shutting off the pipeline bringing renewed infusions of Old World tastes to immigrant communities. The original immigrants had their food habits assaulted from all sides, particularly if they moved out of the immigrant ghettos. Once deprived of the networks of family, marriage, and social ties that helped preserve traditional food habits, they normally found the pressures to Americanize their diets practically irresistible.[27] They would first drop those foods the larger society found most repulsive—Germans their blood sausage, for example—and then push the rest of their distinctive foods to the periphery of their diets.[28]

Those who remained in ghettos were pressured to change by children exposed to the full force of middle-class American food ideas in the schools. A study of the children of Japanese-born parents in Hawaii noted that the schools, which "unflinchingly urge them to change their food habits," were not only successful at this but also affected the parents' diets. "My children tell their mother what foods are good for our health," said one mother. "They say we must eat more vegetables and fruits and less rice. They learn this in school. I believe their teachers are better informed along this line so I do not interfere or ignore their suggestions."[29] The anthropologist Paul Radin discovered in San Francisco that even many Italians—perhaps the most obdurate immigrants when it came to food—took to heart the dietetic lessons their children brought home from school.[30] (Others were not so receptive. When Leonard Covello, who was raised, like many Italo-American children, on breakfasts of bread and coffee, took home some oatmeal as an example of a proper breakfast dish, his father shouted, "What kind of a school is this? They give us the food of animals and send it home with the children!")[31] For those whose parents persisted in the old culinary traditions, lunch times could be particularly excruciating. Some Mexican-American children in California would walk home for lunch rather than face the embarrassment of unpacking tacos at school. Others would try to trade them for "American" food like peanut butter and jelly sandwiches.[32] At home the children would nag their mothers to cook with flour rather than corn meal and to forsake chiles. A Mexican-American mother whose husband had to eat American food all week and came home on weekends yearning for spicy tamales complained that even then her children would not eat them, preferring American food.[33]

By the 1930s many American-born children of immigrants had formed families of their own and were forsaking the cuisines of their parents' homelands and trying to eat like "Americans." Those of European origin tended to find their ancestral ways of eating too heavy and spicy. Stews, sausages, and peasant one-pot dinners were abandoned in favor of fried or grilled dishes served with a starch and vegetable on the side. Garlic was a particular embarrassment in a culture with a real phobia about it. (A 1939 *Life* magazine article on baseball player Joe Dimaggio assured readers that he did not slick his hair back with olive oil and "never reeks of garlic.")[34] Foreign cuisines also seemed too difficult to prepare, particularly since ingredients were not readily available outside of the immigrant neighborhoods and most dishes were not available in prepared form. A Japanese-American high school graduate in Hawaii differed little from millions of second-generation Europeans in this respect: "I don't like Japanese foods," she said. "I don't like rice and fish. Fish smells bad and rice takes too long to cook. Anyway, it's troublesome to prepare Japanese dishes."[35]

One major immigrant group held out against this Americanizing process—Italo-Americans. They not only managed to retain many of their distinctive food tastes, they were able to watch them become part of the mainstream. Thanks in large part to food's centrality in the preservation of their intense family ties, they had managed to evolve a version of the various cuisines of their homeland that had proven remarkably resistant to Americanization. That during the 1920's Americans had begun to accept their "signature dish," spaghetti and tomato sauce, had also helped.[36] The Depression, of course, was made to order for pasta's economies. In 1930 and 1931 the macaroni manufacturers' association, supported by the large flour millers, mounted a $1.3 million campaign to tell the public that spaghetti was healthful and economical.[37] Food editors picked up its themes. The *Chicago Evening American* said spaghetti's high gluten content made it "an ideal food." Its ratio of protein to carbohydrate was much closer to the "ideal" than that of bread, and, when combined with tomato sauce and the additional protein and minerals contributed by cheese, it provided "a complete food as nourishing and hearty as it is delicious."[38] Even Betty Crocker pushed spaghetti as "the most nourishing food for its cost" and distributed recipes for an "Italian Dinner"—spaghetti and tomato sauce.[39] Whether influenced by the campaign or not, Americans began cooking more and more spaghetti with tomato sauce and grated cheese, particularly in the hard winter of 1932–1933.[40]

Of course, the versions served were very much adapted to American tastes. Rarely did recipes for tomato-based sauces call for even a scrap of the dreaded garlic clove. *Good Housekeeping*'s recipe for spaghetti and meatballs called for beef suet, horseradish, and "bottled condiment sauce" in the meatballs, but no garlic in the accompanying tomato sauce. A little bit of chopped green pepper added the only zest in its spaghetti-with-tomato-sauce-recipe. For its version of that dish, *American Cookery* had cooks flavor two cans of tomato soup with a tablespoon of Worcestershire sauce.[41] Often, to avoid a too intense tomato flavor, the tomato sauce was thickened with flour rather than by boiling down the tomatoes. Homemakers were also simply advised, as by *Good Housekeeping*, that "delicious spaghetti with tomato sauce and cheese now comes in cans."[42] *Al dente* was as yet an unknown phrase; recipes generally called for spaghetti to be cooked until soft. Some people (including the author's mother) even used ketchup as the tomato sauce. So did the U.S. Army, whose cookbook listed it as an alternative to tomato pulp in spaghetti and tomato sauce.[43] Nevertheless, when all is said and done, something quite extraordinary was taking place: Italo-Americans were becoming the only one of the "new," post-1880 immigrants who not only retained much of their culinary heritage but substantially influenced that of mainstream America as well. Pasta and tomato sauce, originally a symbol of intransigent resistance to Americanization ("Still eating spaghetti, not yet assimilated," noted prewar social workers), was crossing ethnic, regional, and class lines.[44] Woolworth's distributed recipes for it, and housewives in Oklahoma prided themselves on their versions of it.[45] This newly acquired taste would, as we shall see, play an important role in the reshaping of the American diet in the postwar years.

It is easy to see the processes of homogenization and standardization as the inevitable byproducts of relentless economic forces such as industrialization and urbanization—and to a great extent they were. But other factors were involved as well, for the last place that Adam Smith's "Economic Man" sits peacefully is at his table. Americans were not just responding to the price and availability of foods, they were also reacting to ideas about the social role of food: who should prepare it, how it should be prepared, and what eating it said about them and their society. Here economic forces did come into play, but in the form of the vast amounts producers and processors spent to create the images that sold their products.

The main object of this massive outlay was the abstraction called the "American housewife"—and with good reason. Since the mid-1920s advertisers had been mesmerized by the discovery that she made the crucial decisions in allocating up to 90 percent of the household's disposable income. It was also apparent that, although the economic crisis certainly contributed to the breakup of many families, for the most part the Depression was reinforcing family ties. The Lynds noted that "Middletown itself believes . . . that many families have drawn closer together and 'found' themselves in the depression."[46] Indeed, as Warren Susman noted, rather than evoke demands for radical departures, the Depression stirred deep-rooted conservative instincts among most Americans—in particular a search for the stable verities embodied in the family. The era's movies, magazines, and immensely popular radio soap operas all reaffirmed the traditional family values of the "real" America.[47]

Nowhere was this more visible than at the dinner table. The mother, the preparer of the food, was central to this enterprise.[48] "Home cooking" was placed on the uppermost of the domestic pedestals. The kitchen was "women's sacred domain," said a typical cookbook, the "forbidden realm [to males] of the culinary arts."[49] "The housekeeping job can be as scientific and engrossing as any office job," and marriage could be made into a career, said Eleanor Roosevelt.[50] An article in *Scribner's* entitled "The New Woman Goes Home" explained in meticulous detail how "the average woman who cooks, cans, preserves, bakes, and launders at home, for her own family, produces more wealth than she could produce by earning money."[51] Budgetary pressures also led to more entertaining at home.[52] There was a revival of the middle-class dinner party, a tradition that had been declining in the 1920s. Ads and articles in women's magazines made much of that recurring trauma of an age of diminished expense accounts: having to impress the husband's business associates with a home-cooked dinner. One of the most important things women could do, Mrs. Roosevelt often said, was to teach others of their sex to cook better meals.[53]

Perhaps as a result, much of the unease over the value of cooking that had crept into middle-class kitchens in the 1920s seemed to dissipate during the Depression, as the middle-class housewife once more saw her culinary role as an important and satisfying one.[54] But as she sniffed about the grocery store and contemplated what to do at her stove—making hundreds of decisions about food each week—where could she turn for guidance? The housewife's traditional bedrock of useful household information, her mother's and grandmother's recipes and advice, seemed outdated and irrelevant. Whether she was a middle-class woman, thoroughly apprised of the demands of the New and Newer Nutrition, whose home and marriage differed markedly from her mother's, or a working-class one who could afford a greater variety of foods than her mother ever could, or the child of immigrants wanting to cook the "American" way, she felt that she had to base her food choices on much different principles than those of twenty or thirty years earlier. By the 1930s, then, the mass media had replaced family wisdom as the major source of culinary advice for American housewives.

The circulation of women's magazines had climbed steadily in the 1920s, and, although the proportion of space they gave to household matters varied, all devoted considerable space to cooking. Before World War I, magazines such as *Ladies' Home Journal* and *Good Housekeeping* had cleverly cultivated an upper-middle-class image in order to appeal to a middle-class readership. But secondary education for women, which had been mainly for the middle and upper classes before about 1910, had become almost universal during the 1920s; by the end of the decade more females were graduating from high schools than males. This literate pool continued to increase during the 1930s as high school enrollments soared to new levels.[55] A new breed of magazine, which dropped the class appeal one notch, was able to tap this vast new market. Magazines such as the *Delineator* and *Woman's Home Companion* portrayed their readers as solidly middle-class to exploit this newer market lower on the class ladder. The era's greatest success story, *Better Homes and Gardens,* was founded in 1922 to appeal to "the average family," which the publisher defined as a family with an income over three thousand dollars—hardly a princely sum.[56]

By the mid-1930s the older women's magazines, faced with Depression-induced circulation declines, had dropped their higher-class pretensions and also adopted an essentially classless stance, the main thrust of which was on women's common identity as women. Class differences were almost never mentioned or even alluded to, particularly with regard to food and housekeeping. Their readers were conceived of as the most middling of the middle class, whose husbands might have "bosses" but never foremen, and whose most common crisis was the arrival of the boss for dinner.[57] "Hitler Threatens Europe—but Betty Haven's Boss Is Coming to Dinner, and *That's* What *Really* Counts," said a September 1939 ad for *The American Home* in a food industry magazine.[58]

For better or for worse, readers had often learned the basics of cooking in home economics classes.[59] There, teachers enthralled by the benefits of the industrial revolution in the kitchen taught that simplified processes and efficiency were the keys to culinary success and that canned and processed foods were invaluable tools. The women's media reinforced this by acting as cheerleaders for processed foods and integrating their copy with their advertising to push processors' products. *Woman's Home Companion* assured food processors that its features were "brilliantly edited to focus the attention of more than three million women on advertising pages."[60] *Good Housekeeping*'s "Seal of Approval" meant that its food writers had certified all the foods advertised in it to be of the highest quality. "Go read a book and grow up!" said the *Delineator* to "the fusspots who still think that to say 'She opened a couple of cans and called it a meal' . . . contains the least hint of opprobrium. . . . Canned foods have long since come into their own. They're teeming with vitamins and other scientific discoveries." (For women ready to "search out new and better combinations," it provided a recipe for "Peas Nana": a can of cream of mushroom soup mixed with a can of green peas, sprinkled with "American cheese," dotted with shortening, and baked.)[61] When a group of poor rural Appalachian women manifested extraordinary esteem for expensive processed foods, researchers concluded it must be connected to the fact that they were avid readers of *Woman's Home Companion*.[62]

By the 1930s most daily newspapers had "women's pages" providing an endless stream of recipes and kitchen hints.[63] More often than not, the page was put together by a harried, underpaid woman, isolated in a tiny ghetto on the outskirts of the city room, charged with coming up with an enormous amount of copy each week to fill the spaces that loomed between the ads. Food companies were more than happy to help out with recipes using their foods; many were run without even editing out their brand names. Enterprising home economists persuaded newspapers to sponsor "home institutes" and cooking schools, which drew audiences of up to twelve thousand to watch cooking demonstrations often presided over by the benign women's editor; the spectators were unaware that they were there to watch processors' products being promoted. Home Economics Service, a major organizer of these events, promised processors "a complete service that will SELL food products."[64]

The largest processors did not have to resort to such deception. They had their own traveling road shows, where home economists staged slick demonstrations of their products to hundreds of thousands of women at state fairs, held cooking and baking contests, gave out free samples, and distributed millions of recipes for their products. Demonstrations such as Lever Brothers' two-hour sessions on "How to Use Spry," whose attendees received a free fifty-page cookbook, were particularly popular in smaller communities.[65] The A&P grocery chain was one of the first to use radio to dispense advice on how to use processed foods in menu planning. In mid-1931 it began broadcasting suggested menus, asking listeners to write in for them. Demand was so great that distribution was turned over to local store managers, who were soon giving out four hundred thousand of the four-page menus each week.[66] Then, two of the firm's executives expanded the menu sheets into a cheap magazine selling for two cents at the checkout counters—and *Woman's Day* magazine, the first of its breed, was born.[67]

However, the most successful of the radio purveyors of food advice was not even a real person. "Betty Crocker" was conjured up in 1926 by Marjorie Husted, who worked in the advertising department of the Washburn-Crosby milling company, soon to become General Mills. Husted sensed that, while housewives could no longer turn to traditional family sources for advice on cooking, they still longed for the personal touch. Betty thus encouraged readers to write to her personally for advice on their culinary problems. The attractive, blonde, cartoon Betty was much younger-looking than the more matronly spokeswomen for competitors. She seemed like a friend and contemporary who could understand the modern housewife's problems in feeding her man, rather than an authoritarian aunt with hoary advice from the old days. In 1927 Husted took to the airwaves as Crocker, dispensing recipes and advice on "The Betty Crocker Show." In 1932 she fleshed out her image even more by interspersing the recipes with celebrity interviews portraying Cary Grant, Joan Crawford, Clark Gable, Helen

Hayes, Norma Shearer, and other stars as absolutely devoted to their home lives. By then, each year hundreds of thousands of letters addressed to Crocker were pouring in to the harried home economists at General Mills, which employed forty of them to respond to the deluge.[68] Pillsbury's tried to counter with "Mary Ellis Ames," whose warnings that poor cooking ruined marriages were aimed particularly at the young housewife. Her cartoons featured anxious young ladies like Sue, about to marry handsome Jim, being cautioned by her father that "the best insurance for a happy marriage is to know how to bake."[69]

The women's magazines were only one step behind in trying to turn their food editors, fabricated or real, into celebrities. *Woman's Home Companion* created "Carolyn Price," *eight* of whom were scattered across the nation giving weekly broadcasts on which—the magazine assured advertisers—they "push[ed] *Companion*-advertised products."[70] *Ladies Home Journal* tried to cash in on the fact that its star food editor, the omniscient Ann Batchelder, actually existed. In 1939 it announced that women it surveyed in grocery stores in five cities picked her as their "favorite authority on food."[71] But the survey must have been confined to authorities who actually existed. No one could match Crocker's renown. By then her immensely popular radio show was supplemented with a weekly food column in over four hundred newspapers.

The most striking fact about this outpouring of information about food is the dearth of material on economizing. Yes, there were "economical" or "budget" recipes, along with advice on how to cut corners, but in no greater proportion than in other, more prosperous, times. To a certain extent, of course, this reflects editors' and marketers' perceptions that those most in need of advice on low-cost cooking could not afford the magazines or products in the first place. However, it also reflected a perception that in hard times Americans wanted reassuring food, that they yearned for the stability and wholesomeness traditionally associated with mother's home cooking. Surrounded by economic insecurity, Americans seemed to aspire more than ever to an uncomplicated, straightforward cuisine—one that would reflect the nation's honest past and restore confidence in its abundant future.

Coming as it did at a time when many women aspired to a degree of thinness and sought to restrain their eating, the yearning for abundance contributed to some conflict between the worlds of the two sexes. Men were expected to be hearty eaters who liked no-frills preparation. "Give a Man Man's Food" was a popular theme.[72] What was this "man-pleasing" food? "Meat for the Males" summed up much of it.[73] This meant "rib-sticking" food, particularly beef, served plainly in large portions, normally accompanied by potatoes. Managers of half a dozen leading restaurant chains surveyed in 1934 agreed that businessmen were as "hearty" eaters as ever. They were rarely tempted by vegetables, ate meat even at lunch, and were partial to pastries and pies.[74] They seemed to have a particular loathing for the miracle vegetable of the early 1930s, spinach (and small wonder, since most of it was canned).[75]

Women, on the other hand, were expected to like "dainty" foods. These were normally prepared for women's luncheons and other functions where hearty-eating men were not present. The portions were to be small, and the food was to appear light. "The course of Empire has been changed by a woman's waistline," warned *Thoughts for Food*, a 1938 menu book. "You can lead some women to a whipped cream luncheon, but you cannot make them eat it." Women's food also reflected the persistence of the nineteenth-century double standard and the Victorian ideal of womanhood. It was dainty in taste as well as quantity, for to have lusty tastes in foods seemed to betray a weakness for other pleasures of the flesh as well. It was also expected to display a certain degree of complexity—the "frills" that men disdained or did not notice and women, with their higher aesthetic sense, appreciated. Thus *Thoughts for Food*'s luncheon menus for women featured seafood and chicken in various kinds of mild sauces, mainly variations of white sauce. The suggested stag dinner for men, on the other hand, was oysters Rockefeller, cheese soup, broiled steak, potatoes O'Brien, french-fried onion rings, and salad.[76]

One of the era's more pretentious cookbooks, the *Philadelphia Cook Book of Town and Country*, had two menus for August luncheons. The "Luncheon for the Ladies of the Garden Club" began with sherry and biscuits, and hors d'oeuvres of an unspecified but presumably dainty and visually appealing kind. The main course was squabs jardiniere (that is, cooked with vegetables), surrounded by an additional "Garland of Vegetables" and "Candied Rose Leaves and Violets." This was followed by a "Salad Mimosa" (lettuce hearts with French dressing and nasturtium seed pods garnished with egg white and nasturtium leaves and flowers) with cream cheese and Bar-le-Duc currant preserves on the side. Fresh figs, green almonds, and "Orange Flower Tea" added the final dainty note.

The "Country Luncheon Especially for a Man," on the other hand, began with cold beet soup and proceeded stolidly to beef and kidney pie, Stout, brown bread and fresh butter, new potatoes, and peas, followed by tomato salad with a tangy Roquefort dressing. Raspberries and cream and coffee topped off that hearty meal.[77] A tearoom manager summed up prevailing opinion well: "*She* may be satisfied with a tidbit—a dressy fruit salad, a dainty sandwich. Not He. He wants real food and plenty of it."[78] Crossing gender lines was not encouraged. Men who liked dainty, visually appealing food were regarded as effeminate. "Fancy compositions" in food, said *Harper's* editor Bernard De Voto firmly, were an obsession of the women's press—not fit for men.[79] Conversely, women who were too taken by the taste of food were not regarded as true women. The news that a newly organized New Orleans epicures' society barred women from membership in 1939 because "the consensus of opinion is that there is no such thing as a woman who is an epicure" provoked amusement, not outrage, in a women's cookery magazine.[80]

Routine home cooking, which had to please both sexes, therefore presented a bit of a problem, but the compromise between the two poles of daintiness and heartiness—between the pleasures of the soul and those of the body—almost invariably leaned mightily toward the latter. After all, since a wife's cooking was regarded as central to the maintenance of a happy marriage, it was natural that wives would cater to their husbands' tastes. Betty Crocker wrote that 60 percent of the letters she received were "from women who, *intuitively* noting in their husbands' poor dispositions at the table a desire for a change in the menu, write asking me for something new."[81] "This book is dedicated to the housewife's private inspiration, that man at the head of the table—to the Adam in your house, and mine," said a typical cookbook preface.[82] The media reinforced the idea that cooking caught and kept men. The *Delineator* called its recipe for chocolate cake "Bachelor Bait."[83] Husted-Crocker recalled receiving a letter that said, "I don't make your fudge cake because I like white cake, but my neighbor does. Is there any danger of her capturing my husband?"[84]

The 1930 edition of the *Better Homes and Gardens* cookbook, a mainstay of the 1930s, was properly straightforward about what pleased these "Adams." The number one rule in menu planning, it said, was "Serve simple meals." These were to be built around the traditional pillars: pot roast, pork chops and apple sauce, New England boiled dinner, creamed codfish, baked beans with brown bread, and desserts like apple pie and peach cobbler.[85] Crocker's suggestion for "When a Man Gets 'Off His Feed' " was canned deviled ham mixed with cream and baked in Bisquick shortcakes.[86] Many of the recipes in the 1933 edition of the *Good Housekeeping Cook Book* called for white sauce, the late-nineteenth-century mainstay upon which most middle-class men had been brought up, slathered on fish, vegetables, and even meat. There were recipes for three different kinds of the glutinous concoction (thin, medium, and thick) and for nine other sauces using it as a base, including tomato sauce, which was white sauce mixed with a can of condensed tomato soup and baking soda. "Curry Sauce" was two cups of white sauce with apples and a mere one and a half teaspoons of curry powder, barely enough to color it faint yellow. The old "scalloped" dishes and croquettes (also based on white sauce) retained their historic pride of place. The few "foreign" recipes seemed to be there mainly to emphasize the cookbook's recommendation that canned goods should be used wherever possible.[87] "Veal Goulash" contained no paprika but required one-half cup of sweet bottled chili sauce and an equal amount of grated "American" (i.e., processed) cheese. Beef goulash was also devoid of paprika but did call for sugar.[88]

The taste for sweetness in main courses—often remarked upon by foreign observers—is not surprising, for the Depression saw no letup in the steady growth of the much-vaunted American sweet tooth. Whether or not it was related to the low price of sugar or the expansion of the ice cream industry, the fact is that sugar consumption continued its historic rise unabated. This was reflected in home cooking—or at least in the books that told Americans how to do it. Of the 209 pages in the *Good Housekeeping Cook Book*, 96 were devoted to recipes for cakes, pies, cookies, fruits, candies, sweet breads and rolls, and sweet beverages. In addition, most of the salads were sweet, with canned fruit, bottled mayonnaise (which was sweet; there was no recipe for the real thing), and/or French dressing. One version of the latter was made of oil, vinegar, and sugar, while the other (particularly recommended for salads) contained sugar, ketchup, chili sauce, and "condiment sauce." Garlic, on the other hand, was treated only slightly less cautiously than arsenic.[89]

The newer cookbooks were also enthusiastic about processed foods. Canning was "the magic key which opens food treasure chests from all lands," gushed one. "Thanks to this progressive industry, every single one of us may enjoy foods which even the richest Croesus would have considered luxuries beyond attainment not too many years ago."[90] For those who strove for creativity and innovation in the kitchen, a standard recommendation was to combine foods from different cans. Even in her otherwise rather sophisticated cookbook, Mary Ellsworth wrote that, while canned soups by themselves were very good ("we are experts if we can do as well ourselves"), "we take canned soups, season, combine, supplement or garnish them, and then produce them with a great deal of pride as products of our own ingenuity." The recipes for "making two good soups into one better one" include adding a can of ABC minced clams to one of Hormel vegetable soup to make clam chowder. Another, for "Zuppa," was a can of Campbell's split pea soup mixed with one of Ancora green turtle soup; some sherry was added, and the bowl was topped with whipped cream.[91]

Of course, it is difficult to know to what extent printed recipes were actually translated into kitchen practice. Today, when middle-class kitchens often boast a whole library of specialized cookbooks, publishers believe that readers consider a cookbook a wise purchase if they consistently use but two or three of its recipes. However, we do know that the people like Husted-Crocker who dispensed the advice thought that the housewives of the 1930s seemed practically starved for it and were impressed by the power of their suggestions.

This vast outpouring of menus, recipes, and other cooking advice constituted another force homogenizing the American diet. Most recipes—like the food industry that originated so many of them and the media that carried them—were aimed at the mass market, conceived of as a middle-class one. Betty Crocker could not afford to devise Bisquick recipes especially for unmarried working mothers who could not get home to feed their children until 6:00 p.m. Nor did it pay for her to give recipes for the upper classes' dreaded "cook's day off." Regional differences could hardly be taken into account, except for the occasional warning about oven temperatures for high altitude baking. The recipes had to be prepared for use in vastly different climates, in urban, rural, and suburban settings, by people assumed to be of middling means.

An interesting result of this was a decided shift—to the Midwest—in what was considered the center of gravity of American cookery. This represented quite a turnabout, for hitherto the name "New England" had denoted the mainstream American tradition, evoking images of cooking that traced back directly to the revered "Pilgrim Fathers," the fountainhead of all that was best about America. To a certain extent, the growing reverence for the Midwest was simply the response of mass marketers to the fact that the demographic center of the country was now firmly planted there. It may also have been linked to the changing image of New England: During the 1920s the previously praised Pilgrims came to be derided as pleasure-hating Puritans. The area's pastoral, seafaring, and Protestant image was tainted by the realization that it was heavily industrialized and distressingly Catholic. Indeed, in the 1930s the antebellum South provided a much more congenial setting for romantic pastoral dreams than Massachusetts, Rhode Island, or Connecticut.

Meanwhile, the Midwest—with its still-fresh tradition of hardy family farmers carving out a new frontier—had become the repository of the Protestant virtues of hard work, devotion to family, and plain living. Its frontier was now regarded as the wellspring of democracy, and its small towns were thought of as places where people named Smith, Jones, or something else easily pronounceable lived in ample frame houses on tree-lined streets—close-knit communities where everyone knew and could rely on each other. Hollywood movies were particularly effective in propagating these images, even though (or perhaps because) they were produced by people with names like Cohn, Lasky, and Goldfish (who became Goldwyn), and the most brilliant director of these paeans to Middle America was an Italian-American named Capra. But the cooking of Andy Hardy's family also became the norm because the head offices of many of the processors dispensing advice and recipes were in the Midwest. Their home economists thus worked out of places like Chicago, St. Louis, Minneapolis, Dubuque, and Madison. They themselves were usually from that region, for the large, well-funded home economics programs in midwestern land-grant universities like Illinois and Wisconsin produced women with home economics degrees on a far larger scale than schools in other regions. Women's magazines responded to the shifting center of gravity. *Redbook* magazine's circulation shot up after it deliberately shifted its focus to "the little ladies in Kokomo."[92] The founder of one of the era's most successful women's magazines, *Better Homes and Gardens*, edited in Des Moines, Iowa, would not hire an easterner as editor. The midwestern bias was so pronounced that the rare easterners who did make it onto the staff usually did not last long.[93]

Midwestern cooking could also claim to being the national cuisine because it represented a good compromise among the regional American styles: a hybrid that reflected the culture of the area itself. Its white population had originally migrated from the three main culinary areas—New England, the Mid-Atlantic states (Pennsylvania in particular,) and the Upper south.[94] There was also a hefty infusion of Germans, who left a greater imprint on nineteenth-century American food than any other European immigrant group. It used enough pork and corn meal to accomodate the South, more than enough beef, grain, and dairy products for the Northeast and Mid-Atlantic regions, and the fruits and vegetables of the Far West. It was weak on fresh seafood but did use quite a lot of the canned variety. Perhaps most important, though—as three contemporary observers noted—the Midwest was "the stronghold of conservative American eating." Its cooking had "risen from a rich and indigenous earth" and used "wholesome, flesh-building produce, tender beef, plentiful pork, butter, cream and eggs and milk as a straightforward working fare for a hard-working citizenry. It was food that pleased the palate and stuck to the ribs. . . . Family dishes seldom go out of favor in the Midwest."[95]

Here was the ideal cuisine for a nation searching to reaffirm its roots in the fertile soil and solid families of the American Eden. Throughout the nation, then, those who developed recipes for food processors, edited women's magazines, and published cookbooks assumed a decidedly midwestern posture. The *New England Kitchen Magazine* had long since changed its name to *American Cookery;* the *Boston Cooking-School Cook Book* had metamorphozed into "Fanny Farmer." Now "New England cooking" denoted merely another regional cuisine—one that, like the others, was looking increasingly like an endangered species.

CHAPTER 3

From Burgoo to Howard Johnson's: Eating Out in Depression America

Even before 1930 food lovers had bemoaned the demise of regional American cooking.[1] Yet the forces of nationalization had not been all-conquering; there was much in their path that they had not yet homogenized. Southern cooking in particular retained much of its distinctiveness. Southerners consumed much more corn meal, lard, pork, and sweet potatoes than people in other regions and ate less beef, butter, milk, and potatoes.[2] They had a decided preference for hot raised biscuits and breads and ate only half as much bakery white bread as people in the rest of the country.[3] They were also more partial to the products of the frying pan than other Americans and tended to use a heavier hand with piquant spices and sauces. The groundswell of northern reverence for southern folkways that culminated in the *Gone with the Wind* phenomenon stoked northern appreciation of southern foodways: the Aunt Jemima syndrome. Northern-edited women's magazines and newspapers were full of "authentic" southern recipes—usually "old family" ones from "the Deep South"—and the upholstery in countless northern homes became infused with the smell of southern fried chicken.[4]

One of the things that opened many Americans' eyes to the persistence of culinary regionalism was the automobile, along with the new ribbons of asphalt highway that allowed them to take long car trips across vast stretches of the country.[5] As one motored westward from the East and crossed into Oklahoma, one left the roadside stands dispensing hot dogs and hamburgers and entered the Barbecue Belt. From there to California stands selling barbecued beef, pork, and ham lined the highways. Barbecue here meant meat cooked slowly and basted continually until it was tender and served drenched in a piquant sauce; in the rest of the country it simply meant grilling on an open fire. Asked about the roadside food after a motoring trip through the Southwest, a bronzed New Yorker replied, "Barbecue! Everything was barbecue!"[6] The Kansas newspaper editor William Allen White remarked that, although much of what was served hardly deserved the name, it was its love of barbecue that separated the Southwest from the rest of the United States.[7]

But regional cuisine was preserved best, not in roadside stands or in daily meals at home, but in the communal eating festivals that were still a hallmark of small-town America. More than ever during the insecure days of the Depression Americans sought—in their church dinners, family and school reunions, political barbecues, camp meetings, civic holiday picnics, and ethnic holiday meals—to reinforce the bonds of family and community by preserving their rich culinary traditions. A real barbecue, declared one Texan, was "the noblest demonstration of American neighborhood eating."[8] In 1939, when the Works Progress Administration's Federal Writers' Project began to collect descriptions of these events from all over the country, it justified the project with the notion that "few nations in the world are as devoted as Americans to group eating."[9]

These occasions when Americans ate together exemplified the extent to which culinary regionalism persisted. Southern gatherings usually featured pork, both smoked and fresh, and chicken, augmented by seafood near the coasts. Midwestern rural groups favored freshly butchered pork, with chicken the favorite for Sunday dinners, while city dwellers favored beef. So did people in the Southwest, where beef was the most common meat for barbecues. The communal food traditions of the Northeast were so varied as to defy generalization.[10] Local variations on regional themes made for even more variety. Possum, squirrel, racoon, and muskrat were favorites in some parts of the Southeast; the latter even had its devotees in northern Michigan.

In Kentucky, it was said, no political campaign could be launched, no thoroughbred horse sale could be conducted, without a "burgoo" feast. James T. Looney, "the Burgoo King," could prepare these for upwards of one thousand people. Burgoo, the centerpiece, consisted of a giant pot-au-feu (his term; burgoo's origins were reputedly French) of soup meat, chicken, squirrel, canned corn, potatoes, onions, cabbage, carrots, and "Burgoo seasoning," which seems to have been heavy on the Worcestershire and hot pepper sauce. In 1931, after Looney presided over forty stoves serving nine hundred gallons of it at a charity horse race staged by the horse breeder "Colonel" E. R. Bradley, the burgoo-smitten breeder renamed his finest equine prospect after him. The next year, horse fanciers across much of the nation puzzled at the name of the winner of both the Kentucky Derby and the Preakness: "Burgoo King."[11]

John Saunders, a counterpart of Looney in the hills of Virginia and North Carolina, toured American Legion and church fetes cooking up enormous cauldrons of his legendary "Sergeant Saunders' Brunswick Stew." Although not dissimilar in conception from burgoo, it contained only squirrel meat. ("Brunswick stew ain't fit for hound dogs 'less you got squirrel meat," remarked a local connoisseur.)[12] When blacks wanted to raise money in North Carolina, they would hold "chittlin struts." Guests would gather in a home while the hostess fried up "messes" of chitterlings. At least one host walked around singing the menu:

> Good fried hot chittlins, crisp and brown,
> Ripe hard cider to wash 'em down,
> Cold slaw, cold pickle, sweet 'tater pie,
> And hot corn pone to slap your eye.[13]

The oyster roast was particularly popular in Virginia, the Carolinas, Georgia, and Alabama. In Virginia it was mainly a winter festivity; the oysters were shoveled onto sheet metal laid on top of outdoor brick barbecues stuffed with logs.[14]

The liveliness of the tastes at these fests—the ketchup, horseradish, Worcestershire, and hot pepper sauce that were poured on almost everything in sight—stood in marked contrast to the bland tastes that characterized the food featured in popular cookbooks, cookery columns, and advertisements in the national media. South Carolinians had a soft spot for Pee Dee suppers, where they would try to digest Pee Dee, a spicy fish dish certain never to make the pages of the *Good Housekeeping Cook Book.*[15] Onions were cooked in salt pork fat and stuffed into the fish, and the fish was slathered in more fried onions, generous helpings of ketchup, Worcestershire sauce, Tabasco sauce, or cayenne, and cooked in huge iron cauldrons over hardwood fires. The dish was supposed to be so hot that you could "hiccough and cry at the same time." It was said that pine bark could be substituted for the fish and no one would know the difference. The traditional accompaniment to the dish, "Pee-Dee-type corndoggers," which were cakes made of cornbread batter stuffed with pieces of raw onion and fried in boiling fat, hardly reflected national concerns over dieting.[16] Nor did regional cuisines echo the national media's injunctions to cook simply. A standard New Orleans family recipe for gumbo called for crabs, shrimp, oysters, ham or veal, okra, green pepper, celery, filé, thyme, bay leaf, salt, black pepper, and cayenne pepper.[17]

Even relatively rootless Californians staged communal eating events, although one of the largest, predictably, was for people born outside the state. Twice a year fifty to seventy-five thousand Iowans would gather in a park in Los Angeles and on the oceanside at Long Beach for monster picnics in which they would display their skill at preparing the chicken, ham, beef, potato salad, pickles, breads, cakes, and pies that had been their picnic fare "back home." Barbecue remained a symbol of California's historic attachment to the Southwest, although purists often derided the local version as little more than small chunks of meat served in a greasy sauce. The annual Sheriff's Barbecue, whose profits went to feed those on relief, was reputed to turn out one of the better versions; at least the sixty-thousand-odd people who showed up for it each year seem to have thought so. Using a Missouri recipe that began "Take 24,000 pounds of best Kansas City corn-fed steer meat," the volunteers slowly drenched it in gallons of barbecue sauce while it sizzled over hickory and oak coals. Meanwhile, twenty-four hundred pounds of Mexican red frijoles simmered in onions, garlic, and chili powder.[19]

North of California salmon barbecues and fish fries were popular. At one of the former, the ten thousand people who gathered in Anacortes, Washington, for the annual Mariners' Pageant tucked back four thousand pounds of freshly caught salmon that had been rushed from Puget Sound and barbecued over a huge alder wood fire.[19] Fish fries were popular in Montana, where they were sponsored by sportsmen's clubs. Hundreds of people, attracted only by word of mouth, would seem to appear out of nowhere at the ordained site in the sparsely settled wilderness, arriving in automobiles, buggies, or wagons, on horseback, and even on foot. There they would dig into large platters of trout fried in bacon fat or barbecued over open fires, complemented by huge pots of potato salad, pickles, and sliced onions and platters of sliced white bread.[20] In the small communities that dotted the levees of the lower Mississippi, a much different kind of fish fry was common. On Saturday nights in particular, men would gather in the houses of women famed up and down the levee for their fish fries to gamble and eat piles of fried catfish washed down with bootlegged liquor.[21] Meanwhile, up in Illinois and Iowa, harvest times, family reunions, quilting bees, and church socials would bring the kind of gatherings more usually associated with America's heartland: potluck dinners in which farmers' wives would each bring large portions of hot food to contribute to enormous spreads of fried chicken, stews of all kinds, dumplings, baked beans, vegetables, pickles, cakes, and pies.[22]

Communal eating events also provided ethnic groups with the chance to join together in spirited rearguard battles against the relentless Americanization of their cultures and diets. Many of those held in the traditional centers of the foreign born—such as the orgy of outdoor eating that accompanied the Italian Feast of San Gennaro in New York City's Greenwich Village—became quite well known. But they also occurred in more unlikely parts of the country, where they reflected the ethnic mosaic that was recruited to dig the nation's mines and exploit its farmland, forests, and fisheries. Each year three thousand Greeks would crowd into a church in Spokane, Washington, for three successive nights of seafood and lamb dinners to celebrate a major Greek Orthodox holiday. In Minnesota and Wisconsin, Norwegian *lutesfisk* (lye fish) dinners—featuring cod served with potato crepes—became so popular as church fund-raisers that a group of Norwegians in Dane County, Wisconsin, formed a Lutesfisk Protective Association to guard against the non-Norwegians, particularly the Germans and Irish, who, they declared, were "invading the sacred *lutesfisk* domains."[23] Norwegians in Cheyenne, Wyoming, evinced no such concerns every May 17 when they commemorated their homelands' Independence Day with a feast of *lutesfisk*, *pastejar*, goat cheese, and *kafflebrod*. A month or so later, French-Canadians in (where else?) Frenchtown, Montana, would congregate around roast turkeys stuffed with mashed potato, ground pork, and sage to celebrate St. Jean-Baptiste Day.[24] Basque shepherds in Idaho, Nevada, Montana, and Wyoming gathered at December lamb auctions to wolf down chorizos and garbanzos, and five hundred Serbs would serve suckling pig to streams of visitors in their houses in the remote mining town of Bisbee, Arizona, not far from the Mexican border.[25]

Of course, not all communal eating events were culinary delights. A writer in Arkansas reported that fraternal lodge banquets in small towns often consisted of a ham sandwich, a dill pickle, and a cup of coffee.[26]

But communal eating could not remain the exclusive preserve of religious, ethnic, and voluntary organizations. Inevitably, it was affected by the new world of commercial food marketing. Some of the most impressive outpourings of people came to events sponsored not by organizations to which they had any emotional or other ties but by food producers. At these events the American penchant for being impressed by quantity more than quality was often taken to ridiculous extremes. Take the state of Washington: Each year thousands would gather in the town of Wenatchee to commemorate National Apple Week by eating the world's largest apple pie. Ten feet in diameter, made from one ton of apples, 100 pounds of butter, 160 pounds of sugar, and 100 pounds of flour, the twenty-three-hundred-pound, ten-foot-wide monster had to be dragged out of its specially constructed oven by a tractor. At the annual strawberry festival in Burlington, pitchforks, spades, and hoes were used to manhandle two hundred pounds of ice cream and a thousand pounds of strawberries into what was billed as the world's largest strawberry sundae. Meanwhile, in Bellevue each year townsfolk struggled to top their previous year's strawberry shortcake record. In 1941 the residents of that village on the shores of Lake Washington managed to serve fifteen thousand people a piece of the shortcake and a cup of coffee—using fifteen thousand pounds of strawberries in the process![27] Although nearby Buckley served thousands of portions of peach shortcake at its Peach-a-Reno Day, it could not quite match these numbers. Perhaps the least appetizing of the promotions was the world's largest clam fritter, cooked each year in Long Beach. A truckload of clams was mixed with an enormous amount of flour and fried in oil to make what was likely also the world's *greasiest* fritter.[28]

Other foods and other states were not to be left behind. Over half of the communities in Colorado were estimated to have special events to promote their food products, including Apple Pie Day, Pickle Day, Potato Day, and Melon Day, which drew to Rocky Ford a crowd large enough to chomp its way through twenty-five thousand watermelons.[29] Petaluma, California, had Egg Day, on which in October 1933 fifteen cooks used a thousand eggs to make an omelet in a fifteen-foot frying pan.[30] Creameries sponsored huge picnics, where thousands of people would bring their own lunches of fried chicken and baked beans and line up for free ice cream and milk. Cheesemakers in Monroe County, Wisconsin, managed to attract thirty thousand people to the tiny county seat each year to eat free cheese sandwiches (and buy good Wisconsin beer) at their annual cheese festival.[31] But it was left to Longview, Washington, searching for something different to mark its annual smelt festival, to stage perhaps the most bizarre of these events. While three thousand people gathered around a ten-foot frying pan set over an open fire on the banks of the Cowlitz River in anticipation of the fried smelt breakfast, two girls with bacon rind lashed to their feet skated around the sizzling pan to grease it. Then hundreds of pounds of the fish were dumped into the pan, and flour, salt, and pepper were stirred in with long rakes. Just as the dish turned into a revolting brown mess and first-time visitors began to contemplate making a quick exit, the whole lot was chucked into the river and platters of fresh smelt fried in a crispy corn meal batter were brought down from a nearby hotel for their delectation.[32]

The contrast between the celebration of food at these exuberant public spectacles and the general mediocrity of the American restaurant-going experience is rather striking. Before the 1920s a number of cities had boasted high-quality restaurants, mainly French in inspiration, as well an array of excellent seafood and chop houses. By 1930 the ranks of the French restaurants had been decimated, and many of the others had either passed away or sunk into mediocrity. Moreover, a general prejudice against French cooking seemed to have reemerged, alongside widespread ignorance of what it was. In a 1929 experiment, the Fred Harvey System, which ran the dining cars for a number of railroads, discovered that even the well-heeled patrons of the Santa Fe Railroad's crack California Limited would order much more steak when it was called "Small Tenderloin, Mushrooms" than when it was labeled "Filet Mignon, Champignons."[33]

A number of factors had combined to virtually destroy fine dining. These included dieting, the food and health mania, the cult of simplicity, and Prohibition, which undercut the financial basis of fine restaurants.[34] Gourmets tended to blame the decline solely on Prohibition, but after Prohibition was repealed in 1933, although public drinking rose quickly from the ashes, fine dining did not. So-called heavy dining—men eating large portions of food accompanied by alcohol—had survived to a certain extent in speakeasies and men's clubs, but pre-1920 quality did not.[35] Yes, Boston still had Locke-Ober's (although the Locke part, which had closed during Prohibition, remained closed), Philadelphia had Bookbinder's, and Chicago had two or three good traditional restaurants, but only in New York, New Orleans, and San Francisco did anything like a tradition of fine dining survive. Even there, it was sustained by only a small minority of those able to afford it, for interest in fine dining seemed to have virtually disappeared, even from the upper class. When President Roosevelt, scion of one of the nation's oldest and most truly upper-class families, dined alone with his family, his favorite dishes were creamed chipped beef or corned beef hash with poached eggs, with bread pudding for dessert.[36] The era's embodiment of high culture, Milton Cross, the radio host of the Saturday afternoon broadcasts of the Metropolitan Opera, made much of his down-to-earth tastes in food. He let it be known that he liked to dine on thick broiled lamb chops or steak, two vegetables cooked in water, and fruit or Jell-O for dessert. On special occasions his wife would serve spaghetti ("taught by the Italian singers at the Met") and one of his two favorites: pineapple upside-down cake or apple pie with hard sauce.[37]

In 1936 *Fortune* magazine reckoned that even in the New York metropolitan area there was only a "tiny band of gourmets."[38] The small number of expensive restaurants, such as the Colony and Voisin, that did serve sophisticated food could not survive on customers interested in fine food but had to attract celebrities and their wealthy camp followers, "café society." Their modest gastronomical demands were exemplified by the food of the city's snobbiest restaurant, the Stork Club, whose menu featured such pedestrian dishes as minute steak, ham steak, and lamb chops—all broiled. Its famed after-theater supper offered chicken à la king, Welsh rarebit, and chicken sandwiches. Its most highly regarded dish was the cheeseburger.[39] At its West Coast counterpart, Chasen's in Hollywood, movie moguls made deals over steak and chili.[40]

Perhaps the sorry state of American gastronomy was best typified by Duncan Hines, the nation's best-known restaurant critic, whose *Adventures in Good Eating* and hotel guide had sold 450,000 copies by the end of the decade. A traveling salesman for a Chicago printing concern, Hines's closest brush with the restaurant industry had been the few months in 1905 when he ran the commissary for Americans at a mine in northern Mexico. (Even that experience seems to have affected him hardly at all, for his ignorance of Mexican food was profound.) He had begun taking notes on restaurants he liked during sales trips and weekend drives with his wife. In 1935 he compiled a list of 167 of them and sent it to friends and business associates instead of Christmas cards. He was persuaded to expand and publish it, and the rest, as they say, is history.[41]

Hines struck just the right chord for middle-class America. At the top of his list of priorities was cleanliness. He would not eat in a restaurant until he had inspected its kitchen first, even if this meant sneaking around the back to peer through open doors and windows. Nor would he eat in one that had a strange smell, tablecloths and place mats that were not fresh, dishes and silverware that did not "sparkle," or staff who were not "neat." He was not impressed by fancy decor ("Candlesticks and decorations don't make a restaurant"), suspecting, it seems, that elegant trappings could be used to camouflage dirt and filth. The same concern about camouflage informed his judgments on food, for—like most Americans—he suspected that restaurants used sauces to mask inferior ingredients. "Much of our cooking falls down through the fact that too many cooks are still trying to discover something that will take the place of good butter, fresh eggs, rich milk, and a loving touch," he said. Like most of his compatriots, he thought himself particularly rigorous in his demands for a good cup of coffee; like them, he thought that percolators produced the best. The adventurousness he urged on his readers meant trying local specialities: to eat clams, lobster, and chowder in New England, oysters and soft-shelled crabs in Maryland, okra and shrimp in South Carolina, freshwater fish in the Great Lakes area, and "Spanish dishes" in Texas and California. But as this suggests, his recommended places to dine were almost uniformly "American," Trader Vic's in Oakland being about as "foreign" as the adventure would get.[42] And why not? Although he had never been outside of North America, he remained convinced that American cooking was the world's finest.[43]

Because of his cleanliness fetish, Hines was quite a fan of chain restaurants such as Stouffer's and the Fred Harvey System.[44] Many had been started at the turn of the century to assuage the kind of consumer concern over "pure" food that troubled Hines. Their brightly lit, white tile interiors, glistening tabletops, and simple food had made them popular with the growing number of middle- and lower-middle-class people looking for clean, reasonably priced, respectable places to dine. Because their profits did not depend on alcohol sales, they had survived and expanded during the 1920s, although competition from tearooms, cafeterias, coffee shops, and drugstore lunch counters, most of which offered lighter meals and quicker service, did force them to supplement their basic menus of steak, chops, beef pot roasts, pork roasts, and stews with expanded sandwich and salad selections.[45]

But the Depression hit the chains hard.[46] Schrafft's, the main upscale chain, whose wood paneling, chandeliers, and other elegant touches had helped it carve out a market among better-off women, managed to survive on its high profit margins and candy sales. However, the more down-market chains had to resort to vicious price-cutting. In 1931 John R. Thompson introduced a five-cent sandwich. Childs' New York City restaurants began an "eat all you want for sixty cents" experiment, at a time when a meal in a good restaurant still cost about $1.25. Boos Brothers, their San Francisco subsidiary, offered all one could eat for fifty cents.[47] Yet within a year the novelty had worn off, and falling incomes and prices had made even sixty cents seem like not much of a bargain.[48] Other ploys were hardly more successful. By the end of the decade the chains were looking like the past, rather than the future, of the industry. They did only 14 percent of its total volume and were its least profitable segment.[49]

Among other problems, chains had great difficulty finding restaurant managers able to oversee the thousand and one cost-cutting details necessary to make their restaurants as profitable as independently owned ones. The franchised chain, which would help solve this problem, was still in its infancy. However, it was taking root in the most promising sector of the industry, the roadside restaurant. A Massachusetts businessman named Howard Johnson was one of the first to exploit this form of expansion. In 1925, forced out of the wholesale cigar business by the switch to cigarettes, he bought a soda fountain in Wollaston, Massachusetts, just south of Boston, and began churning out ice cream made from his mother's recipe on a hand-cranked machine. Its unusually high butterfat content gave it a creaminess that made it an overnight success, and he soon opened a number of stands up and down the coast. In 1935, seeing that even in the depths of the Depression Americans were taking to the road in increasing numbers, he opened his first roadside restaurant in Orleans, on Cape Cod. Unable to fund this expansion on his own, he sold the right to run it to what he called an "agent"—a franchisee.[50]

Many others had sensed that there was money to be made on America's burgeoning roadsides, but most had little capital to invest. The result was a highway system lined with dreary roadside stands. Their "glutinous pies, mongrel hot-dogs, and mass production hamburgers have ruined many a tour," said *Business Week.*[51] Bernard De Voto wrote, after a cross-country motor trip in which he was forced to sample some of the fare at the "Maw's Filling Stations" and other stands, "There could be no stronger evidence of the vigor of Americans than the fact that by the hundreds of thousands they eat this garbage and survive."[52] Johnson now led the way in transforming the industry into one that provided the kind of fast but reassuring dining experience families on the road wanted.

To gain their attention, his restaurants' roofs were sheathed in porcelain tile of a special orange color, which scientists assured him would best reflect light and would therefore be visible from far enough away to enable speeding cars to slow down and enter. For that reason, they were also set on long stretches of flat road. "Only the blind," it was said, "can fail to notice them."[53] Because American motorists were usually in a hurry, he streamlined kitchen operations to emphasize speed and prepared much of the restaurants' food in a central commissary, leaving only heating—usually frying—to be done at the restaurant. He also worked at building a reputation for safe, quality food, even refusing to call his frankfurters "hot dogs" or "franks" to avoid associating them with the lesser versions sold by the competition. He tapped the era's yearning for roots, tradition, and community by making his restaurants New England "colonial" style with white clapboard exteriors, homey lamps glowing in fake dormer windows, and roofs topped by prim cupolas and weathervanes. Inside, knotty pine paneling and ruffled curtains filled out a design one architectural historian has called "a beacon of traditional values."[54]

The reassuring feeling was buttressed with "home-style" food, prepared and served only by females. Conservatism was the watchword. The only venturesome dish on the menu was fried clams, and this was so only for non–New Englanders. By 1940 the East Coast chain had grown to 125 restaurants—every one of them profitable—two-thirds of which were franchised. For a fifteen-hundred-dollar fee the company picked locations for franchisees, supervised construction (which had to conform to its architectural guidelines), trained staff, and checked to see that the places were run according to the provisions of the contract with Johnson. The most important of these was that almost all of the food was to be bought from the Johnson company, and it was there that the company made the bulk of its profit.[55]

By 1937 Johnson was by no means alone in tapping the mobile middle-class market. Roadside restaurants of literally all shapes and sizes were springing up in and around the nation's larger cities. Many strove for the mark of distinctiveness, the visual knockout blow that would cause passing traffic to slam on the brakes and wheel into their driveways. Southern California, whose climate and geography seemed ready-made for automobiles and large parking lots, was the unchallenged leader in this. Many drive-ins were round (to minimize the walk from the parking spaces), wrapped in art deco–style shiny metal and glass exteriors, and topped with glittering towers. Others sought to look like anything *but* a restaurant. Hollywood's famed Brown Derby restaurant was a derby-shaped brown shell in a parking lot on Wilshire Boulevard. The Chili Bowl chain featured bowl-shaped white stucco buildings. The Tam O'Shanter Inn in Glendale was wrapped in a huge, green, pointed tam-shaped roof.[56] Many were far more luxurious than anything Howard Johnson would build. Carl's Viewpark, a huge red brick structure overlooking Los Angeles's Leimart Park, was built to resemble an oversized "Old South" mansion. Billed as "the world's finest drive-in restaurant," it had both a drive-in section with curb service and a dining room. The latter featured a "Georgian-colonial" fireplace, valanced green drapes, glistening chandeliers, and murals depicting "colonial" scenes. Guests could eat there or on the great white veranda amidst the massive pillars of the "Old South" colonnade.[57]

Others took the homey theme to great lengths. The parking lot at Charles Mickleberry's log cabin restaurant on Chicago's West Side was packed with autos from the day he opened in 1933. Customers loved the decor: the stuffed owls, hawks, fish, and pheasants, the dried corn, tobacco leaves, oxen yokes, muskets, pistols, American and Confederate flags, moose heads, and old prints that festooned the walls. There were red checkered tablecloths, cane chairs, and huge stone fireplaces. The food, of course, was utterly familiar: Mickleberry's "Famous Southern Dinners" featured steak, lobster, scallops, fried chicken, and "Our Famous Italian Spaghetti."[58]

Having "Italian Spaghetti" on a "Southern" menu was by no means unusual. (Bishop's Drive-In in Tulsa had two specialties: chicken "Mammy Style," cooked by two black "Mammies," and "Spaghetti with Italian Meat Sauce.")[59] Because customers abhorred new or unusual food, gimmicks in the way of decor or service seemed the only recourse for desperate restaurateurs. The result was often a culinary and thematic mishmash. When San Francisco's Hof Brau restaurant decided in 1931 to turn part of its space into a fountain lunch section, it chose a "Mexican or Spanish" theme for the decor, putting red tile behind the bar and strewing sombreros and Mexican pottery about. The menu, however, betrayed nary a hint of either German or Mexican influences, offering ham salad, tuna salad, meat loaf, and melted cheese sandwiches.[60] In 1938 Wesson Oil advertised that Chef Otto Gensch of Casa Mañana (location unidentified) used its product to fry his specialty, "New England Codfish Cakes a L'Aurore."[61] The Mad Hatter restaurant in Hollywood served the usual American food, but served it backwards, starting with dessert and ending with soup.[62]

For most Americans, however, eating out was not a particularly zany experience. Most small towns relied on a drugstore lunch counter that served sandwiches (some of them "triple decker") and whipped cream concoctions. After his long cross-country automobile trip, Bernard De Voto pronounced these the best places for lunch in small towns, far better than any local restaurants. "Throughout rural America," he wrote, "the craft of making sandwiches—and it is a skilled craft—is confined almost exclusively to soda fountains."[63] Diners, which slathered brown gravy over almost everything, were also fixtures in smaller communities, and these two kinds of down-scale restaurant expanded healthily during the Depression, attracting customers with their straightforward, economical food and familiar, friendly service.[64] Self-deprecating signs such as "Don't Growl About Our Coffee, You'll Be Old and Weak Someday Yourself" signaled that the quality of the food often played second fiddle to economy and a genial, relaxed atmosphere.[65] Larger cities had many of these diners and fountain lunches plus restaurants and tea rooms specializing in fifty-cent table d'hôte lunches. The largest cities could also support chains of self-service cafeterias, whose pinnacle lay in New York City, where the glittering chain of coin-gobbling Automats alone attracted a quarter of a million customers a day.[66] But perhaps the fact that one of the nation's most popular restaurants could be named Ptomaine Tommy's (a Los Angeles landmark, of course) signals the low expectations and general lack of seriousness with which most Americans approached their restaurants.[67]

There were also restaurants aimed specifically at working men—not just the kind of down-and-out place with a dreary "EATS" sign portrayed in the social realist pictures of the time, but cheerier ones as well. The White Tower chain, which specialized in five-cent hamburgers ("Buy a Bagful" was the motto), coffee, and doughnuts, expanded steadily during the decade by carefully cultivating a working-class clientele. Early on, its founders, the Saxe brothers, had noted that most working men no longer walked to work but took subways and trolleys, and often wanted a place to stop for a bite to eat on the way to or from work. There were plenty of mom-and-pop luncheonettes around, but their quality was spotty, and their standards of cleanliness often did not keep up with their clientele's rising expectations in this regard. In particular, like Duncan Hines, many workers harbored suspicions regarding the origins of their ground meat. The gleaming White Towers therefore sparkled both inside and out with white porcelain tile, not just to be eye-catching but also to reassure patrons of the absolute cleanliness of the establishments and the purity of their food.

The Saxe brothers had started the chain in Milwaukee in 1926, after studying the architecture and methods of the successful Wichita-based White Castle chain. However, unlike White Castle, Toddle House, and others that sought a varied clientele, White Tower aimed quite specifically at the working man. It opened in Detroit in 1929 and then spread into the industrial heartland of the Midwest and East. By the mid-1930s there were White

Towers open round the clock at the trolley stops opposite many of the nation's largest factories; more than half of the stations on Philadelphia's Broad Street subway line had White Towers at their exits.[68] The kind of reassurance they represented was exemplified in a brochure their model, White Castle, distributed to customers in 1932. "When you sit at a White Castle," it said, "remember that you are one of several thousands; you are sitting on the same kind of stool; you are being served on the same kind of counter; the coffee you drink is made in accordance with a certain formula; the hamburger you eat is prepared in exactly the same way over a gas flame of the same intensity; the cups you drink from are identical with thousands of cups that thousands of other people are using at the same moment; the same standard of cleanliness protects your food."[69]

Some cities—New York in particular—did support some ethnic restaurants that appealed to nonethnics. The large majority were Italo-American ones based in the pasta-and-tomato-sauce syndrome that was already creeping into American homes. Many of these restaurants had originated as boardinghouses. During the 1920s, when immigration restriction cut off their steady supply of temporary boarders, some boardinghouse owners had converted their establishments into informal restaurants. Although most catered to *paesani* in their neighborhood, some developed a clientele of other Italian-Americans and, soon, "Americans." The latter were frequently attracted by booze, for Italian restauranteurs—who regarded Prohibition as some kind of a sick joke—continued to serve their homemade wine, beer, and, for the strong of heart, fiery *grappa*. By the time Prohibition ended in 1933, the cozy little restaurant in Little Italy with checkerboard tablecloths, candles in wine bottles, and reasonably priced food and drink was already on its way to becoming a cliché. The fact that Greenwich Village, New York City's bohemian quarter, was also an Italian district helped burnish the image. Benign restaurant owners like "Mamma Bertolotti" were reputed to have sustained many a talented but struggling artist and writer through his or her apprenticeship with filling food at reasonable prices.[70] The most affordable of the foods was also the one most acceptable to non-Italians: pasta—usually spaghetti—and tomato sauce, sometimes with ground meat, meatballs, or, for the more adventurous, clams in the sauce.

These unprepossessing restaurants soon spilled out of the ghettos. As early as the mid-1920s, a number of cities had "spaghetti houses" in their downtown business districts. Often, to allay suspicions regarding their commitment to American standards of sanitation, the pasta was boiled in huge cauldrons in the window; for thirty-five cents patrons would be presented with a large plate of spaghetti and a choice of tomato-mushroom sauce, meat sauce, or melted butter. Bread and butter and coffee were the only other foods sold.[71] By 1930 menus had expanded and full-service Italian restaurants had proliferated. There were 409 of them in Manhattan alone and another 204 in Brooklyn.[72] The West Forties in New York City was a veritable jungle of "Roman" or "Venetian" or "Italian" "Gardens" or "Grottos," catering mainly to the "American" trade.[73] "Everybody," said a 1935 article in *Good Housekeeping*, "seems to know a Tony's or Joe's where one can get 'the best spaghetti in town.' "[74]

Otherwise, America's restaurants reflected the same homogenizing forces as its kitchens. Greek-Americans, for example, made significant inroads into the restaurant industry in the 1930s by recognizing this. They rarely opened Greek restaurants but instead built their success on coffee shops, steak houses, and other "American"-style restaurants.[75] Whether it was in the Stork Club, Joe's Steak House, Howard Johnson's, or a White Tower, Americans wanted what they considered to be straightforward cooking, with nothing disguising the ingredients. The label "home cooking" summed up this ideal. This meant the meat-potato-and-one-vegetable dinner, cooked simply and presented in an uncomplicated fashion, that had become the Anglo-American standard, or the sandwiches (including hot dogs and hamburgers) that had become popular in the 1920s. Ham and eggs was also very popular; even roadside stands were thought to turn out excellent plates of this simple fare, which in early 1942 was named America's favorite restaurant food.[76]

In 1936 a mother and daughter managed to turn around a failed sandwich, coffee, and doughnuts counter in New York City's cavernous Grand Central Station by converting it to home cooking—New England–style clam chowder, chicken dinners, specialty breads, and muffins. Analyzing the reasons for their success, the mother echoed much of prevailing restaurant wisdom: "I would not have a man in the kitchen under any circumstances," she said. "I like to get a woman cook who has no experience except in a large family."[77] A 1931 analysis of two groups of restaurants, one successful, the other not, summed it up: Most of the successful restaurants served American food cooked by women; the unsuccessful ones served more unusual food cooked by men.[78] Howard Johnson, it seemed, had got it right.

CHAPTER 4

One-third of a Nation Ill Nourished?

In mid-1933, while still struggling to straighten out the mess left by her lackadaisical Republican predecessor, the crisply efficient new secretary of labor, Frances Perkins, began to read alarming reports from her department's Children's Bureau of a marked increase in the number of children suffering from malnutrition. A former social worker whose concern with poorly nourished children stretched back well over twenty years, Perkins was shocked. Convinced that a crisis was imminent, she called a special national conference of public health officials, child hygiene experts, and social workers to develop plans for dealing with a situation she called "truly appalling."[1] The conference duly bemoaned the inadequate attention being given to child health and nutrition, and warned of the disastrous consequences of state cutbacks in funds for public health nursing, but it accomplished little in gaining federal help to feed the hungry.[2] In this sense it was typical of national efforts to deal with hunger during the decade, for as long as malnutrition was regarded mainly as a poor person's problem, national campaigns to combat it seemed to go nowhere. Indeed, while no one could doubt Perkins's deep personal concern, her crisis rhetoric helped camouflage the fact that the Roosevelt administration was taking much the same approach toward hunger as that of his much-reviled "do-nothing" predecessor, Herbert Hoover.

On the surface a solution seemed literally to lie readily at hand: giving the poor the food stocks accumulating under federal programs designed to maintain farm prices by purchasing "surplus" foods. As we have seen, when the surplus purchasing program began in 1931, Congress managed to have some of the flour given to the Red Cross. But this was a temporary, one-shot measure. By 1933 silos were again overflowing with surplus grain while hunger seemed to stalk the land. Radicals began using this as an example of the irrationality and immorality of the capitalist system. The Socialist party leader, Norman Thomas, denounced the system that produced "breadlines knee-deep in wheat."[3] As the New Deal agricultural program—part of which involved the destruction of surplus foods—unfolded, others joined in. Newspapers and magazines commonly featured pictures of food being stored or destroyed side-by-side with ones of men waiting at soup kitchens. Yet until late in the decade almost every effort to give the surplus food to the hungry was crushed or undermined by powerful farm interests and government administrators. To the powerful farm bloc in Congress, every pound of food given out free was one less pound purchased by consumers—something that further reduced demand and sapped pressure holding up farm prices. Other farmers' representatives were distinctly hostile to those who decried hunger in the cities. In Ohio and Illinois, rural legislators voted solidly against tax increases to help feed the hungry in the cities. "We can't keep feeding those bums forever," said Governor Henry Horner of Illinois.[4] In Indiana farmers opposed county money going for urban relief because, said one of their spokesmen, it would be spent on "cigarettes, malt, and other non-necessitous things."[5]

To make matters worse, Roosevelt's close advisor Harry Hopkins, the former social worker who dominated federal relief programs, opposed direct federal relief to the unemployed. Instead, he favored work relief, which would put cash into the needy's pockets in exchange for work on government-funded projects. Only after Roosevelt was shaken by the outcry when the government began to slaughter six million young pigs and dump them into the Mississippi, bury them, cart them to dumps, or otherwise prevent them from coming to market were some surplus commodities distributed. He directly ordered a reluctant Hopkins and an unenthusiastic Secretary of Agriculture Henry Wallace to set up a system for distributing the pork, as well as surplus evaporated milk, cotton, and coal, to the unemployed. They set up the Federal Surplus Relief Corporation in late 1933, but, hampered by Hopkins as its president and forced to distribute goods through a variety of state and local authorities, it accomplished relatively little. Even so, within two years farmers and businessmen who feared it would take away their customers succeeded in gutting it. Its name was changed to eliminate "Relief," and its carcass—the Federal Surplus Commodities Corporation—was moved to the Department of Agriculture, where it could remain under the watchful eyes of those who saw its main task as selling surplus commodities overseas or destroying them.[6]

So it was that on March 4, 1937, when Roosevelt stood on the windswept steps of the Capitol and vowed in his stirring second inaugural address to help the "one-third of the nation" that he declared "ill nourished," the agency stockpiling food would not give it to the needy. Instead, it suggested that Americans change their diets so they would consume more of the foods that were piling up. This was one of Wallace's pet ideas. He thought that the nation's grain farmers were being pushed to the wall not by the temporary Depression but by the century's long-term decline in wheat consumption. Unless Americans could be persuaded to reverse this trend, recovery from the Depression would only make things worse. The poor would spend their additional income on corn-fed beef, dairy products, fruits, and vegetables and spend even less on wheat flour.[7]

Yet paranoid farm interests pounced on his department's most timid steps in the direction of promoting dietary change. In 1935, when the department's top dietitian, Hazel Stiebeling, complied with Wallace's request to come up with food consumption recommendations for Americans at four different income levels that would emphasize liberal use of overproduced foods, she was vilified by farm groups. Some protested that the amounts of sugar, meat, and animal products in her minimum level diet were too small. Wheat farmers and millers condemned her suggestion that at higher income levels other foods might be substituted for flour. Outraged senators attached a rider to the Agriculture Department's appropriation bill mandating the firing of anyone who suggested reducing consumption of a farm commodity or who asserted that any "wholesome agricultural food commodity" was "harmful or undesirable to use." Only after Wallace called in other experts who showed that, if Stiebeling's menus were followed, American farmers would have to produce more, not less, were the senators persuaded to drop the rider.[8]

Even federal food aid for hungry children was blocked, forcing local authorities or volunteers to carry much of this burden. Most urban school districts had school lunch programs of some kind—some partly subsidized, others not—but few could afford to provide lunches for all those in need. In 1930 Philadelphia businessmen had set up the country's first free school breakfast program for destitute children, but it ran out of money in mid-1932.[9] In some smaller places, parent-teacher associations and local service organizations managed to cobble together systems for delivering meals to schoolchildren. In the mainly rural Puget Sound area of the state of Washington, for example, the first hot lunch program was organized by the PTA of one of the school districts in 1930, with families taking turns providing one hot dish, such as baked beans, macaroni and cheese, spaghetti and tomato sauce, or various kinds of soup. This system was copied by other area school districts. In one the hot dish was prepared by volunteers of the Unemployed Citizens League from foods donated to it. In others the program was aided by community clubs, granges, and other unemployed groups. Some charged enough to cover their expenses; others subsidized the costs from donations in cash and kind. Not until 1936, when the state branch of Hopkins's Works Progress Administration finally stepped in, was there any government aid, and then it was only an offer of cooks and helpers, not money or food. Not until 1938 did some surplus food begin to reach the programs.[10] This was made possible because the food was actually distributed by local authorities, not federal ones, who had to agree not to cut back on their normal food purchases and to give the surplus foods only to students officially certified as "needy."[11]

Even Henry Morgenthau, the conservative millionaire who, as secretary of the treasury, normally tried to rein in New Deal programs, was outraged by shenanigans such as this. When Hopkins and Wallace managed to shelve a program to give surplus dried milk and flour to those hit by a new recession in 1938, he told his staff: "It just turns my stomach to hear Henry Wallace want a hundred million dollars to have people grow less wheat

. . . with people not getting enough to eat. Now there's something cockeyed, crackpot, about this administration. I mean it just goes against all decency and human understanding that they should be trying to find ways and means to grow less. . . . There's people going hungry in America, all over America. Now there's just something—the combination of Wallace and Hopkins refusing to do any direct relief. There's simply something ungodly about it."[12]

But how many people really *were* going hungry? Many decried "malnutrition," but what did it mean? How did one measure it? The fact was that there was not even a universally acceptable definition of death by starvation, let alone malnutrition. Hoover repeatedly pointed out that nobody was "actually starving," yet in 1931 a social worker reported seeing numerous cases of starvation in Illinois mining towns, and New York City hospitals reported ninety-four deaths from starvation.[13] Many people were moved when President Roosevelt labeled one-third of the nation "ill nourished," but—as we shall see—the figure was based on surveys using dubious standards of nutritional adequacy and an even more questionable method for determining whether diets conformed to them.[14]

Conflict over definitions and numbers began early. When the first few years of the Depression brought no rise in mortality rates, Hoover's denials that the Depression was adversely affecting the nutrition of Americans gained credence, for one of the few things upon which there was wide agreement was that worsening nutrition brought higher mortality. By late 1932 the idea was even being bruited about that the Depression had been good for the health of the nation. New York City's health commissioner tried to dispute this, saying that the falling mortality rates of the past few years were due more to mild winters than anything else, but his predecessor pointed out that infant mortality rates, which had little to do with cold weather, had been declining as well.[15] In any event, despite a severe winter in 1932–1933, mortality continued to decline. In September 1933 the *New York Times* noted that as the nation entered the fifth winter of the Depression its general death and infant mortality rates were the lowest in history. Many millions of Americans may have been "spiritually under the harrow," it said, but "people have not been allowed to starve and children have not been underfed or at any rate have not felt the effects of undernourishment."[16]

Yet only one week earlier, in calling the national child health conference, Secretary of Labor Perkins had referred to "the steadily increasing millions of children known to be suffering from malnutrition as a result of the depression." One-fifth of all preschool and school-age children were "showing signs of poor nutrition," she said.[17] But what were these signs? Perkins herself admitted that "no universal method of measuring the child's nutritional condition had been used" and that the data were "fragmentary."[18] The experts at the conference were of little help; they bemoaned the fact that there was no commonly accepted way of diagnosing malnutrition. Indeed, most people familiar with school medical examinations realized that malnutrition diagnoses depended more on the examiners than on the examined.

Take, for example, the dietitians at the New York Association for Improving the Condition of the Poor's (NYAICP) nursing station in East Harlem. From 1923 to 1928, when their orientation was toward correcting specific defects thought to result from poor nutrition, they found 26 percent of children examined to be malnourished. After 1929, when they shifted their emphasis to general prevention, the percentage adjudged to be malnourished declined steadily, until in 1934 only 7.4 percent of preschoolers were labeled malnourished.[19] Yet economic conditions had deteriorated markedly, and relief programs had not accomplished much to improve diets. Perhaps the most surprising statistics emerged from the most pitiful slum in the area, a part of East Harlem that was already known as one of the worst districts in the city before the Depression struck. During the Depression it was inundated with desperate Puerto Ricans fleeing even worse conditions there. Yet its infant mortality rate dropped steadily during the 1930s, plummeting by almost 25 percent in the four years from 1934 to 1937 alone.[20] The director of the NYAICP, Bailey Burritt, was forced to conclude that the Depression had brought improved nutrition and health, at least for those with jobs. (He explained that low Depression wages discouraged the "excessive eating, excessive drinking, and unreasonable hours of sleep" that high wages allowed in normal times.)[21] But then the agency was shocked to discover, in its Lower Manhattan Center, that there was less malnutrition among children of the unemployed than among the employed. This is now ascribed to the free cod liver oil it had given to the children of the unemployed.[22]

It was natural for social service agencies to put the worst face on the crisis and the best ones on their solutions. After all, spending on public health and child nutrition in most cities and states actually declined from 1931 to 1934, and public health authorities, welfare administrators, and social workers all found themselves battling for extremely scarce resources.[23] The problem was that, while the needy were everywhere to be seen, statistical evidence to show that they suffered from hunger and its consequences was very hard to come by, because the three main methods for diagnosing and estimating malnutrition—which had hardly changed in thirty years—were deeply flawed.[24] One, used mainly with children, involved physical examination by a doctor, who searched for visible signs of the ailment. These included poor posture, listlessness, swollen mucous membranes, and suspicious-looking skin, eyes, and hair. But these were only vaguely defined and led to extremely subjective judgments. The number of cases reported could be inflated or deflated almost at will.[25] Another method was the height/weight chart. Its main advantage was that it could be used by a school nurse, but it was of little use with adults and ignored hereditary and ethnic variations. The fact that any child who was 13 percent or more under the average weight for his or her height was declared malnourished also contributed to dubious conclusions.[26]

The third approach was by far the greatest producer of alarms regarding the nutrition of the nation. Dietitians estimated the quantity of various nutrients necessary for good health and calculated the minimum amount of money needed to purchase foods to supply these nutrients. Anyone or any family spending less than this was deemed malnourished. Yet scientists were still in the dark about what most nutrients did and still had difficulty measuring them. As a prominent dietitian acknowledged, the scientific data upon which to base standards for "minimum adequacy" were "fragmentary and inadequate." This contributed to a motley patchwork of relief policies. For example, in 1931 nutritionists in Cleveland and Cincinnati calculated that a family of four on relief needed $14.50 a week for food, while in New York City, Baltimore, and Milwaukee less than two-thirds of that amount was deemed enough.[27] Later, more detailed studies of some of the budgets tried to calculate the amount of certain vitamins and minerals actually purchased, but uncertainty regarding the importance of the nutrients, how they were distributed within families, and whether they were actually consumed made the results of little value. Roosevelt's "one-third of a nation ill nourished" claim was based on a somewhat different, but no less flawed, study. Directed by Agriculture Department home economist Hazel Stiebeling, it used records of what families had recalled eating during one week in 1935 or 1936 to calculate the amount of each of eight nutrients this would provide per person. If even one of these—say, phosphorus—was below a certain minimum level, the person's whole diet was labeled "poor" and he or she was included among the "ill-nourished" on the Capitol steps.[28]

Strangely, as the economy improved, nutrition seemed to worsen—at least according to studies using this method. A 1938 study of blue- and white-collar workers' family budgets, also directed by Stiebeling, used even more stringent requirements for judging the adequacy of diets. It concluded that only 50 percent of the families earned enough to purchase diets that were minimally adequate. Of the one-half who did earn enough, only 30 percent—or 15 percent of the total—bought enough of the right foods to provide a "good" diet. The diets of an equal number were judged to be "poor," while the remainder were merely "fair."[29] As if that were not bad enough, a specialist at California Institute of Technology said that the diets deemed fair should really be classified as poor, for they contained only barely enough vitamin B_1 to head off beri-beri, not enough vitamin B_2 to prevent cheilosis, and inadequate protein.[30] This left only 15 percent of the *employed* workers with adequate diets! It all added up to an image of the physical state of the American lower classes that made Friedrich Engels's grim description of the condition of the working class in Victorian England seem positively rosy. Farmers came off little better. Experts calculated that only five out of every thousand heads or wives of low-income farm families were in good health—largely because vitamin deficiencies were "extremely prevalent."[31]

The end of the Depression seemed to help but little. In 1941, well after the war economy had begun pouring extraordinary sums of money into workers' pockets and boosting farm incomes, Paul McNutt, head of the new Federal Security Agency, predecessor to the Department of Health and Welfare, warned that forty-five million Americans "do not have enough to eat of the foods we know are essential to good health." Ironically, this was the number of people who, it was calculated, could not afford to reduce the proportion of their calorie intake made up by flour and grains to less than 25 percent.[32] (Twenty-seven years later the government would begin trying to get Americans to *raise* the proportion of calories consumed as "complex carbohydrates" to 60 percent.)[33]

Another later-to-be-discarded idea that inspired nutritional jeremiads was the belief in the extraordinarily healthful properties of cow's milk. It had always been regarded as important for infants after weaning, but during the 1920s—thanks in part to the milk industry's clever promotional campaigns—the belief spread that it was essential for older children and adults as well. (This despite the evidence that a goodly portion of the transpacific part of the human race survived and multiplied quite well without even being able to tolerate it.) In 1926 experts were recommending that children up to the age of eighteen should drink one pint of it a day. By 1937 this was the recommended intake for adults; the under-eighteens were up to one quart a day.[34] Yet, at about ten cents a quart in the mid-1930s, milk was not cheap. Meeting this requirement would make a big dent in family budgets, particularly for the poor; the twenty-eight quarts of milk a typical family of five were expected to consume each week would normally have taken up about half the food budget of a relief family of that size in 1936.[35] One can see why studies of low-income people so often called consumption of milk their most serious deficiency and warned of impending health problems.[36]

Yet something all-important was missing from surveys such as Stiebeling's: evidence of a link between these deficiencies and poor health among those surveyed. Nor could they prove that the dietary deficiencies they said were so widespread had affected the health of the nation as a whole. They concluded that health problems must be present on an a priori basis—because calculations based on their theories told them so.[37] The repeated failure of the dreaded health problems to arrive did prompt some questioning of their lugubrious assessments of the nation's nutritional state. A Birmingham doctor surveyed colleagues in major hospitals throughout the country in 1936 and found few who saw a significant increase in malnutrition among the population at large.[38] Dr. C. E. Palmer, a consultant to the U.S. Public Health Service, spent much of his time measuring and weighing children to show that the Depression was not adversely affecting them.[39] He was particularly keen on deflating Dr. Martha Eliot, head of the child hygiene section of the U.S. Children's Bureau, whose data had played a major role in alarming Secretary Perkins.[40]

The steady decline in mortality rates continued to buttress the optimists' case. So did the drop in deaths from vitamin deficiency diseases, which, thanks in large part to the conquest of pellagra, had plummeted into relative insignificance by the end of the decade.[41] In 1935 even the Columbia University nutritional scientist Henry Sherman, who normally sided with the pessimists, admitted that declining mortality and other evidence indicated that American health had largely withstood the Depression's hardship. (However, his explanation for this did allow for future deterioration: It was "largely due to the increasingly intelligent attention which has been given to food and nutrition the preceding years. We entered the depression with bodies better stocked, not with fat, but with mineral elements and vitamins than a decade earlier.")[42]

As someone familiar with the close links between mortality rates—particularly of infants—and nutrition, Sherman had little choice but to concede the opposition's main point. There was not, and still is not, any real evidence that, overall, the Depression harmed the health of the nation. If anything, the evidence points to continued general improvement. The well-meaning nay-sayers who warned of national nutritional disaster helped muddy the waters in this regard. But perhaps their greatest failing was that, in exaggerating the degree to which malnutrition affected the employed and other relatively well-off people, they allowed those who really were suffering from hunger and poor diets to practically slip by unnoticed and uncared for. There were indeed many hungry, undernourished people in America, but they were not where Sherman and his home economist and social worker allies usually said they were, among the middle class or even the "Depression poor" of the cities. Rather, they were concentrated among the "long-term poor," the marginal groups of the pre-Depression era who were a permanent fixture of American life.

Southern sharecroppers were one such group. Except in extraordinary circumstances—such as during the drought of 1930–1931—they continued to scrape by on their traditional diet of salt meat, corn bread, and syrup or sorghum, with a bit of milk here and there, some sweet potatoes, and, on a rare Sunday, a chicken. Many rarely ate vegetables, even in the summer.[43] In tobacco country, where 'croppers would often move five times in six years, never having a chance to plant a vegetable garden, there was what researchers called a "a six-month starvation period" between tobacco crops when they would have to live on a twenty-dollar-a-month "furnish" from the landlord (at 10 percent interest) for a family of five. Then they would eat little more than fat back, corn dumplings, and collards, boiled on rickety wood stoves.[44] Yet although grim, these diets were hardly different from pre-Depression ones and, with the exception of some additional milk, from those of their grandparents.[45] Unmarried black mothers in the ghettos of northern cities trying to preserve rural southern foodways in crumbling apartment kitchens were not much better off than their relatives "back home," but neither were they worse off than they or their mothers had been when they moved to the cities of the north earlier in the Great Migration.[46]

In "Their Blood Is Strong," John Steinbeck noted that Depression-battered migrant farm workers in the West were reduced to diets of dandelion greens and boiled potatoes, but that the "good times" diet—"beans, baking powder biscuits, jam, coffee"—was not much better.[47] Nor were hill people living in isolated communities down the swath of Appalachian mountains perceptibly worse off than before. They continued to live not much differently from primitive hunters-and-gatherers, the main difference being that the hunting in more primitive societies was usually better.[48] The cities harbored many old people without families to take care of them, scraping by on dwindling savings or meager relief payments in a society that only began to pay old-age pensions in 1940—and then only to a very few. But they too predated the Depression, as had the thousands of others who were unable to work for other reasons and were bereft of family or friends to provide for them. In other words, the United States had likely entered the Depression with perhaps 15 to 20 percent of its people living in deprived, distressed, or depressed conditions—people whose diets, life expectancy, and general health were much worse than those of the large majority. The country likely emerged from the Depression with essentially the same people in this category, making up perhaps a somewhat smaller proportion of the population.[49]

Although writers John Steinbeck and James Agee and photographers Margaret Bourke-White and Walker Evans tried to focus attention on the plight of these mainly rural poor people, they were bypassed by most government food programs. The principal conflicts over food relief centered not on them but on urbanites who had lost their jobs—the "Depression poor." As we have seen, there is little evidence that their nutrition and health were adversely affected by the Depression.[50] By 1934 New Deal relief programs seemed to have helped forestall a health crisis by putting money in their pockets in the form of emergency relief payments, jobs, or both.[51] Nevertheless, unemployment did take a psychological, if not physical, toll. Social-psychological studies indicate that, whereas middle-class women emerged from the Depression with enhanced feelings of mastery and self-confidence, working-class women were scarred by the opposite feelings.[52]

The strains on the wives of unemployed workers trying to keep the family satisfied at meal time must have played a role in this. Food costs normally constituted from 30 to 45 percent of workers' family budgets and were—up to a point—flexible. Often, the number of meals was cut from three to two a day. Guests were no longer invited for dinner. Milk consumption was reduced, fresh fruit virtually disappeared, eggs took the place of meat, and drippings stood in for butter. Families fell back on their historic staples: pasta and beans for Italian-Americans, corn meal for Southern blacks and whites, and beans and pancakes for northern native-born whites.[53] The most common complaint was about the monotony of these diets: that the fruits, meats and delicacies that added variety had disappeared. "I say, wife, those are to be eaten, aren't they? I've about forgotten," said an unemployed man staring wistfully at some large black cherries in front of a fruit store. An Italian-American man, complaining about having lived for nearly a month on minestrone and pasta and beans, remarked, "You know, that stuff fills you up, but you don't feel right; no pep or fight. You just want to sleep all the time."[54] (Of course, thirty years later many would regard a diet such as this as an admirably healthy one.)

Others linked poor health more specifically to their monotonous diets of cheap foods. "Who wouldn't be sick," asked an unemployed North Carolina mill hand whose husband could not work because of stomach trouble, "if they had to eat fat back and grease three times a day?"[55] Many also thought lower milk consumption was harming their health. "When you give up milk," said an unemployed New Haven factory worker, "as you finally do if you ain't on relief—they make an allowance for it—why then maybe for the first time you think you might be better off on relief."[56] A Gallup poll in early December 1940 indicated that 40 percent of Americans, mainly lower-income, thought that their inability to buy all the foods they wished was adversely affecting their health.[57]

By then the logjam keeping most surplus food from the poor had finally been broken. Surprisingly, the giant food processors played the key role in this. Farmers opposing free handouts of surplus foods to the poor had been supported by food retailers, who feared losing customers to government distribution centers.[58] The ingenious food stamp plan—worked out in early 1939 by Clarence Francis, the president of General Foods, and Harry Hopkins, now secretary of commerce—deftly co-opted this opposition. There were two kinds of stamps, orange and blue. Each week, participants were required to spend as much on the little orange stamps, which could be used to purchase any kind of food or groceries, as they normally spent on food. For every dollar's worth of orange stamps they bought they received fifty cents' worth of blue stamps, which could be used only to buy selected surplus foods. All the stamps would be redeemed from retailers at face value by the government. Farmers, processors, and retailers were thus assured—so it seemed—that the needy would be forced to consume more flour, margarine, butter, and other surplus foods than they normally would, putting upward pressure on prices and profits.[59]

Skewed as it was toward stimulating people to eat what happened to be in surplus rather than what was most healthful, the plan left much to be desired as a program for improving nutrition. Earlier studies had already shown that the best way to improve the diets of the poor was simply to put more money in their pockets: Rising incomes brought more consumption of meat, fruits, vegetables, and other "protective" foods.[60] However, because it seemed to make so much sense, the general public were overwhelmingly favorable.[61] By May 1941 almost four million people were benefiting from the program. That month, President Roosevelt recommended it to the embattled British as a way to ensure the equitable distribution of the food the United States was beginning to ship to them while also providing for adequate nutrition for the poor.[62]

However, this benign conception of the program was not shared by the agency administering it. To the Department of Agriculture its purpose was "to increase the consumption of food in order to expand the farm market." At times, it allowed blue stamps to be redeemed for foods whose prices were merely in danger of falling.[63] Farmers saw it in exactly the same light, and as surpluses dwindled, so did their enthusiasm for the program. They charged that the program was being abused by recipients and by mom-and-pop grocers who were accepting stamps for cash or nonsurplus foods. By 1943 farm bloc support for the program had virtually evaporated, as had the surpluses, so the USDA dispatched it to the scrap-heap.[64]

On the other hand, the USDA continued to support school lunch programs throughout the war. Politics played a major role in this: Whereas food stamps went mainly to welfare cases in the cities, over 80 percent of those benefiting from federal aid to school lunches were rural, and many of them were in the South, whose seniority-laden members of Congress wielded extraordinary power.[65] But one of the arguments for continuing food aid to rural schools was that studies of farm families had shown that they were even more poorly nourished than city ones.[66] If even farmers—living in the horn of the rural cornucopia—were not immune to the baleful effects of the imbalanced diet, was anyone safe? Was not malnutrition indeed what prominent dietitians were calling it by 1939: a "national problem" that afflicted all kinds of Americans?[67] "Nutritional diseases" were the country's "greatest single health problem from the point of view of disability and economic loss," warned the head of the U.S. Public Health Service in September 1940. Widespread deficiencies in vitamin A would cause night blindness, a disaster for national defense. Three out of ten urban Americans did not earn enough to provide a "good, completely adequate diet," and farm families, poorly supplied with fruits and vegetables, were no better off. Worse still, even among those who could afford good food there was a large group whose inadequate diet was the result of "ignorance, improper dieting, food faddism, or carelessness in spending their money."[68]

Ten years after the onset of the Depression, then, experts were back to blaming poor diets not just on inadequate income but on ignorance and improvidence. In early March 1941, nutritionists at the Department of Agriculture released a study called "Is America Well Fed?" which lamented that food deficiencies among poorer families were often as much due to unwise spending as lack of funds.[69] This reflected one of Stiebeling's studies, which concluded that, although one-half of workers' families could afford to buy an adequate diet, most did not do so. Most worrying were indications that they were not simply ignorant of proper nutrition, but that—in the oft-quoted phrase of the time—many of them frankly just didn't give a damn. A study of reasons for food choices undertaken later that year showed that, while health concerns were uppermost in the minds of the high-income people, most of the low- and middle-income respondents were more concerned about cost.[70] Needless to say, such attitudes were profoundly disturbing to those who now set out to rouse the nation to the supreme physical and psychological effort necessary to fight total war. A strong nation had to be composed of strong people, and only healthy diets could provide that. Considerations of economy and taste would have to take a back seat in the this looming life-or-death struggle.

CHAPTER 5

Oh What a Healthy War: Nutrition for National Defense

By July 1940, as the German blitzkrieg drove through and around conventional defenses, it was apparent that weaponry such as tanks and aircraft, which had played peripheral roles in World War I, had now rendered many of that war's tactics obsolete. Similarly, it seemed, the underpinnings of the previous war's food conservation strategy had also been shattered by things that had been hardly present in Great War thinking: vitamins. "In the last war," said M. L. Wilson, the agricultural economist who helped organize the new defense nutrition program, "fighting airplanes were stepping right along if they made 120 miles an hour. This time speeds of 360 miles are commonplace. There has been exactly as great a revolution, in these twenty-five years, in the science of nutrition."[1] Herbert Hoover's United States Food Administration (USFA), the much-vaunted agency that had organized the World War I campaign to conserve white flour and sugar, had been misled by its ignorance of vitamins, said Wilson and the new generation of experts. Its motto had been "Food Will Win the War." "Vitamins Will Win the War" would be better for this one, they said.[2] The previous commanders on the food front had also defined the theater of operations too narrowly. Their concern over conserving food for fighting men abroad led them to neglect the importance of good nutrition on the home front. Instead of the previous war's emphasis on food conservation, said Wilson, the government would get Americans to eat more vitamins and minerals to improve their mental and physical alertness.[3] Today, said one of the new breed, the government had to "make America strong by making Americans stronger."[4] "The scientists know," reported the *New York Times* in January 1941, "that along with national unity and national faith in democracy, there is another most potent morale builder, the name of which is Vitamin B_1."[5]

The United States was at war in all but name by September 1940, as Washington was swept by an all-out drive to mobilize. France had fallen to the Nazis, and the rest of Western Europe was rapidly following. Hitler had unleashed the offensive against British air defenses that was to be a prelude to a cross-Channel invasion. Americans, particularly in government, began to fear that once this last major democratic obstacle was removed their hemisphere would be next on his plate. Whether willing or not to join the war at this stage—and most were not—most Americans agreed that the nation must prepare for the possibility.

Yet as government contracts began to revive depressed industries and sop up unemployment, some Americans began to worry about the superhuman effort necessary to staff both frantically expanding industries and the armed forces. The continuing reports that more than one-third of Americans were malnourished fueled these concerns. "If our workers are malnourished they cannot be efficient in producing what we need for defense," warned the nation's surgeon general. "Yet every study of nutrition, by whatever method conducted, shows that malnourishment in this country is widespread and serious."[6] The beginnings of the draft that month deepened apprehensions. Of the first million men called for induction, fully 40 percent were rejected for general military service on medical grounds.[7] Many of these, an estimated one-third, were for ailments such as tooth decay thought to be caused by poor nutrition.[8] "We are physically in a condition of which nationally we should be thoroughly ashamed," blustered the newly appointed head of Selective Service, General Lewis Hershey.[9]

In January 1941 an alarmed President Roosevelt told Paul McNutt, the head of the new Federal Security Agency, to coordinate efforts to beef up the nation's nutritional status.[10] McNutt's first step seemed eminently sensible. He asked the National Research Council's forty-one-member Committee on Food and Nutrition, then meeting in Washington, how much of each of the known nutrients a person needed to maintain good health.[11] Alas, the question was not as easily answered as laypeople such as McNutt thought. That month, none other than Dr. Russell Wilder, of the Mayo Clinic thiamin experiment fame, had been brought in to chair the committee. Wilder immediately appointed a subcommittee of three home economists—Lydia Roberts of the University of Chicago, Hazel Stiebeling of the federal Bureau of Home Economics, and Helen Mitchell, who for fourteen years had taught home economics at the zany Kellogg "Sanitorium" in Battle Creek, Michigan—and gave them until the next morning to come up with a set of generally accepted nutrition standards! The three women, who knew that experts were at odds with each other on virtually every known nutrient, stewed all night in a hotel room, convinced that Wilder and his male colleagues were out having a good time "seeing the town" while they wrestled with an impossible task. In the morning they reported that, in Roberts's words, "it couldn't be done, that the evidence was too scanty and conflicting." But the buoyant Wilder merely appointed another committee, with Roberts as chairman, and gave it more time.

Over the next months Roberts's committee surveyed the field and finally came up with tentative recommendations for how many calories and the amounts of protein and eight other nutrients people of different ages and conditions should consume.[12] Roberts was surprised that, after some revision, the standards gained quick acceptance by most experts, but this is easily explained: The committee had nimbly sidestepped McNutt's request to provide "standards"—the minimum amounts required for good health—and instead came up with "recommended allowances," amounts that would avoid deficiencies even among people who needed far more than the average requirements of particular nutrients. This "margin of safety," which elevated most recommendations about 30 percent over the experts' average suggestions, made them high enough to accommodate scientists whose estimates were on the high side and yet did not alienate most of those with lower estimates.[13]

Considerable fanfare accompanied the new table of Recommended Daily Allowances (RDAs) when it was made public at a National Nutrition Conference in May 1941. It turned out to be just what a host of public and private agencies were looking for: an official-looking yardstick for judging how well everyone from factory workers in Chicago to GIs in New Guinea were being fed. The subtle term "allowances" was usually ignored, and individuals whose diets did not come up to its high standards were then deemed undernourished.[14]

The three-day conference had been called by President Roosevelt to suggest a program to improve the nation's nutrition. The fact that the nine hundred delegates were from government, the schools, colleges, media, charities, and industry in itself illustrated the complexity of the problem. While the confrerees did produce a number of useful-sounding suggestions—mainly concerned with educating the public about nutrition—the most interesting thing is what Roosevelt did not do. He did not, as was expected, appoint a "food czar" to coordinate all aspects of food, from its production, shipment, and marketing to its consumption by civilians.[15] President Woodrow Wilson intended this when he made Hoover head of the USFA; it was also something FDR did in other areas, such as rubber and gasoline. But Roosevelt, whose dislike for Hoover was perhaps surpassed only by Hoover's distaste for him, knew full well that, were such a position recreated, Hoover would again be the obvious choice.[16] Instead, he put McNutt in charge of a new Office of Defense Health and Welfare Services, with a small Nutrition Division, and left it to squabble with the powerful Department of Agriculture and the wartime planning agencies over food policy, production, and prices.[17]

Any hope of affecting the nation's health or manipulating demand for certain foods by changing eating habits was now quite divorced from questions of supply and price. At first the latter rested with the Department of Agriculture, then with the price-fixing Office of Price Administration. When rationing was introduced, it too was run by the OPA.[18] McNutt's office was left with a bewildering array of petty programs in a mind-boggling bureaucratic maze. The coordinating committee of the Nutrition Division, for example, had representatives from thirty-one different government agencies, including the Fish and Wildlife Service, Sanitary Corps of the War Department, and Office of Indian Affairs.[19] Worse still, it reported to

McNutt, an administrative lightweight with little influence in higher circles.

Despite their conviction that they were light-years ahead of their World War I predecessors in almost everything, the new generation of government nutrition experts ultimately had nothing like the impact on the American diet of their predecessors. In persuading a large part of the middle class to "Hooverize," Hoover's minions had managed to infuse them with the basic concepts of the New Nutrition. McNutt's managed much less, despite (or perhaps because of) the fact that, while Hoover staffed much of his agency with people from the fledgling advertising industry, the New Dealers turned to many of America's greatest nutritional and social scientists. Moreover, rather than working directly for McNutt (as "Czar Hoover" would have insisted), the latter-day experts operated under the umbrella of the National Research Council—the nutritionists on its forty-member Food and Nutrition Board (FNB), the social scientists on its Committee on Food Habits.[20]

The nutritional scientists spent much of the war continuing to decry the national diet. Undaunted by mortality statistics that seemed to show quite the opposite, Wilder warned that "our food ways for some 60 years have been worse than at any previous time in history. They are worse today than they were in 1914." Even well-off families were consuming only one-third as much thiamin as in Civil War times, and there was "grave vitamin starvation among our poorer families."[21] A study of upper-income diets published in the prestigious *New England Journal of Medicine* seemed to confirm that even the wealthy were not getting the full complement of RDAs. Three-quarters of their diets were judged deficient in thiamin and riboflavin, and—most surprising to some—the physicians among the subjects studied had no better diets than the rest.[22] In 1943, with workers basking in full employment and pay packets stuffed with overtime pay, an FNB study said that applying the new RDAs to 1939 and 1940 studies confirmed that "inadequate diets are widespread in the nation." Newer surveys continued to indicate, they said, "that deficiency states are rife throughout the country."[23]

But if deficiencies were still so widespread, why were adverse health consequences not yet apparent? Nutritionists responded with the concept of "latent malnutrition." Dr. Frank Boudreau, who replaced Wilder as head of the FNB at the end of 1941, explained this "latent or subclinical stage" was the earliest stage of a deficiency disease, which the older methods such as height/weight charts could not diagnose.[24] It was, said another expert, "a borderline . . . condition, the signs of which are not detectable by the ordinary methods."[25] The 1943 FNB study said that malnutrition was rife but was not readily apparent because it was not of the "traditional severe acute type." Its symptoms were not easily detectable, for they were "subacute" and "gradual in their course."[26] In other words, there was no way—except through dietary surveys—of knowing that these deficiencies existed, for they had no discernible effects! Scientists did scramble to come up with new diagnostic tests for these undetectable deficiencies: blood tests, skin tests, darkness adaptation tests (for vitamin A), even X-rays.[27] Eye examinations using slit lamp microscopy were used to try to detect riboflavin deficiencies. (Specialists were encouraged by its potential when all 350 persons in a rural Maine community examined with this method were declared deficient.)[28] But in the end malnutrition remained as undetectable—and undefinable—as ever. Yet, said Wilder, "these milder degrees of nutritional deficiency are the nub of the nutritional problem."[29]

This "latent malnutrition" concept helped keep thiamin at center stage into 1943. Wilder was undeterred by reports that doctors who prescribed thiamin to patients suffering from "nervous exhaustion" had had no success. The patients had likely been suffering the deficiency, in latent form, for too long, he explained. The Mayo Clinic subjects had been deprived of thiamin for only a short time and had therefore perked up quickly after it was restored. Those who had been off it longer took longer to recover. If the deprivation went on for too long, recovery might well be impossible. Latent malnutrition of this sort, he added, was also the likely reason for the recent increase in the incidence of diseases such as arteriosclerosis, arthritis, diabetes, and mental illness.[30] While a man could subsist on as little as 0.6 mg of thiamin a day, "with this small intake he is only half alive. To function efficiently, to do the things that must be done now, he needs at least twice this amount, and for a safety factor at least 2 mg." Yet probably less than a third of the adult male population got that much.[31] The others on the FNB found these arguments persuasive, and the board pushed vigorously for mandatory enrichment of flour. (It had only been "endorsed" by the government in May 1941.)[32] In late 1942, with Wilder as vice-chairman, it was able to prod the government into decreeing that all flour used in the armed services and federal institutions be "enriched."[33] As a result, by mid-1943 about 75 percent of the nation's bread was being made with enriched flour.

Wilder was by no means satisfied with having large doses of his favorite vitamin injected into the national diet. He also called for fortification of other foods with other vitamins on a massive scale. He calculated that most of the calories in the American diet came from vitaminless sugar and "processed fats." Sugar—which was, "of all foods, unquestionably the worst"—could be dealt with by mixing it with nutritious milk solids, although he admitted that soda pop fanciers might not take to cloudy beverages. The government should also mandate the addition of vitamins A and D to margarine, lard, vegetable oils, and even butter, which already contained these vitamins. Rice and corn meal should be fortified as well, he said.[34] In late 1942 he even proposed that fruits, vegetables, and most other commonly used foods should be enriched.[35]

There was logic to Wilder's schemes. After all, if vitamins were so important, and it was now so easy to supply them, why not mandate that this be done? Well, the AMA knew why. To the physicians' guild it seemed that Wilder had become carried away with the public health aspects of his position and had neglected the threat vitamin purveyors posed to the medical profession: that people would rely on vitamin supplements and fortified foods instead of visits to the doctor. The organized doctors thus assumed the tough defensive stance they would maintain for the next fifty-odd years.[36] In late 1942 the AMA declared that fortification of flour and milk was fine, but that was as far as the process should be allowed to go. It would not even approve adding vitamins and minerals to canned baby foods.[37] With a large portion of the Food and Nutrition Board consisting of physicians, Wilder sounded increasingly like a lone wolf on fortification.[38] By 1944, not only were they ignoring his calls for more fortification, they were even questioning his views on thiamin and beginning to lower its RDA.[39]

But even Wilder would not support the next logical step: liberally distributing vitamin pills among servicemen, industrial workers, and other groups deemed in need. Although an army nutritionist promised that, if given five thousand men and allowed to give them extra vitamins and minerals for six months, he would turn them into supermen—invincible shock troops who could beat anyone at hand-to-hand combat—the armed forces were not the major issue.[40] It was generally agreed that their normal rations supplied more than enough vitamins, although a few experts did recommend special supplements for certain units, particularly vitamin A (the "vision vitamin") for night fighter pilots and sentries and the vitamin B complex (which apparently made rats less sensitive to noise) for men under fire.[41]

Most of the conflict raged over the question of whether vitamin supplements should be given to factory workers. Hundreds of thousands of them had been drawn to booming industrial centers, where they lived crowded together with little in the way of adequate housekeeping facilities. Plants employing over twenty-five thousand workers had eating facilities for no more than three thousand. Young workers were a particular concern, for they were reputed to shun milk, fruit, and vegetables in favor of bread, potatoes, and other starches. This raised fears that, without an adequate supply of vitamins and minerals, they would not be able to keep up the frenetic wartime pace of production.[42] Although a well-known experiment on athletes and soldiers in Minnesota in early 1942 had apparently shot down the idea that "Vitamin Super-Charging" of healthy men increased their muscular ability or ability to recover from exertion, vitamin manufacturers succeeded in convincing a number of employers that handing out vitamin pills would improve productivity and reduce absenteeism. A New York State legislative committee reported that those employers who gave out these "pep pills" were universally enthusiastic. North American Aviation said it was "convinced they maintain the health and vitality of our employees." Remington Arms wired, "Pep pills extremely helpful. Feel they help to increase resistance to disease, especially respiratory infections." Dugan Brothers Bakers, in Newark, New Jersey, reported, "Vitamins most helpful, especially to our sales force." Yet the FNB would have none of this and supported the AMA in vigorously opposing it.[43] The two remained firmly aligned with the food industries on this score, arguing that improved nutrition must come through Americans choosing from the available foods more wisely, not through "indiscriminate consumption of various vitamin and mineral preparations."[44] The solution for industrial workers was educating them and making nutritious meals and snacks available to them at work.[45]

A second task of the FNB was to make the new RDAs intelligible to the general public. Since laypeople could hardly be expected to count how many units of vitamins and grams of minerals they were consuming each day, in early 1942 the committee struck on the idea of dividing foods into "food groups" high in the same nutrients. People could then concentrate on eating some food from each of these groups every day. The problem was that there were too many groups—eight at first, then seven—and they were too ill defined to be easily recalled. "Oranges, tomatoes, grapefruit . . . or raw cabbage or salad greens" was one group; "green or yellow vegetables, some raw, some cooked" was another; "potatoes, other vegetables or fruits in season" was a third.[46] In March 1943 the committee tried to simplify the rules and make them more memorable. Each of the "Basic Seven" was given a name and number and made into a slice of an illustrated pie, but, despite this improvement, most people could simply not remember seven different groups.[47] The old message that diets had to be somehow "balanced" did filter through, but what this now meant was never very clear.[48]

The Committee on Food Habits, which was supposed to advise on how to get Americans to follow the new rules, was a more interesting if even less effective body. It was launched amidst high hopes for "applied social science." During the Depression many social scientists had chafed at their traditionally detached stance. They wanted to use their insights into society to help those they studied. What better place to do this than with regard to food habits? Cultural anthropologists were beginning to discover that poor Americans were more often moved by considerations of status than by economy in choosing and preparing foods. The anthropologists were also sensitive to the importance of ethnicity and of the role of food in preserving family and ethnic ties. They were therefore much more aware than the nutritionists, economists, and ad men of the USFA of the difficulties involved in changing food habits, particularly those of the poor. Hoover's people had appealed to reason—explaining the savings to be gained through substituting some foods for others—and patriotism, but neither had much impact on the working class. The new social scientists knew that other sociopsychological and cultural factors had to be taken into account, although they were not sure exactly how. "Science has brought about almost a revolution in nutrition and given us a new basis for diet," said M. L. Wilson, but the understanding of human behavior needed to implement these ideas had lagged.[49] To help close this gap, he pressed the National Research Council to set up a Committee on Food Habits (CFH), chaired by the prominent anthropologist Carl Guthe, to help change what Americans ate.

Guthe recruited some of the country's leading social scientists, including the social anthropologist W. Lloyd Warner and the cultural anthropologist Ruth Benedict, to the committee, but it seemed to drift rather aimlessly. Its first contribution to national defense was hardly an impressive demonstration of applied social science's usefulness for the war effort. In March 1941—perhaps after pondering the social significance of the hit Disney movie *Snow White and the Seven Dwarfs*—it suggested that the following names for enriched flours would have great popular appeal: Vito, Vibio, Bermaco, and Bicapt.[50] Finally, on December 6, 1941, the committee gained some direction with the appointment of a full-time executive secretary: the strong-willed, loquacious, thirty-nine-year-old anthropologist Margaret Mead.[51] She was supremely confident that cultural anthropology held the keys to many of society's mysteries, and her appointment cemented its dominance of the committee. She thought that anthropologists' ability to see cultures as a whole and to see "the regularities in human behavior"—demonstrated in her work on Samoa and the "patterns" Benedict had found in other cultures—should help them to discover the interrelationships between food habits and other aspects of American society.[52] If one could find these linchpins, dietary changes could be successfully introduced, and in ways that did not threaten culture.

But finding useful patterns was not easy. When Wilson asked Mead what lessons could be learned from John Bennett's study of foodways among dirt-poor southern Illinois riverbottom folk, her reply seemed to be that such people could be reached only by appeals to status, "and that only when cast in very local terms."[53] Another memo, based on some conversations with midwesterners, warned that nutrition education programs had to face hostility from men, to whom welfare programs were "symbolic of softness, feminine weakness, and passive defensive attitudes." As an example she cited the following interchange with a fifty-year-old Indiana shopkeeper:

Q: I see you are having a Nutrition campaign here.
Man: (speaking with great vigor) Yes, eat more food and KILL MORE JAPS.
Q: (naively inquiring) Oh, is that what it is about?
A: (in a disgusted voice) NAW. Of course it isn't. It's for *children*. But wouldn't you think in a time like this the government would spend its time on something offensive? (very angry voice)[54]

A long memo to Wilson in June 1943 emphasized antiauthoritarianism in America, suggesting that because Americans resented federal orders his officials should be be particularly sensitive to local sensibilities and should just "suggest" programs on an informal basis.[55] Other anthropologists had little more to offer: some tidbits here and there, but nothing approaching the grand schema Mead originally hoped for. The committee was reduced to savoring the most minor successes: agricultural extension agents who showed farmers' wives how to plant more vegetables in their gardens and "put up" more of their produce; social workers who managed to convince some Italian slum dwellers to feed their children milk instead of coffee. A CFH anthropologist helped society women in New York City organize a series of ethnic dinners at which less affluent people from different cultures could swap economical recipes.[56] Inevitably, their methods involved face-to-face confrontations of the most time-consuming kind. The only workers available seemed to be scattered groups of anthropology undergraduates who could be enlisted to do simple one-question surveys as a course requirement.[57] A CFH experiment in trying to increase cheese consumption in Syracuse concluded that the poor were "little affected by newspapers and radio [but were] quite strongly affected by personal contact." The campaign to collar them took eighteen thousand person-hours and still reached only three out of ten of the city's shoppers.[58] Even those who emphasized the importance of working through networks of friends—a favorite Mead technique—had no solution to the essential problem, which was the very low teacher-student ratio.[59] To change the diets of the millions of Americans who were thought to need better nutrition by these methods would require many hundreds of thousands of nutrition workers.

These experiments in applied anthropology also presupposed that the social scientists had a coherent nutritional message to spread, something that, for the most part, was not the case. The government nutritionists and administrators they were supposed to help were quite horrified by their ignorance of nutritional science.[60] Even Mead later conceded that the CFH "was a dreadful trial to government agencies in its early days because most of its members didn't appear to know anything about the content of nutrition programs at all. . . . It wasn't at all clear why people who knew so little about the particular, precise value of Vitamin C had any right to talk about nutrition at all."[61]

But perhaps the greatest impediment lay in American cultural anthropology itself. Most American cultural anthropologists—particularly those associated with Mead, Benedict, and their mentor at Columbia University, Franz Boas—were committed to the idea of the relativity of values. They were therefore reluctant in the extreme to tamper with any aspect of culture, particularly one as important as food. Their lore was full of stories of primitive cultures that had been destroyed by well-meaning people who had done things as seemingly innocuous as introducing them to the steel axe. Mead exemplified how immobilizing these concerns could be in her introduction to a CFH study of Italian-American "food patterns." "Foods have different prestige and different emotional values in different subcultures," she warned.

> Foods may be used to discipline children, to display maternal solicitude, to demonstrate adolescent independence of parental restraints. The food patterns of the Southeast sharecropper, or the New England fisherman, the first generation Italian or the Mexican resident of Arizona, are all coherent parts of a cultural tradition. Changes made in these patterns without reference to the whole tradition may produce unanticipated dislocations of ways of life which

> are deeply entrenched. Partial alterations in a traditional food pattern to meet new nutritional standards may make that food pattern less adequate than before.[62]

This fixation on the close relationship between food habits and other kinds of social interaction tied their hands in many ways. Mead, for example, was particularly reluctant to recommend anything that might undermine parental authority.[63] So the CFH warned that the cafeteria system, which thrived during the war because it was by far the most labor-efficient way of serving meals to large numbers of people, should not be used in family situations because it allowed children to eat with their friends rather than with their families.[64] Another CFH study warned those preparing to feed evacuees from bombed American cities against tampering with ingrained ethnic food habits. Where possible, separate feeding facilities should be set up for each ethnic group. (Since the study also said that everyone should have free choice regarding which cuisine to eat, it inadvertently raised the rather tempting prospect that the evacuation camps might turn into something like ethnic food fairs, with evacuees being able to breakfast on ham and eggs in the "American" tent, lunch on pirogi in the Polish tent, and dine on veal parmigiana in the Italian tent.)[65]

In practice, then, there were usually so many "complexities," as Mead called them, as to make the committee's research and recommendations of little value to the nutrition program.[66] When the Nutrition Division stopped contributing to the CFH's program of original research in mid-1942, Mead seems to have realized that the CFH could never afford the kind of comprehensive study she thought necessary to get to the heart of how to change American food habits.[67] The committee did manage to fund some studies of ethnic eating habits, but they were so bare-boned that they involved only one informant from each ethnic group.[68] The studies' fine regard for the delicacy of any attempt to change food habits made them of little use to policy-makers. The one on Polish-American eating habits noted that, since there was a Polish proverb warning that eating a green apple before Midsummer's Night would cause a frog to grow in one's stomach, those in charge of group feeding in emergency situations should be careful not to offer Poles apples before June 24.[69] Otherwise, much of the work of the committee involved meetings and conferences (particular favorites of the garrulous Mead) at which people would report on things like the food situation in Mexico City and famine relief in Bengal.[70] In July 1944 Wilson bemoaned the committee's failure to evoke much government interest in its work, but he could think of no way to make it more relevant to the war effort.[71] Mead became increasingly involved in projects having little to do with food, such as explaining to Englishmen why GIs seemed to have inordinate success in seducing English women. Much of her remaining concern over food was redirected to world hunger, and the CFH was reduced to assuring the government that Americans would indeed be willing to see some of their food go to alleviate it.[72]

By war's end there was little to show for the CFH's efforts. Its major accomplishment, said Mead rather cryptically, was having its "point of view" embodied in the United Nations Educational, Social, and Cultural Organization (UNESCO). It had failed, she admitted, to find a solution to "the problem of changing food habits."[73] Over thirty years later, the Congressional Research Service would still lament that for government and nutritionists "little methodology exists to effectively modify the diet of large population groups. . . . The so-called food faddists are apparently more successful in gaining acceptance of their ideas than are the professional nutritionists."[74]

Yet people other than faddists *had* been changing the American diet. However, like the faddists, they were motivated by profit, not public service. This was apparent at one of the conferences Mead organized in 1942 at which experts from various fields gave their perspective on "the problem of changing food habits." There, representatives from the meat industry told how they managed to reverse a mid-1930s' decline in meat consumption. First, their trade association, the American Meat Institute, hired the well-known pollster Elmo Roper to find out the causes of the dropoff. He discovered that, although meat was by far Americans' favorite food, they thought it was fattening, hard to digest, expensive, and not a good source of vitamins. The AMI then hired the Leo Burnett advertising agency to combat these canards. It focused first on "thought leaders in the field of nutrition (doctors, dentists, nurses, nutritionists)" and mounted a special campaign "couched in their own language." Then it targeted home economics teachers, retailers, and finally consumers, who were urged to prepare "well-balanced meals" in an advertising blitz in national magazines, radio, and outdoor advertising. This was abetted by a separate "well-balanced meal" campaign by the National Live Stock and Meat Board. Its home economists ran cooking schools and sent out weekly bulletins extolling meat through news syndicates covering 238 newspapers. It also regularly distributed material on the wonders of meat to 240 of the nation's 800 radio stations, maintained a "clip-sheet" service for 102 metropolitan newspapers, sold a cheap textbook called *Ten Lessons on Meat* to high schools, and plied home economists with literature on meat's healthful properties, including a monthly bulletin slyly called *Food and Nutrition News.*[75]

While impressive, the meat story inadvertently demonstrated something quite demoralizing for those in the government nutrition program: Not only were food processors spending millions to promote their particular brand-name foods and vitamin producers pushing their products, but producers were also devoting enormous resources to promote certain *kinds* of foods. The flour millers, dairy producers, citrus fruit growers, grocery manufacturers, can and bottle makers—even the macaroni manufacturers—spent millions engaging in more or less the same kinds of promotions as the meat people. What prospects did the government, with its relatively puny promotional budget, have of gaining a hearing for its nutritional message above the resulting cacophony? One more campaign was not going to have much impact. Hoover's World War I organization had churned out posters, placards, and pledges at a time when food advertising was still in its infancy. His volunteer public speakers criss-crossed a land that had not yet been exposed to the slick teams from Crisco and Spry. The government representatives at the national nutrition conference in May 1941 acknowledged this relative ineffectiveness by getting commitments from the food industries to help spread its nutritional message. The large processors made good on this promise, but in ways that distorted the message to highlight the importance of their own products.[76]

The Nutrition Division also worked out a system patterned on the *Good Housekeeping* and AMA "seals of approval" for giving government benediction to foods deemed nutritious. Processors whose nutrition claims were approved by the Nutrition Division could advertise this fact and put a special insignia on their products. The logo was amateurish-looking and corny—a profile of what looked like a leering Uncle Sam holding a forkful of food to his mouth and the slogan "US Needs US Strong, Eat Nutritional Food"—but producers rushed to flaunt what amounted to official endorsement of the nutritional value of their products. Of course, breakfast cereal makers were particularly zealous, flooding the befuddled bureaucrats in McNutt's office with requests for endorsements for everything from the wholest of whole grain cereals to products that consisted of little more than white flour or rice, air, and sugar.[77] "DRINK—TO YOUR HEALTH!" said an ad for Stokely's tomato juice emblazoned with the symbol and slogan.[78] Birds Eye Frozen Foods gave out over half a million leaflets with the Basic Seven chart on one side and a chart showing how vitamin-packed its products were on the other.[79] The Meat Institute inserted the message that MEAT IS AMONG THE IMPORTANT DAILY FOODS RECOMMENDED IN THE GOVERNMENT'S PROGRAM FOR BETTER NUTRITION in the "US needs US strong" logo itself. "At least two fruits daily," said the smiling Uncle Sam in a 1942 Libby's ad, "and one of them can still be Libby's *field-ripened pineapple from Hawaii!*"[80] Wheatena cereal's smiling Uncle Sam reminded workers that "your government wants you to start each day with a breakfast that promotes health and vitality."[81] Kellogg's told restaurateurs that the "U.S. Official Nutrition Food Rules" would mean bigger breakfast checks for them because they encouraged "coffee and roll" customers to "step up" to its whole grain or "restored" cereals.[82] The Doughnut Corporation of America, however, ran into a roadblock. Its 1941 "Vitamin Donuts" campaign had fallen flat. Nor had a successor campaign built around "the great morale value of donuts" been a success. Now, Nutrition Division nitpickers barred it from switching to the term "Enriched Donuts" and would only approve the unappetizing-sounding phrase "Enriched Flour Donuts."[83] On the other hand, the Nutrition Division was so pleased with Sunkist's booklet "Feeding the Child for Health" that *it* distributed the pamphlet as well.[84]

Advertisers' enthusiasm for the official program soon waned. When the

Food and Nutrition Board, seeing them repeatedly chop up or otherwise distort its carefully crafted nutrition chart, decreed that its Basic Seven list could only be reproduced in toto, most advertisers decided to drop it rather than publicize competing foods.[85] They turned to ways of tying their often-dubious nutritional messages to the war effort without having to gain government approval. Ads for "Nutritious" Nestlé's cocoa said chocolate had been selected by the U.S. Army for its Type D emergency rations "because it is a concentrated energizing food."[86] Candy manufacturers played other notes from the energy theme, sending the schools charts showing how sugar was used in the "combat and emergency foods of U.S. fighting forces."[87] They even persuaded the first lady, Eleanor Roosevelt, and the army quartermaster-general to appear on a nationwide broadcast supporting their campaign to have Americans send tons of confections to loved ones in the service to "fortify energy" and "boost morale."[88] Florida grapefruit became "the Commando Fruit," its symbol a bayonet-wielding soldier with a maniacal glint in his eye; housewives were urged to "fortify the family" with it because it was "Rich in Victory Vitamin C."[89] The H. J. Heinz Company, trying to dissuade restaurateurs from cutting corners on condiments, pointed out that "Uncle Sam" had asked ketchup manufacturers to set aside much of the 1943 pack for soldiers and sailors, *"and Uncle Sam is an expert in nutrition."*[90] Some California food producers took over moribund voluntary local Nutrition Division committees and had them attest to the healthfulness of their products.[91] By mid-1943 the government's fork-wielding Uncle Sam urging Americans to "keep US strong" had disappeared, and with him went the government nutrition experts' confidence that their efforts would surpass Hoover's in effectiveness.

Politics helped push the nutrition program deeper into the bureaucratic morass. In December 1942 FDR tried to mollify farmers resentful over having food prices administered by legions of city-slicker lawyers in the OPA by giving one of their own, Secretary of Agriculture Claude Wickard, the impressive title of national food administrator.[92] Although decisive control over food prices and rationing remained with the OPA, less important food functions were then transferred to his department's new War Food Administration (WFA), including, in April 1943, M. L. Wilson's Nutrition Division. This meant that the nutrition program would thereafter function within a government department constructed almost entirely of stumbling blocks to giving Americans unbiased information about food. Hazel Stiebeling's close call with walking the plank for suggesting that the well-off need not eat as much wheat flour as the poor had testified to the risks run by departmental employees dealing with nutrition. Another sign of the minefields involved was posted in July 1941 when the department's consumer counsel took to the radio in a seemingly innocuous attempt to explain new government regulations for the labeling of margarine enriched with vitamin A. An outraged congressman from dairy-producing Minnesota took the floor of the House to demand a congressional investigation of such "un-American activities" as well as the "subversive work of the 'lavender' lawyers, 'pink' economists, and 'mauve' home economics ladies, operating together on Uncle Sam's payroll to transform our social institutions."[93]

The WFA brought in none other than the Mayo Clinic's Russell Wilder to head its Civilian Distribution Branch. Predictably, given his thiamin-mania, its first official order was a decree that *all* bakers' white bread be enriched.[94] That done, he announced that his agency's first project would be to study war workers' diets and "develop new foods to replace those which require a lot of labor to produce and distribute."[95] This never quite happened. The CDB did amass a pile of data on industrial eating, but to little avail. Unlike the system in Britain, where government-run factory canteens had a clearly beneficial effect on the working-class diet, factory feeding of workers in the United States remained entirely in the hands of private employers, most of whom were not interested in providing nutritious or even palatable meals. Many thousands of armaments workers continued to eat in the most rudimentary facilities in jerry-built plants. There was only one cafeteria for the more than ten thousand workers at Ford's Willow Run, Michigan, bomber plant, and it was too far from the factory floor for workers to make it there and back in their half-hour lunch break. Instead, they ate at lunch wagons, which sold what they regarded as overpriced "lunch meat" sandwiches, macaroni, or beans—slopped out of dishpans onto paper plates—and coffee generally reviled as "dishwater."[96]

An enraged San Diego aircraft worker managed to get through to the authorities by wiring Secretary of the Treasury Morgenthau: "SEVENTY-FIVE THOUSAND WORKERS FOR CONSOLIDATED AIRCRAFT ARE BEING FED SUBSTANDARD FOOD. MORALE IS LOWERED IN AN APPRECIABLE DEGREE BY TYPE COFFEE SERVED. EXAMPLE TONIGHT'S MENU CALLED FOR HAM AND MACARONI. SPENT 35 CENTS FOR HAM AND MACARONI. HAD MY PLATE CHECKED BY SIX CO-WORKERS WHO WERE UNABLE TO FIND FAINTEST TRACE OF HAM ON PLATE. . . . PERSONALLY THINK WE ARE BEING FED WORSE THAN IN CONCENTRATION CAMPS IN GERMANY."[97] However, when the telegram was sucked downward in the bureaucratic vortex to the Nutrition Division's industrial feeding section, its officials declared themselves powerless to do anything but make recommendations to private industry; they could deal with individual employers only on a "personal and random basis."[98] The National Association of Manufacturers lamented that surveys showed aircraft workers were not eating nutritious meals, but it promised to help educate their wives, not their employers, in nutrition.[99]

That the government was unable to do anything about factory food may not have been a great tragedy, for the stratospheric standards used in many of its surveys gave it a rather cockeyed view of the problem. One survey of cafeteria trays in Illinois graded workers' diets "poor" if only *two* of any of the following items were missing from their lunch trays: an 8-ounce glass of milk or two foods made from an equivalent amount of milk; ¾ cup of green or yellow vegetables; one serving of meat, cheese, fish, or eggs; two slices of whole grain or enriched bread; one square of butter or fortified margarine; and one of the following groups of foodstuffs: 4½ ounces of tomato or grapefruit juice, half a grapefruit or an orange, or 4½ to 5 ounces of raw cabbage or green pepper. Not surprisingly, 71 percent were judged "poor" and 21 percent were called "borderline," having missed out on only one category. Only 8 percent were classified as "adequate."[100] It was probably no great loss to the war effort, then, that Wilder's agency spent most of the war grazing unobtrusively in a remote bureaucratic pasture. It had no input into the important decisions about how much food civilians required; nor did the FNB. The real powers-that-be, at the OPA, used as their yardstick not the RDAs but average civilian consumption for the years 1937 to 1941.[101]

But by early 1943 nutritional considerations were being pushed to the background by the hubbub accompanying the introduction of food rationing. In December 1942, faced with the first meat shortages, the Nutrition Division mounted a Share the Meat campaign, which attempted with little success to persuade people to voluntarily limit themselves to two and a half pounds a week.[102] The next month it canceled plans for a National Nutrition Week because the onset of rationing would bring "a high and continuing interest in nutrition which will not require special stimulation."[103] The drift away from exclusive concern over nutrition was exemplified by its new title when it joined the WFA: the Nutrition and Food Conservation Branch. By the end of that year, conservation—particularly the promotion of home canning—had become more important to the unit than nutrition.[104] How ironic it was that the people who, armed with the new nutritional science, had set out to blaze a much different trail than their World War I predecessors ended up following in their footsteps.

Some had thought that support for school lunch programs might provide the government with opportunities to redirect the American diet toward the goals of the Newer Nutrition, but here again the USDA embrace meant the kiss of death. In 1943, with stocks of surplus foods dwindling and transportation snags bottling up many farm products, members of Congress from the farm bloc pushed through an appropriation of fifty million dollars for local school boards to purchase foods that were abundant locally. Meanwhile, the USDA continued to send them items it had purchased to help support prices, such as evaporated milk and canned prunes. In all, almost a third of the nation's schoolchildren—most of them rural—received some food aid.[105] But commodities were bought not because they were needed for lunches but because farmers could not sell them at a good price. School districts were inundated by foods they did not want and could not store. Perishable foods rotted en route to schools or arrived unannounced at schools that could not refrigerate them. Finally, in 1944, as criticism from consumer groups, home economists, and parents reached a crescendo, the appropriation was temporarily blocked by urban representatives trying to break the USDA's grip on the program. But after almost a year of struggle—

during which it became an issue in the 1944 election campaign—all that was achieved was to have the Office of Education, a bureaucratic lightweight, given three million dollars to supervise the actual distribution of the foods and provide a modicum of nutrition education while doing so. In return for allowing this tiny nutritional toe in the door, the USDA allocation for the program was raised to one hundred million dollars.[106]

Clearly, government programs to improve the nutrition of the people—whether aimed at housewives, war workers, or schoolchildren—had little impact. Nevertheless, the nutrition experts did lay some groundwork for the future. The medical establishment's deep suspicion of vitamin sellers had become entrenched in important government agencies. It had also become clear that the American diet could not be changed by science or even government alone, that the food industries had to be on board as well. However, these were long-term considerations. At the time, they were pushed into the background by the government program that certainly had the greatest impact on the American diet: the wartime system of price controls and rationing.

CHAPTER 6

Food Shortages for the People of Plenty

The nation did not greet the declaration of war with a determined commitment to belt-tightening. Instead, it responded with a run on sugar. Thousands of shoppers, recalling World War I sugar shortages, rushed to grocers demanding hundred-pound sacks of it. The government called the unseemly rush the product of unfounded rumormongering. The next month, however, it conceded that sugar would indeed be rationed, and in May 1942 Americans grimly trooped to local schools to be issued their first ration books.[1] Some nutritionists pointed to the brighter side: Compelling what one public health doctor called "the greatest overconsumers of sugar in the world" to cut back on white sugar, devoid of nutrients, would force them to get more of their sweets from nutritious fruits. There would be a salubrious turn toward honey, molasses, and other sweeteners thought (erroneously) to be more nutritious than sugar. (Blackstrap, the lowest grade of molasses, was a particular favorite among health food enthusiasts, even though its minimal mineral content came from the deteriorating metal containers in which it was boiled.)[2] But most Americans were loathe to give up the white stuff. Despite the fact that the rationing (one-half pound per person per week, with additional amounts allowed for those who said they needed it for home canning and preserving) was not particularly onerous, they spent the rest of the war carefully husbanding their supplies of the precious real thing.[3]

The subsequent introduction of rationing for other foods provoked similar hoarding sprees and caused black markets to sprout almost instantly. The government tried to avert pandemonium by announcing impending rationing weeks in advance, simultaneously freezing prices, but to little avail. The announcement in early 1943 that canned meats and fish would be rationed prompted a gigantic hoarding frenzy. Housewives pushing baby carriages and anything else with wheels invaded groceries, accompanied by children pulling toy wagons, and cleaned the shelves of almost everything metallic. "Never saw anything like it," said an astonished New York store manager. "Some women have bought enough this morning to feed a family nothing but processed food for a month," declared another.[4] Sometimes false rumors of impending rationing acted as self-fulfilling prophecies, causing shortages that themselves led to rationing. Shipments of evaporated milk to the domestic market soared by 40 percent between July and October 1942, yet hoarding caused a severe shortage and, ultimately, rationing.[5] Adding foods to the ration list seemed to heighten their desirability. In mid-1942 rumors that coffee was to be rationed prompted frantic shoppers to sweep the shelves clean of it several months before the rationing actually went into effect. Almost overnight it became regarded as a necessity—something experts now called "an essential morale builder." When it was rationed in November 1942, its popularity soared further as many non-coffee drinkers began to use their prized ration stamps to savor its delights. Yet, lo and behold, when coffee rationing was ended in July 1943, sales dropped markedly as it lost much of its allure.[6]

The contrast with how the British, who shared a culinary heritage with America, reacted to much more severe deprivation is striking. While food rationing and shortages seemed to bring out the worst in Americans, it brought out the best in the British. While Americans hoarded and connived at beating the system, the British took considerable pride in equitably sharing their meager food stocks. While the British by and large had faith in their government and never doubted that the shortages were real, many Americans questioned the necessity and even the legitimacy of rationing, refusing to believe that certain foods were really in short supply. The British would later look back on their years of "making do" as their finest hour. In America, William S. White wrote in 1947, there was "a tremendous legacy of resentment at the mere memory of the wartime controls."[7] Even Canadians took food rationing with more equanimity than Americans, despite the fact that one-third of the nation was lukewarm, at best, about the war.[8]

The differences probably stemmed from some deep-rooted American attitudes. Although most Americans supported rationing in principle as a fair method of equalizing sacrifice, in practice it ran afoul of their historic suspicion of government and its impositions, which made them prone to thinking that it was being imposed unwisely and administered unfairly.[9] Government declarations that there were food shortages also ran head on into the idea that the country's true symbol was the cornucopia. The Depression, with its talk of farm surpluses and reverence for hearty home cooking, had reinforced this idea of a land of perpetual agricultural abundance. For many Americans, to ration food—particularly when no invader had touched any of their fabled croplands—seemed to make no sense. "Famine in a land of plenty is a disgrace," said an irate businessman questioning whether there really was a meat shortage.[10] When a shortage of rice and grits hit South Carolina, a Charleston newspaper, suspicious of its origins, warned Washington that "people in a land of plenty cannot live long on hope while their stomachs grow empty."[11]

Americans were therefore easy prey to rumors of conspiracies behind the apparent shortages. South Carolinians thought the rice that was rightfully theirs had been diverted to "Puerta Rica" and Canada, of all places.[12] When coffee was rationed, the word spread that there was really more than enough of it but that it was being destroyed because there was no space to store it. An Idaho senator said he opposed meat rationing because he had been "reliably informed" that in Philadelphia alone over one million pounds of meat had been destroyed for lack of storage space.[13] The House Un-American Activities Committee investigated reports that Japanese interned in camps in Wyoming had hidden away huge caches of food, particularly mayonnaise. A former official of an Arizona camp testified that its internees hoarded enormous amounts of bread—hardly an Oriental staple—to be dried and stored in the desert, ready to supply parachutists and other invasion forces.[14] It was often said that conservation campaigns such as that to conserve fats for nitroglycerine production were intended not to cope with real shortages but to make civilians feel that they were contributing to the war effort.[15] "I suspected the ration system was a patriotic ploy to keep our enthusiasm at a fever pitch," recalled a soldier's wife who bought black-market food. "Almost everybody had a cynical feeling about what we were told was a food shortage."[16] Stories that the armed forces were secretly destroying immense amounts of excess food circulated around large military installations. Critics charged that canned goods were rationed only because the army had bought much more than it needed and that it did not even have the space to store the surplus.[17] In July 1943, when Gallup asked a cross-section of Americans if "the current food situation should be investigated by an impartial committee"—implying that there was something fishy about the shortages—75 percent said yes.[18] It is no wonder that Margaret Mead thought that false rumors posed the knottiest problem facing the government food program.[19]

Shortages of foods that were not rationed, often caused by transportation or storage snafus, also jangled consumers' nerves. By fall 1942 shoppers were accustomed to seeing whole sections of grocery shelves empty. The dried fruit counter might offer only prunes; canned salmon was seldom available; and canned fruit juice would be labeled "2 to a customer."[20] Supplies of popular nonessentials such as whiskey, canned beer, and bubble gum were also intermittent. At times, the government seemed to go out of its way to exacerbate the situation. The January 1943 order that all bakers' bread be enriched was overshadowed in the public mind by another part of the order, intended to reduce wastage, which banned the sale of sliced bread because it went stale faster than the unsliced kind.[21] Outraged restaurateurs were further infuriated by a subsequent order that they buy only unwrapped bread.[22]

Shortages would often see-saw with gluts. In 1942 citrus fruits were in very short supply, but the nation seemed knee-deep in apples.[23] In fall 1943 there was a great shortage of eggs, but by April of the next year an oversupply had home economists racking their brains for new ways to use them.[24] Although the practice was illegal, wholesalers would force food merchants to buy items in oversupply if they were to get those which were scarce.

These "tie-ins" would then be quietly imposed on shoppers. Onions had been in very short supply in the winter of 1943, but by the end of May 1944 there was such an oversupply that people wishing to buy scarce string beans were being forced to load up on onions as well.[25] Housewives had to read newspapers "with serene and minute care," wrote Anita Brenner, to pick up hints about future gluts and shortages. "Impending storms around some particular food gather outside the news for weeks or months: remarks dropped by the grocer ('We're running low on this, better get some more . . .' or 'Sorry, we couldn't get any this week')" and the ever-present rumors.[26] On January 1, 1944, Margaret Mead glumly reported: "Rather than public opinion having gone forward in the understanding of rationing from Stage 1, 'the Government wants us to eat less,' through Stage 2, 'rationing keeps other people from taking an unfair advantage,' to Stage 3, 'rationing is a way by which *we* get the food we need,' it has regressed. Last spring a great many people reached Stage 2; now Stage 1 is again more in evidence."[27]

The most disturbing of the little ration stamps seemed to be those for fresh meat. The failure of the voluntary Share the Meat campaign in the fall of 1942 had contributed to a meat shortage, forcing an OPA price freeze, which in turn fed a flourishing black market.[28] So in March 1943, when a new rationing system was adopted, meat was added to the list. The new system replaced the coupons allocated for individual foods with colored stamps denominated in points. Rationed foods were initially divided into two categories, canned goods and fresh foods, the former with red and the latter with blue stamps. The new system allowed the OPA to charge more points for foods that were in great demand, such as beef steak, and fewer for those that were not so popular, such as organs and other innards, as well as to adjust the amounts as supplies rose and fell. The system was complicated enough for the OPA to precede its introduction by a press and radio campaign lasting several months.[29] Predictably, in the week before the new system went into effect butchers were swamped by panicky buyers. Police had to be called out in Columbus, Ohio, to control a mob of punching, shoving people. In Cleveland fifty thousand people caught up in an apparent carnivorous frenzy milled around three big markets, trying to load up before the fearsome new system took effect.[30] In New York City mounted police had to disperse a mob of two thousand butchers who marched on the wholesale meat market in Brooklyn to protest their own inability to get new supplies.[31] Soon after meat rationing began, Wyoming miners demanded extra meat rations. Lumberjacks in the state of Washington defied labor's no-strike pledge and walked out in support of the same demand. Meat shortages also played a role in the coal strike of 1943, as disgruntled miners accustomed to having three pork chops in their lunch pails were appalled at finding lettuce sandwiches there instead.[32]

Like the sugar ration, the meat ration was not, by Allied standards, particularly meager. While about two and a half pounds per person per week remained the American norm, the British were getting along on one pound, of far inferior quality. Most Russians, of course, rarely saw it. Nor did the people in German-occupied France in 1943–1944 or the Dutch, who in 1944–1945 were deliberately starved by the Germans. As for the enemy, the Germans were limited to twelve ounces, and the Italians were scraping by on three and a half to four ounces.[33] The Japanese, of course, got by on what was by American standards a derisory portion. The contrast is even more striking when one recalls that ration stamps were not needed for poultry or in restaurants or factory cafeterias.

The hundreds of letters that inundated new War Food Administrator Marvin Jones in August 1943 provide some insight into why Americans nevertheless felt so deprived. Cattle producers had bought full-page ads in major big-city dailies to reproduce a newspaper editorial from Jones's hometown, Amarillo, Texas, calling for a raise in the ceiling price of beef. Again, the theme was that there was no real shortage of beef—in this case, price controls had caused the present shortage of slaughter cattle and the flourishing black market in beef. One statistic struck a particularly raw nerve among those who wrote to Jones: Whereas five years earlier sixty-one million cattle were roaming American ranges, now eighty-two million were "dammed up" there.[34] This seemed to prove what so many Americans wanted to believe: that there was no real shortage of food in America. "Why are there record cattle on the Western ranges and empty meat plates thru out the rest of the country?" said a typical letter. "If there were a genuine shortage of meat, there wouldn't be one single American who would dare raise his voice in complaint." "It is and has been for some time a known fact to thousands of us who look the facts in the face that we have sufficient meat on the hoof to supply our people, our armed forces, our lend-lease committment and an additional eighty-five million people," said an angry San Franciscan. A Missouri farmer found it outrageous that "our country is long on cattle while our city friends haven't tasted beef for weeks and some of them months."[35]

Of course, many of the irate citizens found the "would-be college economist professors" of the OPA at fault. Particularly annoying was an OPA suggestion that Americans might substitute other proteins such as soy beans, which were in good supply, for meat. "These New Deal social revolution dreamers . . . just cannot make over the eating habits of the people while the farms and ranges have an oversupply of cattle," said an indignant California realtor. "If those experts in Washington who are trying to force the soy bean upon the American people had brains of croton oil there wouldn't be enough to physic a flea," said the more rustic Missouri farmer. Others attested to the unique place beef occupied in the hearts of Americans. "The propaganda that . . . soy beans are just as nutritious as a T-bone steak, while probably true, is laughable," said a Brooklyn woman. "You can't place a flock of soy beans in front of persons who are accustomed to thick steaks, and then tell them that they're deriving exactly the same nourishment out of the beans as they would from the steak." "I hope that War Food will hurdle the difficulties that have been unnecessarily created," wrote a plaintive New York businessman. "I relatively dislike yeast, soy beans, fish, chicken, and pork in the order named, but I do love roast beef and sirloin steak."[36]

In January 1945, when asked which rationed product they found it hardest to do without, most Americans said sugar, meat, or butter, in about equal proportions. Only 2 percent said canned foods.[37] One reason canned foods were so low on the list was that the war brought a remarkable rise in consumption of fresh vegetables, which were not rationed. Initially, the government had not helped the vegetable supply when it interned almost 120,000 people of Japanese ancestry—many of them the most efficient market gardeners on the West Coast—in camps in the interior, but its revival of the World War I program of Victory Gardens proved eminently successful. By fall 1943 there were twenty million Victory Gardens producing 40 percent of the nation's vegetables in backyards, in vacant lots, or alongside factories.[38]

The parallel campaign to have people do their own preserving was almost equally impressive. The War Production Board diverted steel from the munitions industry to pressure-cooker production. Department stores ran films and displays on canning, society ladies enrolled in classes on it, home economists lectured on it to ladies' clubs, extension agents demonstrated it to farmers' wives, and charities taught it in the slums. Never before in the nation's history had such huge quantities of food been preserved at home: Three-quarters of America's families put up an astounding average of 165 jars a year.[39] Novice canners using shoddy wartime equipment also produced a record number of disasters. The files of state Victory Garden committees were full of stories of "Victory Model" pressure cookers with faulty gauges, leaky valves, and a frightening tendency to erupt, as well as of exploding jars, rusted jar bands, and defective lids. Innumerable stoves were ruined, kitchens were splattered, and victims were hospitalized with severe burns, cuts, and botulism.[40] At war's end a grocery industry analyst concluded that so many women had had "such unhappy experiences" with home canning that a decline was certain.[41]

Many women had little time for canning, for the demands of war encouraged a 50 percent rise in the female labor force. By 1945 37 percent of adult women worked outside the home. Yet although three-quarters of the new female workers were married, the large majority of wives—three out of four—still remained at home.[42] The national ethos that saw food preparation as women's work was therefore hardly shaken. Propaganda urging women to join Rosie the Riveter was a mere trickle compared to that emphasizing women's crucial role in keeping their families well fed and working for the war effort. When the American Public Health Association selected its wartime slogan in May 1942 from among over one thousand suggestions, it chose "They *Are* Rolling; They *Are* Flying; Keep *MEN* FIT TO WORK OR FIGHT!"[43] "Mother, captain of the kitchen, guards the health and strength of the family these difficult days," began a typical article on how to prepare "victory lunches" for working men.[44] Thousands of women wrote

Betty Crocker acclaiming her creation of the Home Legion, devoted to encouraging their contribution as "homemakers" to the war, as "the finest thing that had ever been done for the American homemaker." Over seventy thousand of them signed up, pledging to work at developing full-time "homemaking" as a career.[45] Even those married women who had left hearth and home were regarded as having done so only for the duration—at least by government, industry, and the media.

Most important, men's expectations hardly changed. When pollsters asked servicemen for a "blueprint" of their "dream girl" they described, not someone with Betty Grable's legs or Katharine Hepburn's wit, but a short, healthy housewife, thoroughly devoted to her children, whose cooking mattered much more than her "braininess."[46] So, amidst all the recipes for rationed foods and advice on labor-saving in the kitchen, there was nary a suggestion that perhaps the household tasks of two-income families might best be accomplished by having men take charge of some of the cooking duties. Only one cookbook aimed at men seems to have appeared during the war, and two-thirds of it consisted of recipes for breakfast, snacks, and mixed drinks.[47]

One consequence of the shortage mentality was the virtual disappearance of the mania about dieting. However, the idea that overweight contributed to poor health persisted, albeit at a relatively low level of intensity. In July 1943, at the height of wartime food shortages, when Gallup asked Americans if they thought that "most people you know would be healthier if they ate less," 64 percent said yes and only 21 percent said no. Although the poor and working-class respondents were not as concerned about excess poundage as the upper and middle classes, a sizable majority of the lower-income groups (59 percent) still said yes. Moreover, 77 percent of them thought that most of the people they knew should lose weight.[48] But class differences in attitudes toward food and health did not disappear. One study of a wartime community in the Midwest noted that executives and skilled workers had more "balanced" diets than unskilled workers, not just because they had higher incomes but also because, particularly among the executives, to eat the "correct" foods was the "proper" thing to do—something that denoted higher status.[49]

Although the executives' diets may have been "better balanced," the war did help close the dietary gap between the classes. In fact, it confirmed what some experts had already noted at the end of the 1930s: that higher incomes would do much more than anything else to improve the diets of the poor. The poor themselves had thought this all along. When asked in December 1940 whether their family's health would be better if they had more money to spend on food, 60 percent of a cross-section of Americans said no, but the large majority of the lower-income groups said yes. Asked what they would spend additional income on, most of them said meat, vegetables, fruit, and dairy products, in that order.[50] True to their word, when war work did raise their incomes, lower-income Americans did devote a much greater proportion of their increased earnings to more nutritious food—particularly meat, milk, fruits, and vegetables—than did those with higher incomes. Rationing and price controls furthered the equalization process by keeping what were regarded as "better" foods accessible to lower-income groups.[51] "Ration coupons have given Negro Americans a new equality," said Andrew Brown, a black Agriculture Department official who was otherwise pessimistic about how blacks were faring.[52] A researcher in rural Virginia noted that high tobacco prices had allowed lower-income families to "buy the kinds of food and clothing they have been seeing higher income families buy for a long time . . . the better cuts of meat, cheese, canned fruit and vegetables." The poor also seemed to be eating more of the rationed foods—the full allotment of two and a half pounds of meat, and so on—because they now thought these were officially dictated healthy amounts.[53]

The many variations on these themes were reflected in evidence that low-income diets improved quite remarkably, much more so than those of the better-off.[54] Government nutrition surveys, although questionable as accurate portrayals of what people were really eating, are still useful for showing trends. They indicated that from spring 1942 to fall 1944 consumption of meat and poultry by the poorest third of the population rose by almost 17 percent, while that of the middle- and upper-income groups declined by 4 and 3 percent.[55] From 1936 to 1948 the poorest third of city dwellers increased their consumption of protein by 30 percent, of calcium by 60 percent, and of iron by 61 percent—much more than the increases recorded for the middle, and particularly the richest, thirds. Lower-class consumption of other vitamins and nutrients—such as thiamin and niacin—also appears to have increased more than that of the better-off. Indeed, by 1948 differences among the three income levels seemed to have narrowed to insignificance, while in subsequent years they again widened.[56]

Meat shortages contributed to all-time highs in consumption of eggs and milk, as Americans turned to these as sources of animal protein.[57] Macaroni and cheese, cooked in milk, which had been a popular Depression economy dish, now became a kind of patriotic dish—a healthy, meatless source of protein whose appeal crossed class and regional lines. Americans also rediscovered beans and legumes as sources of proteins and became somewhat more amenable to one-pot dishes and meals. So, as in the First World War, interest in foreign cooking increased, as at least some Americans realized that foreigners had developed tasty ways to cook nutritious meals with a good variety of vegetables and little meat. The popular novelist Pearl S. Buck commended a Chinese cookbook to American housewives as of "inestimable value to the war effort" because there they could "learn to use meat for its taste in a dish of something else, instead of using it chiefly for its substance."[58] A *Journal of Home Economics* reviewer heartily agreed.[59] Even *American Cookery* magazine, which called American cooking "the finest in the world," admitted that "our allies in this great struggle have much to teach us about the proper preparation of food." (Swedes, whose cuisine it featured, might have been surprised to discover that they had lost their neutral status.)[60]

Meanwhile, the canned goods shortage encouraged canners and home gardeners to try relatively unpopular vegetables such as eggplant and broccoli or revive almost-forgotten members of the easy-to-grow squash family. These reinforced the longer-term forces impelling Americans to eat more green and yellow vegetables. A 1942 study of New York City schoolchildren showed them consuming much more milk, fresh fruit, and especially fresh vegetables (particularly, to the researchers' amazement, spinach) than a comparable group studied in 1917.[61] A survey in rural Texas that same year turned up similar results. Four hundred rural families—white, black, and Mexican; owners, tenants, and wage laborers—in three different parts of the state were found to have diets that were very varied and, by the lights of the day, healthy, especially compared to those in a 1927–1929 study. They consumed generous amounts of milk, eggs, and butter and twice as many green vegetables as in the previous study. They now also prepared their vegetables in the best manner: baking their sweet potatoes, cooking cabbage and greens only a short time, and using the pot liquor of greens.[62] As a result of all of these salubrious trends, in 1945 American consumption of fresh vegetables hit an all-time high.[63] So did consumption of vitamin C in the food supply, which never again reached its 1944–1945 peak.[64]

Of course, there remained the hard nut that never seemed to crack: the 15 to 20 percent mired on the bottom of the ladder. Not only did they seem little affected by wartime improvements, quite the reverse often seemed to be the case. Despite the OPA's efforts, food prices rose substantially during the war, causing hardship for those on fixed incomes or unable to find half-decent jobs in the war economy. In October 1943 Dr. Frank Boudreau, the head of the philanthropic Milbank Memorial Fund and chairman of the NRC's Food and Nutrition Board, pointed out that, unlike in Britain, where food distribution among the needy was based on physiological need, programs in America had been based on the existence of surpluses; once these had disappeared, so had the programs. Yet "many millions have not shared in wartime prosperity and are being squeezed between fixed incomes and higher prices for food and other necessities."[65]

Later, Labor Department statistics confirmed this gloomy view. Whereas in 1942 an individual or family living on a thousand dollars a year had to spend five hundred dollars a year on food, by 1945 that figure had risen to seven hundred dollars, a proportion that today is associated mainly with Third World poverty.[66] Andrew Brown, the black USDA official, noted in 1943 that at least poor urban blacks used to have a monopoly on pigs' ears, snouts, and tails. Now they had to compete for them with higher income groups.[67] The complaint was by no means frivolous. Before the war, for instance, most innards were unobtainable in the well-off western parts of Detroit. In 1943, however, one neighborhood butcher alone was reported to be selling three hundred pounds of tripe a week. For the first time in living memory, stores in Grosse Pointe, the suburb favored by wealthy white auto executives, were selling pigs' tails, snouts, ears, and feet.[68]

Still, for the rest of Americans the war represented a kind of high-water mark for equality in food habits. Not only did it help equalize the sharing of food and nutrients, it also represented a low point for class distinctions in how food was prepared and eaten. Throughout the 1920s and 1930s, as we have seen, food tastes were becoming ever more minor weapons in the social armory. Where the food was consumed—opulent mansions, "exclusive" restaurants, men's clubs, country clubs, restrictive resorts, and so on—was more important than what was eaten. ("You are *where* you eat," a distinguished-looking patriarch counseled a younger man in a 1991 *New Yorker* cartoon set in one of these establishments.) During World War II, even the few who clung to the older traditions of fine dining felt the pinch, for rationing and the labor shortage made many of the old ways of dining and entertaining unpatriotic or impossible. Even when expensive ingredients were available without resort to "Mr. Black," there was usually no one left in the kitchen who could cook and serve them; female servants in particular had turned in their aprons for more desirable war work.

In Washington, a city whose social life revolved around private entertaining, only the grandest of the city's grande dames managed to continue the tradition of formal entertaining. Most society hostesses abandoned the formal dinner party in favor of soirées at the Chevy Chase Country Club, known for its exclusive membership list, not its cuisine.[69] Nine months after the war's end, the aging Mrs. Cornelius Vanderbilt, still guarding the family mantle as the grandest entertainers of fin-de-siècle America, hosted thirty-odd United Nations delegates at a formal dinner party. Although she had recently moved from her fifty-four-room Fifth Avenue mansion to a more modest twenty-eight-room one further uptown, she was reported "not to have altered her style of entertaining noticeably." The table gleamed with silver and crystal, shone with white napery, and was laden with a gorgeous profusion of flowers and decorative fruit bowls. The food was served on beautiful gold-rimmed china. The menu, though, was much less impressive: consommé, grilled shad, turkey, potatoes, peas, asparagus hollandaise, pineapple and ice cream, cake, and coffee.[70]

"There is nothing wrong with the American army," British soldiers liked to say during the war, "except that they're overpaid, oversexed, and over here." What astounded the Brits most, however, was the extent to which the GIs seemed overfed. Their mutual enemy, the Germans, felt the same. Those who overran American positions often expressed amazement at discovering the quantity and variety of food GIs had at their disposal, as did those fortunate enough to be taken prisoner by the Americans. Never in the history of warfare, it was often said, had an army been as abundantly supplied with food as this one. While the average male civilian in 1942 ate 125 pounds of meat, the average soldier was alotted 360 pounds, most of it beef.[71] In early 1943 William Morgan, New York City's commissioner of markets, calculated that each of the bombardiers training at a Texas base was alloted eleven pounds of food per day. Yet the average civilian consumed only about four pounds. While declaring himself "as proud as anyone over the fact that our boys in the service are being better fed than any others in the history of warfare," he suggested that the ensuing waste was "partly responsible for our present food shortages."[72] Here is a typical day's menu at another U.S. Army Air Force base, at Randolph Field, Texas, in 1942:

> Breakfast: Assorted fruit, dry cereals, broiled bacon, two eggs, French toast and syrup, toast and butter, coffee or milk.
>
> Dinner (midday): Hearts of celery, green olives, head of lettuce, roast turkey and cranberry jam, mashed potatoes, raisin dressing, giblet gravy, buttered jumbo asparagus tips, creamed cauliflower, lemon custard or ice cream, rolls and butter, layer cake, preserves, coffee or tea.
>
> Supper: Fresh celery, smothered round steak, escalloped potatoes, frosted peas, strawberry ice cream, layer cake, bread and butter, coffee or milk.

No wonder that most of the men stationed there gained from ten to twenty pounds per month.[73] A study of one group of infantrymen showed them gaining an average of twelve pounds during and shortly after basic training, despite it being a time of peak physical effort.[74]

But there was more to the armed forces food experience than abundance. Rare is the person who has experienced service life who does not think that it affected his or her food tastes. By August 11, 1945, when the Japanese finally surrendered, 16,354,000 Americans, most of them quite young, had served in the wartime armed forces. For most, it meant exposure to a kaleidoscope of strange people with different values and food tastes. But these were mainly other Americans. Some observers thought that the foreign travel war involved would give these young soldiers a taste for exotic foods. In 1944 a *New York Times* writer speculated that the GIs in North Africa would return with a taste for couscous, those in the South Pacific would introduce America to the delights of breadfruit and soursop, and those stationed in India would bring back curries.[75] Some years after the war, observers such as James Beard credited an apparent broadening of American food tastes to wartime exposure to the delights of foreign cuisines. Troops in Italy, for example, were thought to have brought back a taste for pizza.[76]

But they were far off the mark. Few of the troops in invading armies had the opportunity to dine in the areas through which they marched. Soon after liberating a relatively unscathed part of Normandy, GIs did watch in amazement as townsfolk rapidly reopened the cafés and produced steaks, veal, eggs, artichokes, bread, butter, and wine from nowhere, so it seemed. But this was for the locals, not the infantrymen.[77] More often, as in Italy, civilians were starving, restaurants were destroyed, and the better food items were reserved for the pricey black market: officers' fare, perhaps, but not for ordinary GIs. By the time Italy and France were getting their culinary houses back in order, most GIs had boarded troopships for home. The occupation troops in Germany lived among a populace facing starvation—guaranteed only fifteen hundred calories per person per day—to whom they surreptitiously gave chocolate bars and other American foods. On the other hand, places such as Tunisia, the Philippines, New Guinea, and Okinawa never had much to tempt American palates. The only Asian cuisine that seemed to interest American troops in India was the fare served by enterprising Chinese chefs who set up American-style "chop suey" houses near their bases.[78] The real mark was left by the "square meals" American service personnel sat down to in their mess halls. The millions of meat, potatoes, and one-or-two-vegetable meals, accompanied by salad and dessert and washed down with cold milk, played a major role in speeding the process of nationalizing and homogenizing American food tastes.

The structure of the U.S. Army contributed to this. Unlike many armies, built around combat regiments whose members were recruited from the same localities, it threw recruits from various regions together. New Yorkers and Georgians ate side by side, dining on food cooked by Vermonters or Texans. One of the cardinal rules of military caterers throughout the world is to give the troops familiar foods. The last thing the provincial people who comprise the bulk of the enlisted ranks want—the thinking goes—is to experiment with food. In the American armed forces, this meant eschewing regional and foreign dishes and sticking to the kind of basically midwestern, "All-American" cooking that had become the national norm. The Georgian would have to do without grits for breakfast, the New Yorker would eat frankfurters with beans, not sauerkraut and onions. But both could look forward to roast beef and potatoes with peas and carrots. The "foreign" tastes would be foreign in name only: spaghetti with three cloves of garlic in enough sauce for one hundred soldiers; "chow mein" made of beef, celery, and Worcestershire sauce; "chop suey" made with beef, bacon, onions, turnips, corn, tomatoes, celery, chili powder, and Worcestershire or barbecue sauce.[79] Navy cooking followed the same principle, even down to having a similar recipe for the Worcestershire sauce–flavored beef chow mein.[80]

In 1944 the director of nutrition of the army's surgeon general's office announced that, after inspecting army kitchens around the world—England, Italy, North Africa, Egypt, Persia, India, and China—he was happy to find that "American fighting men wanted and were getting American food." Although cooks might be forced to use local fuels and even stoves, "they turned out our kind of food . . . good, plain cooking as a steady fare."[81] After the war, the army quartermaster general admitted that the determination to use American foods was a prime factor in the shipping shortages that plagued the military effort in the first years of the war. Food was second only to petroleum products among the items shipped overseas.[82]

The men in white T-shirts standing over enormous pots and pans cooking essentially the same foods in mess kitchens throughout the world did more than undermine regionalism; they helped mute class differences as well. Many young recruits were children of urban working-class immigrants, brought up in homes where ethnic food habits still prevailed. For them, the mess halls provided the final step in the process of adopting middle-class American food habits that had begun in school lunchrooms. Many a youth of Italian or Eastern European parentage, brought up to regard soft-crusted "American" white bread with contempt and think of milk as baby food, actually came to like both. The system also broke down barriers at the higher end of the scale. The officer/enlisted personnel distinction was maintained by segregating where officers ate and how they were served; but both groups' rations were essentially the same, except for some "extras" purchased for officers' messes.

All of this was largely the result of the modern turn the "ration" system had taken over the past fifty years. Early in the century, military leaders had been shaken by nutritional studies indicating that the lower classes—their raw material—were underfed and malnourished. The British were shocked at the runtiness of recruits in the Boer War; the French and German general staffs were troubled by pessimistic assessments of the diets of their lower classes; in 1918 the Americans called the physical condition of their draftees a national disgrace. As a result, modernizing military men had revamped their military rations in line with the ideas of the New Nutrition. Mess officers were enjoined to use their food allowances to supply certain amounts of protein, carbohydrate, fat, and calories per person per day. There were "garrison" rations, designed to provide the optimum ratio of proteins to carbohydrates, and "field" rations, which were richer in fats and calories. (The U.S. Army also had a smaller "Filipino ration," which it maintained into World War II.)[83] Still, how the foods were prepared was left up in the air. Only in 1896 had the U.S. Army finally come out with a cookbook, and it was simply a collection of some recipes gathered by the Quartermaster Corps from some experienced army cooks. Formal training for army cooks did not begin until 1905.[84]

The informal older system had its advantages, for there was considerable leeway for building meals around cheap local ingredients and using the money saved to purchase more luxurious foods. It also meant that troops stationed in any area would eat, and often get to like, local foods. The 1896 cookbook, for example, had a separate "Spanish" section, consisting mainly of Mexican- and "Tex-Mex"-style recipes, reflecting the large number of troops stationed along the Mexican border. By the time a new cookbook was issued in 1916, the new nutritional concerns had shoved this kind of diversity aside. It pontificated on the proper "balance" of proteins, carbohydrates, and fats and provided mainline "American" recipes for achieving it.[85] Still, the recipes were suggested ones, and there remained considerable leeway for local variation. After 1932 the amount of money provided to mess officers to feed the troops was based on the cost of a list of thirty-nine basic foods, but they were free to substitute nutritionally equivalent foods, as long as they kept a proper "balance" of nutrients, and could cook them as they wanted.[86] In the seaborne part of the navy, having to buy a ship's fresh provisions from what happened to be available at a particular time in Boston, San Francisco, Hawaii, Manila, or Shanghai made adaptation to local foods a necessity, if not a pleasure.

At the outbreak of World War II, however, both armed services worked toward obliterating these vestiges of heterogeneity. Cookbooks were no longer compilations of suggested recipes; they were manuals from which cooks learned a repertoire of set recipes. In 1941 the army, alarmed by the nutritional surveys lamenting Americans' poor nutritional status, began moving toward an extraordinary system of "master menus" whose goal was to ensure proper nutrition by serving soldiers exactly the same dietitian-formulated menus no matter where in the world they happened to be.[87] In early 1942 the quartermaster general hired Mary Barber, the Kellogg Company's director of home economics, as a "dollar-a-year woman" to furnish every U.S. Army command with the same monthly menus—prepared seven months in advance—detailing recommended menus for three meals a day for each day of the month. These would provide each soldier with all the necessary nutrients and at least five thousand calories a day.[88]

Even though it was assumed that some of the food would be lost, destroyed, or wasted before it was eaten, this still represented an extraordinary amount of food, and the fifty-four cents per person per day with which mess officers were expected to provide it was, by the standards of other armies, a small fortune.[89] As a result, although griping about food continued to be a hallowed tradition, most service personnel not only ate much more than they ever had in civilian life, they also ate a much greater variety of higher-status foods. The meat ration, for example, was heavily weighted toward the national favorite, beef. The official army ration apportioned ten ounces of beef per person per day versus only four ounces of pork and two ounces each of chicken and bacon.[90] Meal time was also one of the most pleasurable periods of the day, a respite from the harassment and the tedium of much of military life. Of course, there were still complaints about meager servings. Large stateside bases in particular were prey to rumors that meat portions were not what they should be because cooks were selling food to black marketeers. "Chief cooks are the richest men in the navy" was a common saying in naval shore installations.[91]

The services' concern over proper nutrition was manifested in how hard they worked at spreading the gospel of the balanced diet. The army Quartermaster Corps set out to obliterate soldiers' dislike for vegetables—fostered, it was thought, by indulgent mothers and regional prejudices. It made sure that meals always included a large proportion of vegetables, so that, according to the chief of its Subsistence Branch, "if a man wants to satisfy his hunger, he must eat some of them and thus he gradually acquires a taste for vegetables."[92] Troops could have seconds and even thirds of beef or other desirable main courses, but only as long as they came up with a "clean plate."[93] While they were allocated three times as much meat as the average male civilian ate, they were also provided with more than twice as much fruit and vegetables.[94] A naval doctor who studied seventy-five hundred sailors' meals in 1944 was satisfied that they were learning to eat well-balanced meals. Even tastes in vegetables were in transition, he said, as a new appreciation for those cooked for short periods in little water replaced the older preference for long-boiled ones.[95] Service dietitians often pointed out proudly that, while coffee had been the favorite beverage in World War I, in this war it was fresh milk.[96]

Of course, after eight or ten days at sea, sailors no longer had fresh milk to drink and fresh vegetables became scarce. But dry powdered milk provided them with plenty of creamed dishes, and dried and canned vegetables still allowed for considerable variety. (Submarines on long voyages posed more difficult problems—constipation plagued submariners.) In the army, troops away from their bases had to make do with various levels of field rations. Yet even here the army went to extraordinary lengths to keep diets abundant and adequate. Mobile field kitchens followed the troops into battle, committed to providing them with at least one hot meal per day. Chefs baked fresh bread in special portable ovens and cooked hot soups and stews with specially cut meat. Quartermaster Corps NCOs learned how to reduce a ton of beef from 134 cubic feet to a mere 32 cubic feet by clever trimming and boning. They even parachuted fresh eggs to troops on maneuvers in the United States.[97] Ice cream machines were standard equipment; their cooling products were particular favorites in the South Pacific.[98] The Coca-Cola Company prided itself on quickly erecting bottling plants in newly liberated areas to keep an ample supply of its product flowing to the troops at the front. For those unfortunates under fire, the famous (or infamous) C-rations had to do: two cans for each meal, one of which contained a meat and vegetable combo, the other instant coffee, sugar, and nine biscuits.

The emergency ration—for when soldiers were stuck in places where they could not even be supplied with C-rations—was an "improved" version of the World War I emergency ration: a chocolate bar with oat flour added to inhibit melting. (The main improvement was in the packaging, which was impermeable to gas attack, and the addition of 150 units of presumably morale-enhancing thiamin.)[99] Chocolate bar manufacturers who trumpeted this use of chocolate in their advertising did not mention that, according to the officer who perfected the new bar, the chocolate bar form was chosen because, at twenty-four hundred calories to the pound, it was "the nearest approach to straight fat (the most concentrated form of calories known) that we could make edible."[100]

In 1942 the emergency ration was more or less replaced by the K-ration, which was closer to the cutting edge of food science and technology. The new marvel weighed only two and a half pounds, yet it contained separate packages for breakfast, lunch, and dinner. Breakfast was a four-ounce can of veal luncheon meat, a packet of twelve malted milk tablets, foil-wrapped instant coffee, and three lumps of sugar. Lunch was a can of pork luncheon meat, a packet of twelve glucose tablets, and a tube of bouillon. For dinner there was a canned cervelat sausage, a chocolate bar, and two foil-wrapped discs of lemon powder with three cubes of sugar to turn it into lemonade.

Each meal packet also included four "defense" biscuits, four compressed graham biscuits, and a stick of chewing gum.[101]

By the end of the war, much more variety had been added to the much-derided C-ration, which now offered a choice of ten different main course meat cans. The "10-in-1" ration, for those on the front lines who were not directly under fire, included roast beef, canned fruits and vegetables, and even canned hamburgers. Yet although the army called these developments "almost revolutionary," it never pretended that the quality approximated the national benchmark for fine food—"home cooking."[102] Indeed, many of the innovations were distinctly unpopular. Dehydrated soups and vegetables, developed at tremendous cost and effort, were almost universally reviled. So were the newfangled biscuits and the powdered lemon drink, which was supposed to provide vitamin C.[103] Ironically, though, Spam, one of the most commonly derided luncheon meats ("ham that didn't pass its physical") became a postwar favorite—for reasons which remain a mystery.[104]

By the end of the war, there was much confidence that, whatever they thought of the food itself, the service experience had dramatically improved men's eating habits. The tin trays or plastic plates in which most service meals were served—with a section for meat flanked by smaller ones for potatoes and vegetables—symbolized the triumph of the idea of the balanced meal. At war's end an upbeat naval doctor told the American Public Health Association that the eleven million young men who had learned to eat balanced meals would apply that knowledge to the families, totaling forty to fifty million persons, they would soon be heading.[105] The president of the American Dietetic Association was confident that America's returning servicemen would become "the apostles of good eating . . . who would save the country's undernourished from themselves."[106] The hopes of mothers who had spent years teaching little Joe to drink milk and eat vegetables and fruit had been fulfilled, glowed a nutritionists' journal in December 1945. "The men in training camps all over the country were served meals featuring fresh milk, milk products, and ice cream, fresh fruits and vegetables, different kinds of meat and fish, whole grain and enriched bread and cereals. And now in the redistribution centers the kinds of foods chosen by men waiting for discharge indicate that their food habits include even more milk and ice cream, fresh salads, fruits and cereals."[107] A war correspondent on a troopship packed with GIs returning from Europe in June 1945 wrote that "on the first night out each man was served a glass of fresh milk, and if it had been liquid gold it could not have been more welcome. For the next two days, 'milk' was the chief subject of conversation."[108]

That the war's impact on civilian attitudes was rather more ambiguous was illustrated during the food crises of the first postwar years. While their bodies may have emerged from the war in better shape than when they entered, civilians' psyches seemed to have been scarred by the shortages, many of which continued into early 1946. Black markets flourished in foods such as sugar, butter, and the better cuts of meat, which were still rationed or price-controlled.[109] Even bubble gum sold at enormously inflated prices in schoolyards across the nation.[110] A welter of interest groups wrangled over rationing and price controls in Washington, attacking the menagerie of New Deal bêtes noires in the new Truman administration. Confused, hesistant, and transfixed by the domestic situation, the government seemed paralyzed as a new, devastating food crisis loomed abroad. In August 1945 the administration had cut off Lend-Lease shipments of food to Britain and other allies, forcing the British into even more severe rationing than during the war. The next month an international conference warned that the world faced a severe shortage of wheat and only the United States and Canada could help avert disaster. The Americans, however, cited ridiculously optimistic forecasts about the rapid recovery of world agriculture to justify doing nothing.[111] Within months, as crop failure stalked country after country, headlines were telling of 600,000 Chinese dying of starvation, millions in South Asia facing hunger and starvation, and 125 million Europeans in grave nutritional peril. Meanwhile, the United States seemed to be wallowing in nature's bounty. Its wheat crop was astounding—the largest in its history. Milk, meat, and rice production were at all-time highs. Total food production was one-third more than in the last prewar years. It was now producing fully one-tenth of all the world's food.[112]

Yet any attempt to redirect their food toward a needy world ran head on into Americans' expectations that, with the war over, restrictions on their eating should end as well. So while other nations, including Canada, maintained food rationing, in November the U.S. government abandoned rationing of everything but sugar and declared that food prices would be controlled only until July 1946. Through December and into January, it shut its ears to warnings of the spread of famine. Only in late January, when it realized that Americans were consuming wheat at such a rapid rate that it might not be able to meet its minimal foreign commitments, did the administration begin to act.[113] On February 6, 1946, Truman took to the air to warn that "more people face starvation and even actual death for want of food today than in any war year and perhaps more than in all the war years combined."[114] A confidential government report supported this assessment: One hundred and forty million Europeans faced severe deprivation, it said, and one hundred million of them would have to live on fewer than fifteen hundred calories a day. A failed wheat harvest in India, short rice crops in South Asia, and the flooding of the fertile Yellow River Valley in China due to Japanese destruction of its dike system would cause famine in Asia. The world was face to face, it concluded, with "the threat of the greatest famine of modern times."[115]

Yet Truman proposed only the most timid measures to meet this crisis. He asked flour millers to use more of the whole grain when making wheat flour and ordered that whiskey and beer production be reduced for sixty days.[116] Consumers were asked to voluntarily cut back on wheat, fat, and oil consumption so that more could be exported. To lead this voluntary conservation effort, Truman chose the man who had done it during World

War I, Herbert Hoover. He was made head of a Famine Emergency Committee, composed mainly of business leaders and media owners, which was to persuade Americans to cut wheat consumption by one-quarter. Hoover initially toyed with the idea of calling for a 40 percent reduction in order to reach the 25 percent mark, but when the committee realized that anything more than a 15 percent reduction would require rationing, it lowered its sights. A more modest campaign was mounted to persuade each American to forego what amounted to three slices of bread and one tablespoon of fat per day.[117] "This is not a difficult program," said the Richmond, Virginia, *News-Leader*. "The entire household must cooperate. If there is a cook, pains must be taken to acquaint her with the hunger of the world and the part the family must have in relieving misery. The average Negro cook is sympathetic."[118]

But the problem ran deeper than the sympathies of cooks, for the wheat conservation campaign opened up the Pandora's box of American meat-mania. In fact, wheat supplies were not nearly as high as the administration had first estimated, and they were dropping steadily because an enormous amount of grain was being fed to livestock.[119] Ceilings on wheat prices had made it much more profitable for farmers to use grain to fatten their hogs and cattle than to sell it to millers.[120] To change the system and allow a rise in grain prices would involve a mare's nest of problems—breaking the commitment to avoid a wage-price spiral, going back on the vow to farmers who had already sold their wheat that prices would not rise, and so on. The hope was that Hoover's blue ribbon committee, which included virtuoso molders of public opinion such as Henry Luce, the head of Time-Life, Inc., ex–New York mayor Fiorello La Guardia, Eugene Meyer, the publisher of the *Washington Post*, and the pollster George Gallup, would be able to convince the public to cut back.[121]

But while the public's heartstrings were tugged by an outpouring of stories about overseas famine, their belts were not tightened. Gallup's pollsters reported that Americans agreed overwhelmingly with sending food aid abroad and that 60 percent said they had cut back on their consumption so this could be done. Yet there was little evidence that in fact they had.[122] Indeed, in April the disheartened secretary of agriculture reported to the president that bread consumption was actually running 15 percent ahead of the previous year's level.[123] Moreover, grain continued to pour down the gullets of hogs, cattle, and chickens rather than into the holds of cargo ships headed overseas. As Luce's magazine *Fortune* pointed out, the livestock population had risen from 510 million in 1939 to over 640 million; Americans who before the war ate a yearly average of 127 pounds of meat were now downing it at the rate of 150 pounds apiece.[124] Meanwhile, the secretary of agriculture told Truman that the nation was simply not meeting its overseas commitments.[125]

Hoover was hardly much help. He spent most of his time abroad, studying the crisis firsthand on a five-week round-the-world trip that took him to Vienna, Cairo, Baghdad, India, Bangkok, and other Far Eastern points.[126] He had barely returned from that trip when he was off to South America, saying he wanted to check on food supplies there. Then he journeyed north to Canada, where he praised that nation's impressive efforts to ship food overseas without mentioning that this was made possible by what he would not countenance—rationing.[127]

Meanwhile, although Americans continued to tell Gallup's pollsters that they were supporting the campaign by cutting down waste, reusing fats, eating less bread, and even replanting Victory Gardens, the food consumption figures showed otherwise. Chester Davis, the committee chairman, reported that in a swing through the South he found almost no interest in the campaign and that food-producing areas were particularly unsympathetic. La Guardia grumbled that the committee had made a mistake by endorsing the food and fund-raising campaigns of CARE and the United Nations Relief and Rehabilitation Administration, which hardly made a dent in world hunger yet distracted attention from the more important task of food conservation. (Ironically, he would soon be appointed to head the latter organization.)[128] Business support was also disappointing. The retail merchants' subcommittee, a powerful-looking group led by the head of Macy's, drifted aimlessly. Eugene Meyer resigned from the larger committee, pleading the pressure of new duties at the World Bank. As school summer holidays approached, the committee realized that all of the effort it had expended in the schools would soon evaporate. By the end of June, Davis reported that the situation in the Washington office "can only be described as one of personnel disintegration."[129] Again, voluntary conservation had been a dismal failure; the FEC was soon wound up.

Although the beginning of the 1946 harvest brought an easing of the immediate threat of famine in Europe, the crisis was far from over. Indeed, the winter of 1946–1947 was one of the coldest and grimmest in European history—conditions in Britain were worse than during any of the war years—and famine again stalked much of the non-European world. But Americans seemed concerned only by the food situation in their own backyard. In November 1946 they went to the polls in what was labeled "the beefsteak election"—a name derived not from concern over hunger abroad but from an extraordinary fit of meat-mania at home.[130] The buoyant postwar job market had put even more dollars in the pockets of meat-loving workers. When the OPA removed controls on meat prices on July 1, 1946, farmers flooded the market with their overstuffed livestock, but so great was the pent-up demand, and so widespread was the ability to pay, that meat prices still rose relentlessly. On September 1, with prices having risen 70 percent in only two months, the OPA reimposed price controls on meat. OPA director Paul Porter warned that farmers who held their livestock back from the market in the hope of returning to the "Alice-in-Wonderland days" of no price controls were indulging in wishful thinking. Sooner or later they would have to sell. Those who warned that price controls would cause a "meat famine" were dead wrong.[131]

But livestock farmers were not to be intimidated. Encouraged by another bumper grain crop, they again held their animals back from the market, stuffed them with feed, and demanded a return to "free enterprise" in food. The political opposition smelled blood. Republican candidates charged that price controls were destroying the meat industry and helping to deliver the country over to communism. The media, largely Republican-controlled, helped create an atmosphere the writer A. J. Liebling likened to a great "gouamba"; this, he said, was an African word for a "meat hunger," the craving of the meat-starved for meat. Headlines shrieked of a "MEAT FAMINE," a "MEAT CRISIS," "MEAT SCARCITY," and "MEAT LACK." "AMERICANS DINE WELL—IN CANADA" said the headline of an Associated Press picture essay. It featured a rotund Detroiter gorging himself on T-bone steak across the river in Windsor, Ontario, but, like Hoover, did not mention that Canada was maintaining food rationing. Even the staid *New York Times* joined in with a piece headlined "QUEENS RESTAURATEUR, WORRIED OVER MEAT, DIVES OFF BROOKLYN BRIDGE AND SURVIVES."[132] A Newark butcher began selling horsemeat; a laboratory in Columbus reported that its rabbits had been stolen for food. "The only thing people will talk about is meat," four soon-to-be-defeated Democratic congressmen wrote Truman. "Party workers canvassing the voters are being told by Democrats 'No meat—no votes.' "[133]

Truman could withstand the cries of a meat-crazed public no longer. On October 14, 1946, barely four weeks after Porter had warned farmers about Alice in Wonderland, he caved in. He removed price controls on meat and began to dismantle the OPA. ("You've deserted your president for a mess of pottage, a piece of beef—a side of bacon," the embittered president wrote in a speech that, mercifully, he never delivered.)[134] Farmers and finishers soon began releasing their penned-up stock, and the panic over meat supplies subsided. However, within six months—after the Republicans had swept Congress and the extra livestock withheld from the market had worked their way through the system—meat prices were again soaring, putting the GOP media in a bit of a bind. They tried to make the administration seem responsible for the "soaring," "skyrocketing," and even "gyrating" prices, but they had to skirt any suggestion that the removal of price controls, which they had so vociferously demanded, had played a role.[135]

Meanwhile, the situation overseas remained critical. As the winter of 1947–1948 approached, Truman warned that millions in Europe and Asia faced starvation, but again he called only for voluntary conservation. "From now on," he told the nation on the night of October 5, "we shall be testing at each meal the degree to which each of us is willing to exercise self-control for the good of all." What did this amount to? That no meat be eaten on Tuesdays and no poultry and eggs on Thursdays, that everyone eat one less slice of bread a day, and that restaurants serve bread and butter only on request—hardly a starvation diet.[136] Again, the business community was given the leadership. This time, the handsome thirty-eight-year-old head of Lever Brothers, Charles Luckman, was given "his biggest selling job."[137]

Again, a majority of Americans—albeit a dwindling one—swore that they planned to comply.[138] Again, they did not. Even on the first Meatless Tuesday, butchers reported demand near normal.[139] Catholics, already committed to meatless Fridays, complained that they were bearing an unfair burden and asked, unsuccessfully, that the meatless day be switched to Friday. Within less than two months, Luckman had resigned, claiming, quite incredibly, that enormous savings had already been achieved.

Truman, calling for a "more intensified" effort, passed the job on to a three-man cabinet committee, but it was even less successful.[140] Its most important member, Secretary of Agriculture Clinton Anderson, had already said that the administration attached little importance to the Tuesday and Thursday food savings; they were mere "symbols of sacrifice."[141] On January 5, 1947, restaurant and hotel associations declared that the program was "bogging down" and advised those few members who had not already done so to abandon the meatless and eggless days as "impractical." The program's director called this "a real kick in the pants," but in fact its death knell had already rung.[142] By then food aid to Europe was slated to become part of the proposed Marshall Plan, and political and military concerns were taking precedence over hunger with regard to Asia.

Eventually Truman was able to turn the tables on his critics: One of the factors in his upset victory in the presidential election of 1948 was that he was able to blame high beef prices on their demand that price controls be removed. But by then the wartime and postwar meat shortage had taken its toll, reinforcing and perpetuating the beef-centeredness of a very carnivorous country. The idea that food was the measure of America's abundance had played a prominent role in family life during the Depression. Wartime shortages had perpetuated it. The "great gouamba" helped it hit the postwar era at full stride. It would hardly miss a beat for at least ten more years.

CHAPTER 7

The Golden Age of Food Processing: Miracle Whip *über Alles*

"Of all the violent upheavals that have shaken and transformed the American market," said *Fortune* magazine in October 1953, "none had been bigger, or more baffling, than those affecting food." One of the few seemingly immutable laws of economics had been contravened, that formulated by the German Ernst Engels in the late nineteenth century. This simple proposition said that as family income rises the proportion spent on food will tend to fall. Yet since 1941 Americans' real incomes had risen steadily, and so had the proportion of their budgets spent on food. In 1941 the average American family spent 22 percent of its income on food; in 1953 the figure was 26 percent. At first, from 1941 to 1947, the main propulsion was increased spending for better "basic" foods such as meat, dairy products, eggs, fruits, and vegetables, particularly among low-income groups. Since 1947, however, the increased spending had been concentrated among the higher-income groups and had gone to purchase, not better food, but more foods with "built-in-service"—that is, preprepared and otherwise processed foods. "There are few jokes these days about young brides whose talents are limited to a knowledge of the can opener," said *Fortune;* "16 billion pounds of canned goods are now going down the national gullet every year. . . . One out of five home-made cups of coffee drunk in the U.S. today is made from a soluble preparation. In many supermarkets you can now buy a complete turkey dinner, frozen, apportioned, packaged. Just heat and serve." The food industry, like the auto industry, was expanding by selling more and more "extras."[1]

What *Fortune* labeled this "relentless pursuit of convenience" derived much of its initial steam from returning veterans' relentless pursuit of the American Dream.[2] With the end of the war, millions of them turned to the delayed task of family-building; one of the smallest child-bearing generations in the country's history began marrying and producing children at an impressive rate. The "baby boom" generation was on its way, and almost immediately it began to shape and distort the national agenda. Like a python that had swallowed a pig, the United States would spend the next eighty-odd years with many of its most important priorities dictated by the steady course of this massive bulge through its system.[3]

One factor encouraging the bulge was the failure of the expected post-war depression to arrive. Job opportunities continued to expand, as did real incomes, particularly among the middle class, which grew by leaps and bounds. Veterans' assistance programs such as the GI Bill of Rights spurred upward mobility by subsidizing education and skills training. In only six years, from 1947 to 1953, the number of families in the solidly middle-income group ($4000–7500 in 1953 dollars) grew from 12.5 million to 18 million, until they comprised 35 percent of all families. From 1947 to 1959, the proportion of families living on under $3000 fell from 46 percent to 20 percent, while the percentage of those earning between $7000 and $10,000, a high middle income in those days, rose from 5 to 20 percent.[4] Low-interest federal home mortgages for four million veterans helped Americans move into new homes at the rate of over one million a year, a pace that was sustained through the 1950s. Since only new homes were eligible, most of them sprawled out into suburbs, where the automobile was a lifeline to everything, including the new supermarket. In the four years after the war, Americans bought 21.4 million cars and 20 million refrigerators; consequently, they could buy more food less frequently.[5]

Food and appliance producers hardly missed a beat in switching from patriotic wartime themes to extolling the virtues of the middle-class American home and family, the new core of the mass market. Betty Crocker disbanded her wartime homemakers' legion and began a new radio program—"Design for Happiness"—on "how to create happy new homes."[6] Women were now "entrusted with the biggest morale job in history," said *House Beautiful:* redomesticating the (presumably brutalized) returnee. At first this would mean "understanding why he wants it this way, forgetting your own preferences. After all, it is the boss who has come home."[7] There was good reason to treat the boss with kid gloves, for his income was expected to provide the material basis for the family's dream—good housing, transportation, and home comforts.[8]

Women, on the other hand, were warned against asserting any war-inspired independence. Their contribution would continue to be the creative task of presiding over a happy home.[9] Marjorie Husted, Betty Crocker's creator, told advertising executives that women must be made to feel that "a homemaking heart gives her more appeal than cosmetics, that good things baked in the kitchen will keep romance far longer than bright lipstick."[10] Polls indicated that a large majority of both men and women opposed women working outside the home and believed that women who stayed home had "a more interesting time" than those who worked.[11] The few oddballs who questioned the conventional domestic division of labor were given short shrift. When a woman at a 1948 U.S. Women's Bureau conference in Washington suggested (à la Charlotte Perkins Gilman) that new apartments be constructed without private kitchens, instead being linked to central kitchens via dumbwaiters, and that other forms of hot meal delivery service be encouraged, the response was generally hostile. "I'm agin it," said Richardson Wright, editor of *House and Garden* and head of the Wine and Food Society. "The women wouldn't like it. Taking cooking out of their hands is like telling them that they can't have children. Most women find cooking gives them a chance to use their imagination. Some of them even feel it's a way of holding a husband and making the kids happy." Mrs. Roger Straus of the New York State Food Commission was also dubious. "Isn't it one more step in the direction of standardization?" she asked. "Most of us like our homes because they are individual, personal."[12]

Popular culture reflected this notion that food preparation was central to women's role in binding family ties. Wherever one looked, whatever one heard, the competent housewife in her kitchen seemed well-nigh ubiquitous. The popular radio soap opera "Ma Perkins" revolved around wise old Ma dispensing wisdom to her small-town family and neighbors amidst the clunking of her mixing bowls. Much of the action in TV's most popular family sitcom, "Father Knows Best," also took place in the kitchen and dining room. There wise and gentle Karen Anderson (Jane Wyatt) would prepare and serve properly balanced meals to husband Jim (Robert Young) and the other good-looking Andersons and become mildly exasperated when the children tried to rush through their meals to attend to the crises disrupting their lives. Harriet Nelson, co-star of "Ozzie and Harriet," seemed to hardly ever leave her kitchen. She spent much of each program surrounded by gleaming new electric appliances manufactured by the sponsor, Hotpoint. Wearing a particularly frilly apron to make the point that this involved no hard work at all, she would confidently bake cookies or prepare lunch while calmly holding conversations with manic children or commenting on her husband's latest harebrained scheme—no mean feat in the days before videotape. Many of Jackie Gleason's "The Honeymooners" TV comedy skits also took place in the kitchen, but since the lead characters were urban working class (a TV rarity), it was bare, gray, and depressing—not at all like the ideal ones in the suburban or small-town sitcoms. Alice Kramden, the sensible, all-suffering wife of Gleason's buffoonish bus driver, was forced to prepare her meals on an old gas stove and serve them in the dismal kitchen on a plain table covered with oilcloth. Nevertheless, she too was a competent cook who sported a frilly apron.[13]

The other media echoed that competent cooking was central to women's role. In the 1952 Hollywood movie "My Son John," when Helen Hayes begins to suspect that her intellectual son ("He has more degrees than a thermometer," she says) may be a Communist spy, she confronts him with her credo: "I've always gotten *my* strength from two books," she says, holding them up, "my cookbook and the Bible." Women might occasionally admit that they are not good cooks, wrote the author of a 1950 cookbook, "but they no longer boast about it. It's the *thing* to be knowledgeable about food; it's smart—and smarter still to be able to produce it, of gourmet quality, out of your own kitchen, especially if you can do it seemingly without effort, and be gay and carefree about it."[14] That same year *Vogue* magazine insinuated that women who did not cook well were "nervous, unstable types," who would probably end up on the psychiatrist's couch.[15] "At first I found it hard to believe that being a woman is something in itself," a redeemed ex–working woman confessed to *Good Housekeeping* readers in 1951. "I had always felt that a woman had to do something more than manage a household to prove her worth. Later, when I understood the role better, it took on unexpected glamor."[16]

Dad's only culinary responsibility in all of this seemed to be to carve large joints on festive and ceremonial occasions. Although the bumbler's annual battle with the Thanksgiving turkey invariably evoked hilarity in sitcoms, it was also serious business, for it was a symbol that in the end—after Mom's alchemy in the kitchen—it was still the man of the house who would apportion the meal's centerpiece. Carving was supposed to be a skill passed down from father to son, but the fact that most men had not learned it led many to their only confrontation with their wives' cookbooks, in search of chapters such as the one in the *Good Housekeeping Cookbook* entitled "When He Carves." These featured detailed instructions for slicing and dismembering roasted meats that would have given the finest of French maîtres d'hôtel pause.[17] So inadequate did most men feel when faced with the ceremonial task that they shrank from carving almost anything else. (*Better Homes and Gardens*, thoughtfully recognizing this, recommended meat loaf meals on that account; they were "inexpensive, simple, delicious—and Dad won't have to carve!")[18]

The only other kitchen chore expected of men was dish-drying. This simple task, which required even less skill than dish-washing, was more of a social function than anything else, providing, as it did, an opportunity for couples to discuss the day's events or other important topics. (Indeed, when the Oscar Mayer meat packing company decided to get their salesmen's wives involved in a new sales campaign, it gave them free dishtowels emblazoned with the campaign's slogan to remind them to bring the topic up during the dish-drying conversation.)[19] It was also as remote from preparing and serving food as one could get without leaving the kitchen altogether, so it posed no threat to women's presumed mastery of the mysteries of food.

So complete was women's supposed monopoly over family food preparation that male food writers for women's magazines were often forced to assume female pseudonyms. Marshall Adams, one of *McCall's* best food writers and editor of its *McCall's Food Service Bulletin*, wrote under the pen name "Marsha Roberts."[20] But food prepared outside the house was a different story. It was taken for granted that the finest restaurant chefs were men. When *Better Homes and Gardens* featured seafood dishes from four famous restaurants, the recipes were credited to the male chefs of each and accompanied by the comment that this kind of "perfection in seasoning and cooking could be achieved [only] by a knowing chef."[21] Moreover, there was no question that the creative geniuses behind the much-ballyhooed "revolution in the kitchen" were men. The trade journal *Food Processing* billed itself as "the Magazine of Applied Technology for Men Who Manage." There was nary a female face at gatherings of top food processing executives nor, of course, at the annual meetings of food chemists and food engineers. *All* of the food industry leaders surveyed about future food trends by *Food Engineering* in 1960 were men.[22]

Virtually unnoticed behind the idealized image of men "bringing home the bacon" for full-time housewives tending efficiently to their homes was the fact that many men could not pay for the new homes, cars, and appliances without a financial contribution from their wives. By 1953 30 percent of housewives were working, compared to 24 percent in 1941.[23] In 1957 twenty-two million women were working full-time—32 percent of the labor force—and over half of them, twelve million, were married.[24] By 1960 there were twice as many working wives as there had been in 1950, and the number of working mothers had increased by 400 percent. Most important, whereas before the war the vast majority of working mothers had been working class, by the mid-1950s about one-half were middle class.[25]

Food processors recognized that these women represented an excellent market for convenience foods; they did not have the time to prepare "balanced" family meals but could afford to have industry do some of it. In 1957 *Food Engineering* cited the rising percentage of working women who were married, widowed, or divorced to show that "everything favors convenience. . . . Working wives and mothers are great buyers of convenience foods." Later, it highlighted a 1960 survey indicating that almost 25 percent of supermarket shoppers were working women to stress the importance of packaging in selling to these women "shopping on the run."[26] The industrial designer Egmont Arens told industry leaders that the large proportion of the convenience food market made up by working women made it particularly important to have simple recipes on their packages: "When Mary Smith rushes home from work late in the afternoon she wants to buy food that not only will look pretty on the table but is something she can get ready in the half hour before her husband comes home to dinner."[27]

Yet working women were invisible in food advertising. They were also ignored by the recipe writers for the women's media and regarded as a negligible market for cookbooks. Instead, "Karen Anderson," who spent her day cooking and raising her family, remained their target. Top food executives who spoke patronizingly of "our boss—Mrs. Consumer" thought of her as a housewife and nothing but a housewife.[28] At the 1962 Grocery Manufacturers Association convention, the group's president (using very dubious figures) credited convenience foods with having cut the average housewife's daily kitchen time from five and a half to one and a half hours a day in ten years. There was no mention of the obvious fact that the major factor in lowering kitchen time was the growing number of working women, who had no choice in the matter. An executive of the American Can Company told the assemblage that "the packaging revolution" had helped give the American family not more time for women to work but "more time for cultural and community activities."[29] Charles Mortimer, head of General Foods, boasted that "built-in chef service" had now been added to "built-in maid service," implying that housewives could now lead the lives of the leisured upper class.[30] Even in 1969, when it had become the norm for married women to work, the chairman of the board of the Corn Products Company saw the "social revolution" convenience foods had brought only in terms of the full-time housewife. "We—that is, the food industry—have given her the gift of *time*," he said, "which she may reinvest in bridge, canasta, garden club, and other perhaps more soul-satisfying pursuits."[31]

The repeated assurances that convenience food would indeed make for free time to play canasta were based on the postwar era's unbounded faith in the American genius for labor-saving technology. "You'll Eat Better with Less Work," *House Beautiful* had assured its readers in January 1946. It predicted that within two years they would be spending 50 to 75 percent less time on feeding their families. The bulk of food shopping would be done ten or twelve times a year rather than several times a week, and there would be more home entertaining because an elaborate company dinner would be prepared in half an hour.[32] *Better Homes and Gardens* said that canned whole meals, using technology developed by the army, would soon be commonplace on home shelves, along with canned hamburgers and frozen coffee and frozen grapes. Both magazines were enthusiastic about the savings that would accrue to consumers through prepackaging vegetables in cellophane, which would eliminate the wastage that came from dealers' mishandling and shoppers' poking.[33] The prospects of exotic or out-of-season foods flown in from the far corners of the world by giant cargo planes also excited optimism, as did the possibilities of dehydration and anhydration. "The day is coming," one writer enthused during the war, "when a woman can buy a boiled dinner and carry it home in her purse . . . when a well-stocked pantry will be reduced to a few boxes . . . when you'll serve the girls a bridge luncheon of dehydrated meat and potatoes with powdered potatoes and powdered onions, a dehydrated cabbage salad, and custard made with powdered eggs and powdered milk for dessert."[34] But above all, almost everyone agreed, the surest new path to liberation was the one carved out by Clarence Birdseye, the Gloucester, Massachussetts, businessman-inventor who developed "quick frozen" food.

Although he is often called the "inventor" of frozen foods, in fact Birdseye's most creative invention was that myth itself, which he fostered with a tale of how the idea had come to him on a fur-trading trip to Labrador, where he ate delicious fish and meat that had frozen almost instantly in the subzero weather. There were already a number of methods for freezing foods when his inspiration first struck in the mid-1920s, but they were used primarily to preserve foods that were already going bad from deteriorating further; this had fostered a connotation between freezing and low quality in the public mind. Birdseye's major contribution was quintessentially American: new packaging to overcome this poor image. Working with the DuPont chemical company, he developed a moisture-proof cellophane wrapping that allowed foods to be frozen more quickly. Making the wrapper around it out of waxed cardboard prevented it from disintegrating into an ugly mess when the food thawed. The rotting-food connotation was combated by using the freshest possible foods and calling his foods "frosted" rather than "frozen." After a rather promising start, in 1929, only months before the stock market crash, Birdseye managed to lure megalomaniacal E. F. Hutton, head of the General Foods conglomerate, into entering a bidding war for the company with archrival Junius P. Morgan of Standard Brands—a war that, to Birdseye's enormous profit, Hutton won.[35]

This "marvellous invention" by the man who "wrought a miracle . . . *may change the whole course of food history*," said General Foods in its first advertisement for what it now called Birds Eye Frosted Foods. But the Depression economy dampened these high hopes. Home freezers were beyond the reach of almost everyone, and most grocers were unwilling to lay out upwards of a thousand dollars for the relatively small commercial chests.[36] The wartime OPA gave frozen fruits and vegetables a shot in the arm by taking them off the ration list seventeen months before canned goods were derationed, but production of home freezers had ceased.[37] When it resumed in 1946 and refrigerators with small frozen food compartments began to appear, industry optimists predicted that by 1955 one-quarter of the nation's food expenditures would go for frozen foods. Others went further and said that most of the nation's food would soon be prefrozen.[38] But then another setback occurred, as small entrepreneurs rushed frozen products of dubious quality into the market. General Foods eventually crushed most of them (in what the industry labeled the "Great Blood Bath"), but frozen food's reputation was again besmirched.[39] Who would sell frozen foods was not clear either, as special shops selling only frozen food vied with department stores, ice companies, and home delivery services for dominance. Macy's invested heavily in a line of individually frozen dinners that "went flat on their face," said its chief food taster.[40] To almost everyone's surprise, it was not food but orange juice that became the postwar era's first major frozen success story. By 1949 more of the frozen concentrate was being sold than the two previous frozen food leaders, peas and strawberries, combined.[41] In 1953 orange juice comprised fully 20 percent of all frozen product sales.[42]

Retailers' display cabinets had remained small and expensive into the 1950s, but the development of large open-top freezers for self-service establishments paved the way for supermarkets to dominate frozen food sales. By then Swanson's had come up with frozen individual meals, which, although they excited few palates, were serendipitously named "TV dinners." This tie-in with the era's TV boom allowed consumers to rationalize the obvious lowering of dining standards with the excuse that they were intended to be eaten in untypical circumstances—in front of the TV set—even though this was rarely the case.[43] "Advances" such as this and the development of main course treats like chicken à la king (an early popular favorite) led frozen foods' relentless march onto the nation's dinner table.

Frozen vegetable sales—led by brightly colored green peas and less dazzling string beans—boomed in mid-decade, particularly as large institutional kitchens switched to them. While sales of canned goods continued to outstrip them, by the end of that decade the gap was narrowing.[44] By 1959 Americans were buying $2.7 billion worth of frozen foods a year, 2700 percent more than in 1949. Over half a billion of these dollars were spent on "heat and serve" prepared dishes.[45]

Freezing was by no means the only process transforming American food production. Food producers talked of facing what they called the "fixed stomach": Americans could not be persuaded to eat more food.[46] Increased profits would therefore have to come mainly from two sources—economies in production and more value added to what they produced. The first was achieved in the conventional American fashion, by replacing humans with machines. For more than fifteen years after the war, most of the new investment in the industry went not to expand production but to mechanize it. "Continuous" operations replaced "batch" processing, eliminating human handling at various stages in everything from butter-making to bread-baking.[47] At the La Rosa Company's giant new pasta factory, noodle-making was turned into a continuous hands-off operation. Automatic equipment sucked up the semolina from freight cars, filtered it, dumped it into mixers, and then extruded, dried, and cut it. Frozen food producers installed machines that took foods from a conveyor belt into a breading contraption, plopped them into hot fat to be automatically fried, drained them, and slipped them into trays for freezing. Bakeries were "robotized" with new precision instruments regulating dough-mixing, fermenting, proofing, and cooling.[48] As a result, small processing companies fell by the wayside, bought out or driven out by larger ones seeking diversification or economies of scale. Even canning companies, historically a haven for small rural entrepreneurs, were affected. From 1947 to 1954, their number declined by almost one-quarter even though production increased. "This shows how rapidly food is becoming an industry of large-scale manufacturing operations," said *Food Engineering*.[49]

The other aspect of the postwar industry's structure—the drive for more value added—was based on the idea that processors had something almost as valuable as sustenance to sell to busy postwar housewives: time. Longer shelf-lives, more processing, precooking, and packaging all had one great justification: to liberate "Mrs. Consumer" from the drudgery of the kitchen. Indeed, there is a paradoxical contrast between the processors' advertisements, which, like the women's media, portrayed cooking as an interesting, nurturing, and creative pursuit, and their claims that new processing techniques and packaging would free women from this boring, unpleasant task. Asked in 1957 why people wanted things so "highly packaged," the president of Campbell's Soup replied: "To save trouble. The average housewife isn't interested in making a slave of herself. When you do it day after day [cooking] tends to get a little tiresome, and the young housewife is really less interested in her reputation as a home cook today. . . . She doesn't regard slaving in the kitchen as an essential of a good wife and mother."[50] The American Can Company claimed that in 1951–1952 frozen orange juice had saved housewives the equivalent of fourteen thousand years of "drudgery."[51]

By 1954 the value added to the cost of food by manufacturing was already 45 percent higher than in 1939. It continued to rise, until by 1959 it had accounted for most of the decade's increased spending on food.[52] By then chemical producers were among the main beneficiaries of the trend. Immediately after the war, chemists had set about putting wartime innovations to profitable peacetime use. Many schemes, such as those hoping to sell upgraded C-rations in supermarkets or to use army-developed dehydration techniques to reduce almost all foods to powders, did not work out. The inevitable predictions of meals-in-a-capsule were also, as usual, far-fetched. But the chemists' major opening came from the fact that most of the current techniques for processing, preserving, precooking, and packaging had one thing in common: They made foods lose their taste, texture, and normal appearance.

The result was a kind of Golden Age for American food chemistry. From 1949 to 1959, chemists came up with over four hundred new additives to aid in processing and preserving food.[53] Preservatives such as calcium propionate extended the shelf-life of bread and seemed to promise virtual immortality for some kinds of baked goods. A whole array of chemicals prevented foods with fat from going rancid. Others stopped their color from fading.[54] Frozen and dehydrated foods were sprayed with sulphur dioxide gas or dipped in solutions containing sulfites to achieve the same ends.[55] "Smootheners" such as hydrolized starch, which could withstand the heating, freezing, and mechanical manipulation of modern food-processing, were a boon to manufacturers of baby foods, pie fillings, puddings, gravies, and stews.[56] New food colorings provided scope for creativity in conjuring up illusions of freshness. Sophisticated flavoring agents liberated manufacturers from dependence on natural fruits and flavors. ("There are not sufficient strawberries grown in the world to supply the demand for strawberry flavor," said a researcher for General Foods, makers of Jell-O, a bit defensively.)[57] One miraculous substance, monosodium glutamate, was even found to enhance desirable flavors and suppress undesirable ones![58]

Food processors got agricultural scientists to pitch in by developing foods more suitable for processing. In 1945, at the urging of A&P Stores' poultry research director, representatives of ten poultry organizations, the USDA, and two poultry magazine editors organized a national Chicken-of-Tomorrow Contest to underwrite the development of a better—that is, cheaper to produce and easier to market—chicken. The cross-bred finalists assembled at the University of Delaware in 1948 formed the basis for new breeds of battery chickens that were meatier, with broader breasts, thicker drumsticks, and fewer blemishes than their scrawnier, barnyard-scratching (albeit tastier) ancestors. The synthesis of Vitamin B_{12} in 1949 allowed that growth vitamin to be added to feed. This helped chickens grow faster while lessening their need for protein, making chicken feed cheaper. The next year, researchers discovered that the contagious diseases and depressed growth caused by the stress of crowded batteries could be alleviated by adding antibiotics to the feed. Now, the number of broilers in the same "house" could be increased from three thousand to more than twenty thousand and even forty thousand. As a result of these and other innovations, the feed conversion ratio—that is, the number of pounds of feed it takes to produce one pound of chicken—was slashed dramatically, and the length of time it took to produce a fully developed four-pound chicken plummeted as well. In the words of proud poultry scientists, "chicken on Sunday became an everyday treat."

Another expert marveled at the "spectacular results" antibiotics and vaccination had achieved with livestock, allowing the number of cattle crowded together on feed lots to be increased from hundreds to tens of thousands.[59] Scientists working on their relatives in dairy herds were no slouches either, particularly as "hormonizing" techniques grew more sophisticated. By 1974 there were only half as many dairy cows as in 1950, but they were producing just as much milk. In agriculture as in food-manufacturing, however, more efficient production meant the demise of the small operator. Eighty-five percent of the dairy farms working in 1950 had gone out of business by 1974.[60]

Yet, as in food-manufacturing, the so-called advances were in the economics of production, not in taste. It was widely acknowledged that in practically all spheres taste had been a casualty of processing. Food trade journals were full of articles about and advertisements for flavoring agents, all of which assumed that they were to be used to replace tastes lost during processing. Even *Fortune* magazine—no critic of the food industries—acknowledged in 1952 that "it is hardly surprising that, in the opinion of many, the flavor of American food and drink—in jars, cartons, cans, fifths, and pints—leaves something to be desired."[61] At the producers' level, everyone involved in breeding, whether of animals or plants, understood that there had to be trade-offs for gains in economy, appearance, or shippability, and taste was the most easily traded-off quality.

This was particularly so in America, where food industry moguls had a generally low opinion of consumers' taste buds. How else explain the 1959 interview in which the president of Campbell's Soup, with a straight face, told an interviewer that the "biggest improvement" in food production in the past twenty-five years had been "in the breeding of plants to get better flavor in vegetables and fruits"? Even "the tomatoes are better," he said. "They have higher color and higher flavor. And that is typical of many vegetables."[62] The manager of the frozen foods division of Marshall Field & Co. harbored no such illusions. In 1956 he told the Grocery Manufacturers Association that most frozen vegetables and fruits were tasteless, "with absolutely no comparison to the fresh product except in appearance." This was because hotels, restaurants, and other institutional users had complained that the first packs of frozen peas shipped to them, which were young, sweet and fresh, broke on steam tables and produced unacceptable amounts of waste. As a result, packers now let vegetables mature beyond their prime or switched to hardier but less tasty varieties.[63] Ordinary consumers apparently did not notice the difference.

Nutrition seemed of even less concern than taste. "Every woman likes to say she thinks a lot about the health of her household," said an industry analyst in late 1945, "but nutritional considerations are more of an undertone than anything else in the planning of family meals."[64] Other industry leaders dismissed out of hand those who continued to harbor prewar concerns that nutrients were lost in processing. "Today's processed foods have a food value at least equal, and often superior to, raw produce," said Paul Willis, president of the GMA, "but many housewives are still spending countless hours preparing raw produce in the erroneous belief that they are feeding their families more 'healthfully.' "[65] A Pillsbury vice-president noted rather smugly that market research into the effectiveness of advertising the protein content of cereals revealed that housewives were abysmally ignorant of what protein was and what it did. They knew only that it was a good thing and that their families should have some of it.[66]

A remarkable aspect of the postwar transformation of food-processing was the minimal extent to which it affected traditional American tastes in food. One reason was that most of the effort went into "improving" familiar foods and products. It took about five years and a large investment to introduce a completely new product, while "new and improved" foods such as Heinz "hot" ketchup (one of the big busts of 1959–1960) or new flavors of Jell-O could be whipped up in a year or so. Indeed, despite their self-congratulatory back-patting over their innovativeness, the food-processing industries consistently ranked near the bottom in the proportion of sales invested in research and development—a "mediocre" performance at best, according to one study. In 1962 the large processors employed only 10 to 15 percent more scientists than in 1939.[67]

The "new foods" columns, in which women's magazines regularly waxed ecstatic over their advertisers' innovations, were in fact dreary recitations of minor variations on ancient themes. *Better Homes and Gardens*'s "These Foods Are News!" column in 1959 and 1961, for example, was full of distinctly unnewsworthy products. "Potato salad from a package!" hardly seemed a miraculous labor-saver. A six-ounce box of dehydrated potato slices had to be boiled, seasoned with a packaged mix, and chilled before mayonnaise and hard-cooked eggs were added. New packaged soup mixes were the old standbys: tomato, mushroom, and chicken and rice. Canned condensed cheese soup was hailed as delicious on its own and as a sauce for vegetables or an ingredient in casseroles, hardly different roles than canned mushroom soup had been playing since the 1920s.[68] When Charles Mortimer, head of General Foods, chafed at the limited number of vegetables there were to process and ordered his experts to find a "new" one, they came up with—not finnochio, chayote, or Chinese eggplant—Rolletes, a mixture of pureed carrots and peas frozen on a stick.[69] Innovative packaging enveloped the same old foods. Alcoa's new boil-in-the-bag aluminum packages were used mainly for macaroni and cheese. It developed an ingenious package for steaming frozen foods on the stove—a three-shelved aluminum tray with ice on the bottom shelf—but its main customer, the Gunsberg Company of Detroit, used it only for corned beef.[70] Even those radically new products that did strike it rich, such as cake mixes (only ten years after hitting the market in 1947 they were being used for over half the country's home-baked cakes), were usually new ways of preparing old foods.[71]

While some scientists worked in corporate labs to improve processing, others were out front defending it. The Nutrition Foundation, set up in 1941 by the major processors to fund research on how to revitalize foods, was used to marshal scientific opinions to correct "superficial and faddish ideas" and to combat those questioning any of the 704 chemicals that by 1958 were commonly used in foods.[72] When a lone congressman, James Delaney of Brooklyn, managed to parlay some political debts into permission to head a special committee to investigate the use of chemicals in food in 1950 and 1951, scientists with the food and chemical industries at first refused even to defend their practices. The National Agricultural Chemicals Association denounced testimony that DDT was present in cow's milk and seemed to accumulate in human body fat as "careless and unsubstantiated criticism" that threatened "to injure large segments of agriculture" and created "an unjustified fear" among consumers.[73] In late 1951, after testimony that chemicals used to make bread softer and whiter might be harmful received some publicity, the Food Protection Committee of the National Research Council reported that, "contrary to some ideas that had been circulated, reliable food processors had not reduced the nutritional quality of foods or created inferior products through use of chemical additives."[74]

In 1959 the H. J. Heinz Company hired a slew of the nation's most prominent nutrition experts to oversee the *Heinz Handbook of Nutrition*, a comprehensive reference manual intended to cover the entire field of nutrition. Subsequently translated into Spanish and Arabic, it warned that "discussions of modern methods of food manufacture inevitably highlight partial losses of a number of valuable nutrients during processing . . . while the large number of advantages are ignored or taken for granted." The advantages included providing more balanced diets for urban populations and the development of processed infant foods that provided "essential nutrients seldom supplied before when they were needed most."[75]

Some scientists went further, hailing processing as adding nutrition. "Often the availability of certain nutrients in natural foods can be improved by a proper degree of processing," said the well-known biochemist Conrad Elvehjem, soon to become president of the University of Wisconsin.[76] The head of food science at the University of California at Davis admitted that nutrition came last in the order of the food qualities that food scientists worked for, but this was because "if food isn't safe, convenient, good to eat and resistant to spoilage most people would throw it out regardless of its nutritive value."[77]

Government agencies also provided solid backing for the new food technology. In 1951 the FDA, jogged by the Delaney committee, had begun to demand more power to police food additives, but the Eisenhower administration, which took office in early 1953, was unsympathetic to this kind of government regulation of industry. In 1958 Congress mandated that it play a more active role, but its deputy commissioner saw the new powers as allowing more reassurance, not enforcement. Perhaps, he told the New York State Bar convention, they would "allay public concern over its food supply" caused by "incorrect" reports that carcinogenic chemicals were being added to food. The speaker who followed him, Dr. Phillip White, secretary of the AMA's Council on Food and Nutrition, agreed, assuring the lawyers that the quality of American food had never been better.[78] When some renewed concerns about "overprocessing" of foods arose, the FDA issued a pamphlet—thousands of which were distributed by processors—saying there was no such thing. "By patronizing all departments of a food store we can easily supply all of our nutritional needs," it said. "The American food supply is unsurpassed in volume, variety, and nutritional value."[79] Experts at the USDA and the state agricultural experiment stations, who devoted considerable effort to developing new ways of processing foods, were also supportive. In 1953 they hosted a "research luncheon" at the USDA laboratories in Beltsville, Maryland, for President Dwight D. Eisenhower, which included powdered orange juice, potato chip bars, a whey cheese spread, "dehydrofrozen peas," beef and pork raised on new (hormone- and antibiotic-added) feeding methods, and lowfat milk.[80] USDA home economists set about proving that not only did processed "convenience foods" save the housewife time, they also saved her money.[81]

This enthusiasm about the new ways of processing foods was paralleled by appreciation for that most visible new way of buying it, the supermarket. These had first come on the scene in the early 1930s, but their development was held up by Depression economics and wartime shortages. Chains such as A&P, which were the most dynamic segment of the grocery network, consisted mainly of relatively small stores, often little larger than the thousands of independent corner grocery stores they drove out of business. Their great advantage lay in their enormous purchasing power and centralized warehousing systems, which allowed them to get price concessions from manufacturers and save on distribution costs. After the war their capital resources gave them a head start in the rush to construct new suburban supermarkets and allowed them to shed most of their smaller inner-city stores and convert the rest to supermarkets. As a result, from 1948 to 1963 large chains increased their share of the nation's grocery business from 35 percent to almost half.[82] As early as 1956 the independent corner grocery store, while still visible, was a relic of the past. Full-fledged supermarkets accounted for 62 percent of the nation's grocery sales, while smaller, self-service "superettes" took in another 28 percent of the food dollar, leaving the 212,000 small food stores to share 10 percent of the market.[83]

Supermarkets, with their dizzying arrays of processed foods, came to be regarded as quintessential symbols of the triumph of American capitalism. In 1957, when the U.S. government wanted to display "the high standard of living achieved under the American economic system" at the Zagreb Trade Fair, it reproduced a supermarket stocked with American processed foods and produce.[84] Simultaneously, across the Adriatic, where, according to William B. Murphy, president of Campbell's Soup, Western Europeans were "twenty-five years behind us [in] the kitchen revolution," the government mounted an exhibit at Rome's Levant Trade Fair that revolved almost exclusively around American food-processing equipment. Thousands of Italians walked under an eye-catching six-foot-high "U.S.A." sign to watch, presumably in awe, while a machine halved local peaches, removed their pits, peeled, washed and then refrigerated them in readiness for freezing.[85] Two years later, when Vice-President Richard Nixon stood amidst the glittering white kitchen appliances at the American exhibition in Moscow ("a lavish testimonial to abundance," the *New York Times* called it) and engaged General Secretary Nikita Khrushchev in the famous "Kitchen Debate" over the merits of their two systems, he pointed to the number of choices it provided in consumer goods as evidence of capitalism's superiority. (When he remarked that the appliances were intended "to make the life of our housewives easier," Khrushchev replied, "Don't you have a machine that puts food into the mouth and pushes it down? Many things you have shown us are interesting but are not needed in life.")[86] The next year, when Khrushchev visited America, he and his party were taken into a San Francisco supermarket. "The expression on their faces was something to behold," one of the hosts, Henry Cabot Lodge, told the Grocery Manufacturers of America.[87]

Not all foreigners were impressed. The bountiful Thanksgiving dinner displayed at the 1957 Dijon Food Fair elicited typically Gallic skepticism. "Who has an oven big enough to cook something like that?" said one woman, contemplating the monstrous turkey. Another, reflecting a common European conception of American food, wondered why the Americans had a gastronomic exhibit at all when everything they ate came from cans.[88]

The ascendancy of the supermarket played a major role in reshaping the marketing of processed foods. A 1960 DuPont company study indicated that there had been an unprecedented rise in "unplanned"—that is, impulse—purchases since a previous study in 1949. Most of it was attributable to supermarket shopping, for close to three-quarters of all supermarket food purchases were unplanned.[89] This kind of buying was not welcomed by the established food manufacturers, for it threatened their bread-and-butter, brand loyalty.[90] It also undercut the incentives producers' salespeople had traditionally offered to grocery store owners and their clerks. As the head of a major advertising agency explained, selling food products in self-service stores eliminated entirely "the possibility of substitution by a friendly or inimical clerk."[91] Now it was one-on-one, as it were, between the processor and the housewife, with no referee. As a result, packaging and marketing became much more important than ever. Psychologists with horn-rimmed glasses now prowled supermarket aisles, clipboards in hand, to determine why consumers picked certain foods off of shelves and not others, trying to turn packaging, hitherto a seat-of-the-pants affair, into a science. Raymond Loewy, perhaps the most brilliant American industrial designer of the century, was hired by Armour and Company to redesign its entire family of over four hundred meat and dairy products.

"In the modern super market women are no longer cajoled into buying a particular brand," wrote the prominent industrial designer Egmont Arens. "As a result, an entirely new kind of package design has developed. Instead of a package which was merely a poster, attractive at a distance, today we design a package for 'readership' "—a quality that would entice "the ladies who trundle their little shopping wagons among the shelves and tables" to pick up the package and read about its contents. "High impact colors" were essential for this first, "stop-traffic" part of the process, he said. When he redesigned A&P's coffee bags, Arens convinced company executives that vibrant reds and yellows were the way to go by taking them to the top of the Empire State Building and noting that the only autos that could be discerned were of those two colors.[92] Later "color studies" claimed to be much more sophisticated, telling packagers such things as that women reached most readily for red packages, while men were more attracted to blue.[93]

The new importance of marketing was reflected in the lofiest corridors of processors' power. When General Foods, the largest food conglomerate, selected a new top officer in 1954, it chose not a production or financial specialist but a marketing expert. This acknowledged, said *Time*, "that the emphasis in the food business has moved more and more from manufacturing to marketing."[94] In 1956 Unilever hired W. Gardner Barker, a market researcher, from Simoniz Wax to be in charge of new products at its Lipton subsidiary. Three years later he was chosen to head the company, where his distinguished career was highlighted by the successful introduction of Cup-a-Soup.[95]

With the ascendancy of marketing, spending on advertising soared. Much of it still echoed prewar themes. Wesson Oil gave recipes for "Man-Winning Tomato Salad"; Pillsbury promised that its new pie crust mix would "put a loving look in your husband's eye."[96] Its immensely successful Pillsbury Bake-Off, begun in 1949, was a national version of the cooking and baking contests that had been transfixing women at state and county fairs for many years. General Mills stuck with ageless Betty Crocker—in whom by 1954 it had invested thirty years and over a hundred million dollars—allowing only a streak of what may have been gray in her cartoon image's hair. Although her new radio personification, singer and actress Adelaide Cummings, was a dazzling blonde ex–fashion reporter who lived in a Park Avenue apartment, her radio persona was distinctly down-to-earth. Her scripts remained what *Sponsor* magazine called "models for the integration of selling and programming . . . an example of how to tie product and program together so naturally they seem like one unit."[97] Advances in color photogravure that reduced the cost of color advertisements in magazines helped reinforce the old home economics lessons about the importance of presentation. Bright pictures of canned peas, corn, pineapple, and tomato sauces, arranged in perfect circles, squares, or triangles, virtually leapt from the pages.[98]

But by 1950 one theme had come to dominate all else: convenience. "Quick 'n' easy," "heat and serve," and "ready in a jiffy" beat tattoos on the pages of the magazines and echoed on the radio and TV. Among the most memorable were the commercials on an immensely popular live drama show of the 1950's, "Kraft Television Theater." While disembodied hands effortlessly mixed Miracle Whip, Kraft marshmallows, Kraft caramels, and Velveeta into some rather bizarre concoctions, the soothing voice of an off-camera male announcer assured housewives that these "easy to make" recipes were "bound to please" everyone in the family. A typical one demonstrated a "speedy way to put together a tray of good-eating snacks . . . in a jiffy" with four different kinds of Kraft "cheese food."[99] Another suggested "Cheese Rabbit" for a quick one-dish dinner: a jar of Cheese Whiz mixed with a can of kidney beans, some onion, pepper, margarine, ketchup, and Worcestershire sauce.[100]

General Mills, grounded in the faltering flour industry, managed to emerge from the decade stronger than ever thanks to its timely development of convenience foods. Until the late 1940s, Betty Crocker had devoted much more effort to promoting Gold Medal flour than to Bisquick, a premixed biscuit and batter mix.[101] By 1950, however, Bisquick, which saved some preparation and baking time, was getting much more play. Then, in mid-decade, both the flour and Bisquick took back seats to her cake mixes, which became one of the great marketing success stories of the time. When originally developed, the mixes had demanded the addition of nothing more than water. However, marketers soon realized that cake-baking was still too important a part of the housewife's self-image to eliminate her contribution completely. They therefore had the directions changed slightly to require the addition of one egg. "Betty" also encouraged minor additions to the basic cake mixes to foster the illusion of individuality. Yet they did not stray from the formula that made her the most successful recipe dispenser of the era: Keep them "simple, quick, and right."[102] Other processors marched to the same simple beat. Arens, the revered package designer, told food producers that recipes on packages must be simple and quick and that the results should be simple as well. "Overelaborate dishes usually are not the kinds that women want for everyday meals," he warned.[103]

Whatever might be said about the gastronomic or nutritional merits of the processors' products, their ascendancy helped buttress the American ideal of a classless society. The war, as we have seen, had had a leveling effect, and some of this egalitarian thrust persisted—at least with regard to food—

through the 1950s. Although class differences certainly persisted, Americans were not especially conscious of them. This was particularly true of the middle-class suburbanites who now set the nation's cultural tone. William H. Whyte noted in his study of Park Forest, Illinois, a middle-class suburb of Chicago: "It is classless, or at least, its people want it to be."[104] Studies of suburban life such as his indicated that social status was based much more on social activities than on family, occupation, property, or consumption habits—including food.[105]

The nature of the food-processing innovations of the 1950s, which mainly brought old familiar foods in different packages, also inhibited food from becoming an important mark of class and status. They were aimed not (as would later be the case) at particular "upscale" niches of the market, but at what *Life* magazine called in 1957 the new "mass-class market" of middle-income families, earning between three and ten thousand dollars annually, who now comprised 63 percent of the population and accounted for 72 percent of consumer purchases. This was the market for convenience foods, noted *Food Engineering*.[106] The weekday dinner table at a corporate lawyer's household in upper-middle-class Flossmoor, Illinois, looked little different from an insurance company clerk's in Levittown, New York: Campbell's canned or Lipton's dried soup, broiled meat, frozen french fries, and a frozen green vegetable, with supermarket ice cream or a Jell-O concoction for dessert—an All-American "square meal." Popular dishes such as tuna and noodle casseroles transcended class lines. The recipes upper-middle-class Vassar College alumnae in New Haven, Connecticut, contributed to their fund-raising *Vassar Cook Book* differed little from those in similar books produced by women considerably below them on the social scale, particularly in reliance on the same processed foods. "Spaghetti West Texas" had a sauce of ground beef, canned tomato soup, and canned corn. The meat loaf was a bit unusual in that it contained a can of Campbell's Vegetarian Vegetable soup, but its ketchup sauce was a familiar sight on tables from Palo Alto to the Bronx.[107]

Nor was there much to be expected in terms of distinctive food tastes from the old upper class. They now feared another wave of war-profits-bloated nouveaux riches would invade their sprawling Westchester mansions and rugged Kennebunkport "cottages." But while their Gilded Age forebears had tried to outdo the parvenus in lavish entertaining and dining, they adopted a strategy of "conspicuous underconsumption," which meant serving more or less the same food as everyone else.[108]

In 1962 Charles Mortimer, chairman of the board of General Foods, credited the food industry's research and development effort with "making possible the enormous processing plants and their time-and-labor-saving output of the best eating the world has ever seen."[109] The kind of food he meant was almost certainly reflected in the list of the most popular TV dinners at that time: fried chicken, roast turkey, Salisbury steak (hamburger), and roast beef—simple food that exemplified the straightforward nature of America's dominant position in the world economy.[110] Perhaps it was natural that, in an era when Americans brimmed with confidence in the superiority of their political, economic, military, and even cultural institutions, they should feel similarly about their food and those who produced it. In any event, this certainly seemed to be the case. That same year, when Elmo Roper's pollsters interviewed 1173 shoppers, almost all female, outside of supermarkets, only 4 percent of them had any suggestions for improvements to be made by food manufacturers—and these dealt mainly with easier-to-read labels.[111]

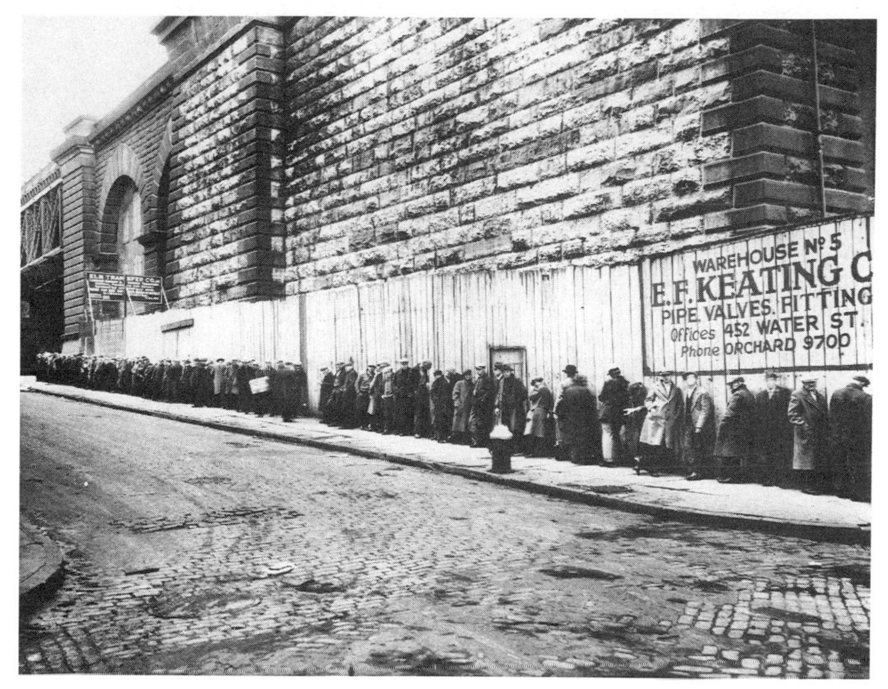

The line-up at a soup kitchen under the Brooklyn Bridge in the early 1930s. Mobster Al Capone's organization was among the many that sponsored breadlines and soup kitchens during the Great Depression. (Library of Congress)

A 1933 addition to the White Tower chain of restaurants, whose sparkling appearance helped assuage the traditional suspicion of restaurants' ground meat. They were often strategically placed along mass-transit lines to attract a working-class clientele. (Paul Hershorn and Steve Izenour, White Towers, copyright © 1979 by the Massachussets Institute of Technology)

Christmas dinner of potatoes, cabbage, and pie at a farm on submarginal land in Iowa, 1936. Until the 1960s, the most serious problems with hunger tended to be concentrated in rural areas. (Library of Congress)

The Depression stimulated a renewed emphasis on women as homemakers, as well as pressure to save money by entertaining with dinner parties at home, as evidenced in this advertisement for American Home *magazine, published on September 18, 1939, shortly after Germany invaded Poland.*

One of the many restaurants which sprang up along the nation's highways during the 1930s which, like Howard Johnson's, tried to lure motorists with their eye-catching appearance and the promise of a clean alternative to the thousands of run-down hot dog stands, barbecue joints, and other food shacks which also lined the highways. (Library of Congress)

Women and children lining up in Detroit, Michigan, in the spring of 1942, to be issued the first ration books, for sugar. Although it was hardly onerous by the standards of other belligerents, food rationing turned out to be quite a trauma for the self-described "people of plenty." (Library of Congress)

Soldiers in "the best-fed army in history" pass the milk at noon "chow" at Fort Belvoir, Va., in January 1943. By any standards, the amounts of food allocated to the armed services were truly enormous. (Library of Congress)

A War Food Aministration poster trying to persuade Americans to eat foods which were not rationed. Government attempts to promote voluntary conservation of food were notably unsuccessful.

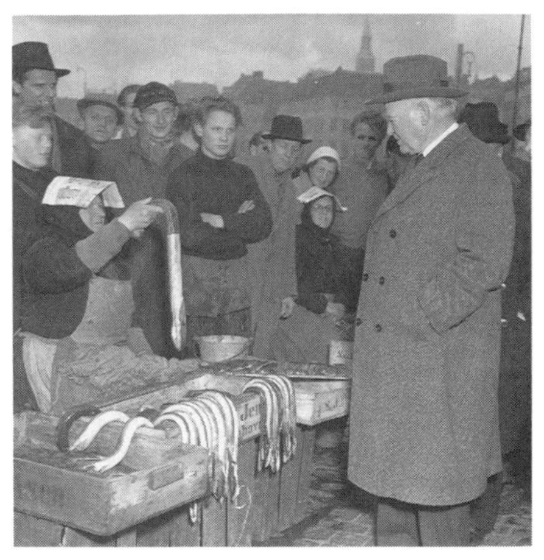

A Copenhagen fishmonger offers Herbert Hoover an eel during the ex-president's world tour in 1946 as head of the Famine Relief Committee, which tried, with little success, to persuade Americans to cut down their food consumption to provide provisions to send abroad. (National Archives)

Photograph distributed abroad by the U.S. Information Agency in 1958 to impress foreigners with the achievements of American industry. The housewife stands amidst her gleaming appliances displaying some of the processed, packaged, and plastic-wrapped foods she bought at the supermarket. (National Archives)

An enormous new machine for drying eggs for use in the new methods for manufacturing foods developed in the 1950s, "The Golden Age of American Food Processing." (National Archives)

A 1959 USIA photograph of a family praying before breakfast. Wonder Bread is about to be pushed down into the toaster, and each bowl contains a piece of shredded wheat. Four children per family was not untypical during the "Baby Boom," which produced the generation which would set the rest of the century's cultural tone. (National Archives)

Julia Child, whose television show and cookbooks played a major role in popularizing French cooking among the middle class in the 1960s, showing how to gut a fish. (Schlesinger Library, Radcliffe College)

Making alfalfa sprouts sandwiches on whole wheat bread at a Washington, D.C., health food co-op, January 1979. The slogan on the T-shirt, which reads "Food for People, Not for Profit," was the title of a popular book of essays attacking the nefarious effects of corporate greed on the health and safety of the food supply. (Copyright *Washington Post*; reprinted by permission of D.C. Public Library)

April 28, 1969: Secretary of Health, Education, and Welfare Robert Finch explains that the law forces him to ban the non-caloric sweetener cyclamate because it caused cancer in rats, even though he thinks it perfectly safe for humans. The government's ambivalent stance spurred fears of chemicals and other additives in the food supply. (Library of Congress)

Two people who arrived too late to apply for food stamps at a local Washington office in 1969. A nationwide outcry against hunger in America in the late 1960s forced the expansion of the food stamp program, even though the relatively high incidence of obesity among the poor seemed to play into the program's opponents' hands. By 1992, one in ten Americans was receiving the stamps. (Copyright *Washington Post*; reprinted by permission of D.C. Public Library)

"Rubbing Elbows with the Ritzy," the Washington Star's *caption for this photo of a McDonald's which opened beside one of the capital's fancier French restaurants in January 1979, implied that the two were incongruous. Yet the era saw a boom in eating out at both ends of the scale which continued until the late 1980s.* (Copyright *Washington Post;* reprinted by permission of D.C. Public Library)

This 1984 New Yorker *cartoon reflects one of the paradoxes of the 1980s: a mania for weight-loss swept the middle and upper classes while what one ate remained a sign of distinction.* (Drawing by Lorenz; copyright © 1984 by The New Yorker Magazine Inc.)

CHAPTER 8

The Best-fed People the World Has Ever Seen?

Its editors readily admitted that the dishes suggested in the new, improved, 1947 edition of the *New York Herald-Tribune* cookbook differed little from those in the 1937 edition. Why? "We believe that today, just as in 1937, good, plain food, discriminatingly cooked and served, is what the American family wants." What had changed, they said, were the methods, processes, and products that could deliver this kind of food.[1] This was not far off the mark, for, as we have seen, one of the striking aspects of the new food technology was how little it altered basic American food tastes. When asked by Gallup in January 1947 what they would choose as the their perfect meal if money were no object, Americans responded little differently than they would have twenty years earlier:[2]

Fruit or shrimp cocktail
Vegetable soup or chicken broth
Steak
Mashed or french fried potatoes
Peas
Vegetable salad
Rolls and butter
Apple pie à la mode
Coffee

Six years later, when *Better Homes and Gardens* came out with the revised version of its 1930 cookbook, the section on menu planning was little changed from that of twenty-three years earlier: Dinner should be meat, a starch, a vegetable, a salad, and dessert. Roast beef, steak, pork, veal or lamb chops, or roasted or fried chicken anchored the meals. Potatoes, carrots, and peas were the favored accompaniments. The "salads" were mainly jellied or otherwise sweetened.[3] Ten years after that, in 1963, when the army polled its enlisted ranks on their food likes and dislikes, the foods that topped each category made up a menu that differed little from Gallup's repast in 1947: steak, french fries, corn on the cob, sliced tomatoes, hot rolls or biscuits, and strawberry shortcake (which just nosed out apple pie.)[4]

Their commanders-in-chief shared these tastes. When Franklin Roosevelt's death elevated Harry Truman to the presidency in 1945, Truman's wife, Bess, brought their housekeeper from Independence, Missouri, to ensure that he could still susbsist on the hearty but simple midwestern food—roast beef and mashed potatoes were among his favorites—to which he was accustomed. His successor, Dwight Eisenhower, had a chef trained in the French tradition, but French food was served only at formal state occasions, which Eisenhower detested and kept to a minimum. Usually he dined on "standard American cooking," such as hash, often from a tray in front of the TV set. His favorite food was beef steak, an inch and a half to two inches thick.[5] The august senators up on Capitol Hill seemed even less demanding. The food in the famed Senate dining room was generally regarded "not as unexciting, but as unspeakable." The standard fare, aside from the famed bean soup and cornpone, leaned heavily on hamburger, cold cuts, ketchup, and gravy-and-fried-bread-crumb items.[6]

National eating times, which practically precluded preparing any but the simplest meals, demonstrated that Truman, Ike, and the senators were in step with the rest of the nation. In 1950 the typical American ate breakfast at 7:00 a.m. or slightly earlier, lunch at 12:00, and dinner at 6:00 p.m. The fact that only one person in six ate dinner as late as 7:00 p.m. was particularly telling, for it meant that most women preparing dinner were very pressed for time.[7] While obviously the case with working women, this was also true of stay-at-home suburbanites with young children, for the hectic after-school hours were often crammed with chauffeuring and other duties that competed with cooking. It is no wonder that most women stuck with the tried and true, particularly if it could be had in "quick 'n' easy" form. Americans' natural suspicion of experimentation in food—"Ugh! He'll eat *anything!*" was the usual response to adventurous eaters—reinforced this culinary conservatism by constricting the housewife's choice of what to make. It is no wonder that, given the narrow range of acceptable foods and preparation methods, most of the seventy-five hundred readers interviewed by *Good Housekeeping* in late 1945 declared that "the most bothersome problem in cooking" was "getting more variety into meals." This was not for lack of forethought; 84 percent said they planned meals one day or longer ahead.[8]

An obvious solution to the problem of variety was to venture down some culinary byways into ethnic cuisines. Certainly the time should have been ripe for this. Not only had wartime shortages helped make some kinds of European one-dish cookery acceptable, but the assimilation of the European immigrant population into American life made their food seem less threatening. Wartime movies featuring roll calls of infantry platoons and bomber crews composed of genial representatives of the major white ethnic groups ("O'Reilly . . . Yo . . . DeAngelo . . . Yo . . . Goldberg . . . Yo . . . Jones") had helped popularize romantic notions of the "melting pot." It now seemed, said the historian Richard Polenberg, that "the melting pot bubbled cheerfully, a far cry from the pressure cooker it had at times resembled in the past."[9] Occasionally this was reflected in food attitudes. Home economists, who in the 1910s and 1920s had urged ridding immigrants of their nasty food habits, now took a more benign view of foreign foods, but mainly to promote tolerance. An "interchange of food habits and customs" leads to "bringing about a greater understanding and friendship among both foreign and native groups," said the head of home economics at Western Reserve University in 1948. The preface to the American Home Economics Association's *The World's Favorite Recipes*, published in 1951 to aid the United Nations Association, said, "What people eat and the way they eat has long been one of the factors in prejudice. . . . Appreciation and adoption of the foods of many peoples is progress."[10] The middle-class women in the Cambridge, Massachussetts, Home Information Center devoted their 1953–1954 meetings to learning how to prepare dishes such as Italian lasagna, Russian kasha and borscht, Armenian shish kebab, and French cassoulet "to familiarize ourselves somewhat with the pattern of cultures other than our own in the matter of tastes, and to enjoy a broadening experience."[11]

But the idea of fostering tolerance through food was hardly widespread. At the very same time as the Cambridge women were earnestly trying to broaden themselves, the 1951 cookbook of the Mothers' Club of the Congregational Church in neighboring Middletown was featuring a recipe for coconut and chocolate candy called "Nigger Heads or Half-Breeds."[12] Moreover, although home economists might remark on the valuable lessons to be learned from foreigners, this was not reflected in what they taught or did. Their cookery texts were still relentlessly mid-American, full of uses for white sauce but with hardly a reference to anything foreign- or ethnic-sounding.[13] Of course, delving into foreign foods often involved doing something most American housewives could not or would not do: going to ethnic enclaves to search out ingredients unavailable in supermarkets. This seemed by far the most exciting part of the adventurous Cambridge women's experience, but there were no indications that they intended to repeat it. Most important, though, it involved overcoming the deep-rooted distate for most foreign foods that was still an integral part of American culture. Heirs to a tradition that for centuries had derided foreigners for their eating habits—"frogs," "krauts," "mackerel-snappers"—and suspected that high seasoning and sauces other than white sauce and brown gravy were camouflage for inferior foreign ingredients, they could not take to savoring Ethiopian cooking overnight.[14] This was even the case in polyglot Manhattan, which harbored an impressive variety of ethnic groups. A 1961 survey concluded that most of the borough's inhabitants ate meals that were "uninspired and monotonous." The typical dinner menu was steak or chops, a green or yellow vegetable, salad, and gelatin or canned or fresh fruit for dessert. Middle- and upper-income groups leaned heavily on the traditional pot roasts, beef stews, and roast beef, lamb, and chicken.[15]

As a result, while some food manufacturers and cookery experts did try to expand the home cook's repertoire by adding international flavors, they were careful not to offend native taste buds. Clementine Paddleford, food editor of the *New York Herald Tribune* and *This Week* magazine, wrote in 1960 that "even with the increasingly popular trend towards foreign foods the dishes come to the table with an American accent." American dishes had often originated in other countries, she noted, "but over the years they have been mixed and Americanized."[16] Betty Crocker's new ethnic recipes, developed for the 1961 edition of her cookbook, exemplified what this meant: Except for their shape, the only thing about the tortillas in her "Mexican Enchiladas" that resembled the Mexican ones was the name. They were mainly Gold Medal flour, beaten up with some corn meal (Mexicans use *masa*, a much different product), egg, and water and poured onto a griddle like pancakes. A mania for authenticity was hardly evident in her other "dishes with a foreign flavor." Beef curry was described as "a typical Chinese curry." It called for beef, mushrooms, tomato, and onion seasoned with curry powder and sugar. Perhaps it was called "Chinese" curry because it was thickened with cornstarch.[17]

Italo-American food continued to be exempt from the general reluctance to venture down foreign byways. Despite the fact that Italy had fought on the enemy side, World War II had enhanced its food's reputation in America. The popularity of spaghetti and tomato sauce, and its ease of preparation, made it a standard in armed forces mess halls. Except for the previously discussed recipes for "chop suey" and "chow mein," spaghetti and meat sauce was practically the only "foreign" recipe in the 1942 edition of the U.S. Army cookbook.[18] But by then spaghetti and canned tomato sauce had become commonplace in American kitchens. In the postwar years, millions of housewives who had never even met an Italo-American had their "own" recipes for spaghetti and tomato sauce with meatballs or ground meat. It was often the first substantial dish that newlyweds or singles living alone learned to cook. Their children graduated from soft, bland baby foods to only slightly less soft and bland canned spaghetti preparations. Had it not been for the inventiveness of the technicians at Kraft Foods, who in 1937 came up with the phenomenally successful Kraft Macaroni Dinner, the old-style elbow macaroni might well have become a kind of relic, used mainly for macaroni salad.

Pizza also captured many American hearts in the 1950s, but it could rarely be reproduced at home. Food manufacturers explored many avenues in trying to turn it into a homemade item. In 1955 Hunt's began suggesting that English muffins be smeared with its tomato sauce and covered with cheese, salad oil, and (optional) oregano. A renegade *pizzaiuolo*—pizza maker—in Florida picked up the idea and began doing it with bagels.[19] But acceptable as the substitutes were on one level, they could not match the real thing.

Another "dish with a foreign flavor" that found favor in American homes in the 1950s was chow mein. Unlike the wartime versions—usually called chop suey—which used ketchup and Worcestershire sauce, it introduced a somewhat different flavoring, soy sauce, to the home kitchen, as well as a new vegetable, canned bean sprouts, and canned fried noodles. It is easy to pooh-pooh the fad for its egregious inauthenticity: Chinese regarded chop suey and chow mein as "a culinary joke at the expense of the foreigner," said one of them.[20] Nor did it represent a breathtaking leap into the unknown, for making it at home out of certifiably clean ingredients and readily identifiable meats took care of ever-present fears of what was really taking place in inscrutable Oriental kitchens. Heinz even had a recipe for it using its cream of mushroom soup.[21] Nevertheless, for many Americans, eating and liking homemade chow mein still represented a distinct step toward breaking out of their culinary straightjacket.

By the end of the decade, rising sales of soy sauce, garlic powder, and other "foreign" ingredients were evoking speculation in the food industry that the nation might be developing a new interest in ethnic food. However, the giants in the industry were unimpressed. Some linked rising spice sales to the ascendancy of processed foods, which themselves "must be kept bland to suit the widest taste."[22] When William Murphy, the no-frills chemical engineer who was president of the Campbell's Soup Company, was asked in 1957 whether there weren't now a great many people who liked French, Chinese, and Italian food, he replied that although there were "quite a few . . . it isn't a very big percentage. The most popular things that people eat are the traditionally American things. The French things, the Chinese things, percentage-wise are small. . . . People tend to eat things that they are accustomed to."[23]

The women's media played a major supportive role in selling the "things they are accustomed to." Panicked by the growing competition from television for the food-advertising dollar, they were more shameless then ever in promoting food-processor advertisers' wares in their editorial copy. Inevitably, a veteran of the women's magazine business wrote, their recipes "featured ready-made food, mixes, gravies, syrups, anything in a bottle or jar or box."[24] This constraint often combined with the need to create an immense number of "new" recipes out of old foods to produce rather bizarre results. One wonders how many people could really have tried the *Good Housekeeping* suggestion that they freeze a stack of bologna slices interspersed with alternating layers of "pasteurized process sharp-cheese spread" and "relish-cheese spread."[25] Popular cookbooks were also filled with recipes for "new" ways of using processed foods. One of the colored cards in the *COOKINDEX* file of Tested Recipe Institute, Inc., is a recipe for "Cheese Apples": balls of mashed processed cheese, chopped nuts, and chopped pickled onions rolled in paprika, with a clove stuck on top as the blossom. Another is for "Frankaroni Loaf," which was boiled frankfurters resting on a bed of elbow macaroni, bread crumbs, processed cheese, and milk; after it was baked and sliced, a sauce of canned tomato soup and chopped stuffed olives was poured on it.[26] Processors' fanciful attempts to think up new uses for their foods were also reflected on the recipes they developed for their labels—which housewives seemed to use more than ever—and the short recipe books they distributed.[27] Kraft's told how to use its cream cheese in "Chocolate Topped Coconut 'Philly' Fudge" and "Baked Chive Cheese

Omelette" and suggested that Miracle Whip be used in "Miracle Whip Spice Cake" and "Punjabi Appetizer Dip." Wise's potato chips provided a recipe for butterscotch soufflé with a crust of chips.[28]

At least the Rombauers' *Joy of Cooking*, another of the era's best-sellers, had a split personality. It was an outgrowth of a small 1931 cookbook by Irma Rombauer, which tried to adapt some of her family's old-country German and traditional midwestern methods and recipes to the modern American kitchen. In 1939 she and her daughter had tried to tap the "career girl" market with *Streamlined Cooking*, a cookbook that saw the newly invented pressure cooker ("It permits cooks to scoff at time") and frozen fruits as salvation for cooks pressed for time. When it fizzled, they incorporated much of it into a greatly enlarged 1943 edition of *Joy*. The 1951 edition attempted to deal with the "continuing shrinkage of housekeeping time" and the disappearance of the servant by striking a balance between traditional cooking and "quick 'n' easy" recipes. However, by the late 1950s Marion Rombauer Becker, the daughter, who had effectively taken charge of the project, could see little ground for compromise between the two. "American cooking," she said, "had turned into a practice divided against itself. One group of cooks—if they can be dignified by that title—followed the grey-flanneled pied-pipers who offered TV dinners, an infinity of packaged mixes, and frozen pie-in-the-sky." The others were those who would "spend the small additional time required to make food wholesome and delicious."[29]

Despite the recent wave of nostalgia for it, it is still amazing to see how often American cooking was included in the postwar era's celebration of the country's greatness.[30] One can understand how midcentury Americans, knowing that their farmers were churning out huge surpluses of food, bedazzled by the immense number of choices they faced in supermarket aisles, could believe themselves to be "the best-fed people on earth" with regard to quantity. But most Americans also seemed to buy the more dubious proposition that the dishes prepared from this abundance represented "the best eating the world has ever seen." In a 1952 article on the excellence of midwestern food in the *Saturday Evening Post*, the popular author McKinley Kantor rhapsodized about the way his aunt and grandmother made pan gravy, describing how, after frying pork chops in fat, they stirred flour into the grease and then slowly added milk to the pan:

> That's where I learned to make gravy—wholesome flour and milk gravy, to go along with pork chops and mashed potatoes. I would rather eat my gravy than any esoteric *sauce supreme* contrived by some third-generation hireling of the original Escoffier, who may be able to enchant sophisticates with his *Foie de Veau Pôelé à la Bourgeoise*, but who wouldn't know how to cook a meat loaf with browned potatoes if the entire fate of the Thursday night church supper of the Berean Guild depended on it.[31]

Unfortunately, most American women did not seem to feel that they were up to Kantor's standards. When Gallup asked married women in 1951 whether they were better cooks than their mothers, 58 percent said no and only 23 percent said yes.[32] It was not, however, for lack of trying. Two years later 67 percent disagreed with the statement that not one woman in ten made a real effort to improve her cooking. Which single dish was the great test of a woman's cooking? In order of frequency, women listed pie, cake, roasts, bread, and steak. Men listed the same traditional things but in different order: pie, roasts, bread, steak, and cake.[33]

American restaurants hardly picked up the culinary slack. "You have never heard of a European visiting America for its food," wrote one observer in 1960.[34] The national provincialism in this regard was perhaps best symbolized when Duncan Hines made his first trip to Europe in 1954. By then a national celebrity, he now lived in a one-story colonial-style home outside of Bowling Green, Kentucky, whose rooftop weathervane gave it a distinct resemblance to the Howard Johnson's restaurants he so admired.[35] Some of the profits from annual editions of *Adventures in Good Eating* and his other guidebooks (sold by the listed establishments, who also paid an annual fee to hang his "recommended" sign on their doorsteps) were now channeled through the Duncan Hines Foundation to support his favorite cause, the Sanitary Foundation, created to promote cleanliness in restaurants. His first encounter with European food at its source had been the "tiny French peas" he had been served on the liner *Liberté.* These were superb, he acknowledged, as were the noodles in Rome. However, he seemed more impressed by the frequency with which French waiters changed ashtrays than by French food. They "use too much sauce on things," he said. "They'd douse it over meat in a way that would contaminate everything else on the plate." He was genuinely shocked to find that Italians actually ate "baby octopus" ("Couldn't choke the damn thing down") and studiously avoided "ground-up meats" ("I wasn't going to take a chance on the raw meat of some varmint I didn't even know the name of") and outdoor dining ("Too many flies and gnats"). The place whose cuisine he admired the most was London, where one could get good roast beef and steak-and-kidney pie. How did American cooking stack up against that of Europe? he was asked. "I still claim it's the best in the world" was the reply.[36]

Needless to say, the vast majority of the restaurants recommended in Hines's guides served American-style food, but the same was true of those compiled by other experts. Almost all the places chosen by a distinguished advisory committee for inclusion in Ruth Noble's *A Guide to Distinctive Dining* in 1954 and 1955 specialized in American cooking, with the emphasis on steak, lobster, roast beef, and roast or fried fowl. Only two French restaurants were listed, both in New York City, along with the usual New Orleans duo of Antoine's and Arnaud's. There were two Italian restaurants (in Detroit and San Francisco) and three listed as "specializing in Mexican and American foods." (One recommended place, La Posada in Winslow, Arizona, was built "in the style of a real Spanish rancho" but featured blueberry muffins as its specialty.)[37] The most renowned of the advisors for Noble's guide was Lucius Beebe, a portly and rather pretentious writer, traveler, and presumed bon vivant who assumed for himself the mantle of America's greatest epicure. Yet, though he was the major food writer for *Holiday* magazine, which itself projected a cosmopolitan image, that magazine's annual restaurant awards also stuck to the safe, sound, and usually American. There were more French restaurants than in Noble's compendium (including the Old Warsaw in Dallas—"Fine French cuisine; expensive"), but the steak, lobster, and roast beef syndrome still reigned supreme.[38]

In early 1961 another writer for *Holiday* and his wife surveyed many of the country's better restaurants on a long road trip that took them from New York City to Florida and then across the South and "the scenically impressive but gastronomically retarded regions of the Southwest." By the time they rolled into southern California they were so sick of beef and steak that ("un-American as this may sound") they could barely look at it. "Whatever it is called," he wrote, "beef invariably arrives flanked by 'Famous Big Russet Idaho Potato' and saucers of sour cream, chives, and chopped bacon. The prologue to this red-blooded all-Western binge is a bowl of finely chopped salad greens wallowing in a viscous dressing of unknown origin."[39] They might have done better to heed Charles Einstein, who some months before had warned *Harper's* readers that there were only five great "eating towns" in America—Boston, New York, Baltimore, New Orleans, and San Francisco—and all were located on a seacoast.[40] Craig Claiborne did not even share this benign view of the food on the coasts. When he began reviewing restaurants for the *Times* in 1959, he later recalled, "I realized that New York was not a great restaurant town. Although the best in America, the food was terrible; it was awful . . . ugh."[41] In fact, even Hines thought most American restaurant food was quite disgusting. "I've run more risk eating my way across the country than driving the highways, dangerous as the latter has become," he wrote in 1947.

> The way many eating places manhandle chicken, then palm it off on the public, is a crime. What is this so-called Maryland or Southern fried chicken they serve at most eating places? Nothing but an old hen parboiled the day before, hacked with a cleaver into the most pieces possible, then fried in stale, thick batter when you order it. What you get is fried batter flavored with chicken. Why don't they list it on the menu as "chicken fried batter?" . . . The gravy one eats in the average restaurant is about as appetizing as warmed over calcimine [whitewash] and the sauces are usually the same stuff so highly seasoned it burns a hole in your stomach lining.[42]

Another traveler, the novelist John Steinbeck, writing of a road trip across the country in 1961, said, "In the eating places along the roads the food has been clean, tasteless, colorless, and of a complete sameness." Only the breakfasts were excellent—if one stuck to bacon and eggs and pan-fried potatoes.[43]

It is not hard to find the reasons for the mediocrity of most American restaurants; they were more or less the same as those undermining home cooking. One was a conservative clientele. "Ask the average person 'What do you associate most with eating out?' " said a restaurant industry magazine in 1952, "and he'll probably answer 'broiled steak.' "[44] Guido Mori, the French-trained chef who headed a staff of sixty-seven at Chicago's famed Pump Room, said in 1951 that his customers went in mainly for steaks, chops, and chicken à la king. "Foreign dishes take too long for most of them," he said, with a touch of melancholy.[45] Customers' conservatism was also the product of the timidity that came from an ignorance about food. "From observing thousands of travelers in eating places," said Hines, "I have concluded that the average American fails to get good food either because he doesn't know what good food is or because he is too timid to insist upon good food."[46] "It is almost as if the customers had no interest in what they ate as long as it had no character to embarrass them," said Steinbeck. They suspected or disliked strong, pungent, or exotic flavors because they put cleanliness ahead of taste. As a result, he said, their sense of taste had disappeared through lack of stimulation.[47]

Mediocritizing influences also came from within the industry. Restaurants had done well during the war. Their exemption from rationing, the tripling of the number of women workers, and massive amounts of overtime work and pay made them reasonable alternatives to eating at home. From 1939 to 1946, restaurant sales almost quadrupled. But the postwar return to normal working hours and the growing devotion to home and family-building brought them face-to-face with the old reality: Restaurant-going was simply not an important part of American family culture. After 1948 sales began to slip, particularly as television—which mesmerized families for practically the whole of the next decade—came to anchor Americans to their homes.[48] The glistening chrome diners, that most beloved symbol of Americana, which had boomed during the Depression and war by serving home-style food in congenial surroundings, were skewered by the surge of suburban family-building. Urban and working-class in origin, they tried to follow their clientele to the suburbs and cater to the family trade. But expanding the table service areas to accommodate children meant reducing counter space and hiding the cooks in annexes, forsaking the genial give-and-take with regular customers that had been the backbone of the business.[49] Some restaurateurs tried to follow their customers home with "take-home" foods. Others emphasized that restaurants offered an alternative to the kind of food one got at home. "In the old days food operators attracted customers to their eating places with the slogan 'Real Home Cooking,' " said one industry magazine in the dark days of 1952. Now it was "the sameness of home meals" that "drives them from dining table to counter or booth. . . . The operator who features 'Home Cooking' should look into his kitchen and see if it can stand some of the good restaurant cooking." But the main thing restaurants had over home cooking, it said, was that customers preferred restaurants' deep fried foods, particularly potatoes.[50] It was hardly the stuff upon which a great culinary revival could be built.

Restaurants also battled doggedly with rising food prices and the old bugaboo of American industry, high labor costs.[51] Increasingly they turned to frozen foods. Frozen vegetables, for example, cost about 15 to 20 percent less than fresh ones, and there was hardly any waste: They did not need peeling or trimming, and what was not used could be kept frozen. Not only did these and other frozen foods cut preparation time, but the most unskilled people could "heat and serve," reducing the need for well-paid and often highly independent help. As a result, even the best eating places used them. In the early 1950s, the Chicago commissary of the Rock Island Line, famed for its excellent food, began to mass-produce, precook, and freeze thousands of portions of swordfish, Swiss steak, rolls, vegetables, and pies for mere reheating on the train. Dining-car staffs were cut from nine to six. Yet, because warming took less time than cooking, customers were served faster and tables turned over more quickly. The Chesapeake and Ohio went the Rock Island Line one better: It purchased complete frozen dinners from Frigidinner, Inc., doing away with most of its talented black cooks, who, famed as they might have been, could not make fine railroad dining profitable.[52]

A gastronomically insensitive public made it easy to switch to lower-cost processed foods. The Rock Island Line plunged ahead with the change after an introductory experiment in which it became clear that passengers did not know they were eating frozen foods. Chains like Howard Johnson's and Schrafft's, which had previously been known for straightforward cooking of mainly fresh ingredients, took maximum advantage of frozen foods. Much of their food was now prepared and frozen in central commissaries and then shipped to the restaurants, where unskilled local help would heat, grill, or fry them. By the early 1960s, 80 percent of the the food served at Holiday Inns consisted of frozen prepared items sent out nationwide from its Chicago commissary. Thanks to this system, its restaurants were able to service two dining areas with only one cook and one dishwasher. "The frozen prepared items have achieved amazing acceptability," said the chain's vice-president of operations. "It will have to come to this," he added, "because of a lack of skilled people."[53] Tad's, the high-volume, low-priced steak chain, was so positive about customer acceptance of frozen foods that it developed an entirely new chain of restaurants, Tad's 30 Varieties of Meals, serving complete frozen dinners—fried chicken, scallops, and so on, wrapped in plastic with potato and vegetable—which the customers defrosted at tableside microwave ovens.[54] Even upscale restaurants, whose high prices enabled them to hire skilled chefs, could not resist the economies of frozen foods. Boston's venerated Locke-Ober's, which Lucius Beebe called "a temple of gastronomy," switched to frozen vegetables in 1958. Apparently, the gastronomes either did not notice or did not think that the temple was being defiled. "People now come up and compliment us on our vegetables," said the manager. "The beans are out of this world is what they say. . . . I've never had such beans."[55]

Other cost-cutters contributed to an institutional sameness in much of restaurant cooking across the nation. Stockpots became relics of the past as the use of powdered soup bases became almost universal. Kraft Foods emphasized their economy ("one step preparation—time saving, no waste. No expensive labor preparing soup stock") and offered restaurants four kinds: "Onion Soup Base, Soup Base Flavored with Beef Extract, Chicken Base, and CS Base with Chicken Fat."[56] Carson Gulley, a soft-spoken black chef who regularly dispensed cooking tips to housewives on radio and television and in the press, recommended them highly to restauranteurs. "For many years we went along without soup bases," he granted, "and still did a commendable job of cooking, but we also had many cooking failures and mishaps that could have been avoided if soup bases had been available. . . . In fact, there is hardly a cook living who can prepare French onion soup at its best without adding a little soup base." Using them was particularly good for profits, he noted, because they had a "stabilizing effect" that extended the life span of soups and "meat extender" dishes.[57]

Perhaps the popularity of soup bases explains why many restaurateurs felt no need to disguise the fact they used canned soups. Indeed, many lunch counters featured racks displaying single-serving cans of soups; customers could watch their choices being opened and heated.[58] Even fresh potatoes went by the boards, replaced not just by frozen french fries but also by "instant" dehydrated ones whose preparation was idiot-proof. All one had to do to make mashed potatoes was add water, mix, add milk, and mix some more. "Simplicity itself!" said Kraft when it introduced the new product to the restaurant trade in 1958. "Even inexperienced kitchen personnel can mix a perfect batch in a few minutes." Processed potatoes also promised strict portion control and "no more seasonal fluctuations in price or product to worry about."[59] By 1959 Idaho potato producers had even stopped promoting fresh potatoes in the restaurant trade; all of their efforts were devoted to the frozen and dehydrated kinds.[60] The elusive vocabulary surrounding canned and frozen foods helped blur concepts of "fresh." In 1952 *Fountain and Fast Food* told of the system Mawby's restaurant in Cleveland used to bring canned vegetables "to serving temperature in about one minute, to ensure freshness."[61] Everyone seemed to assume that "fresh frozen" fruits and vegetables had never been cooked, whereas most were parboiled. They were then "defrosted," never "thawed."

Restaurants also tried to cut labor costs by adopting other industries' mass production techniques. The Varsity restaurant in Atlanta, which billed itself as the world's largest drive-in, sold two million hot dogs and a million and a half hamburgers a year (largely to Georgia Tech students) with a system of conveyor belts that brought orders to countermen in the kitchen and returned the food to the front.[62] The Fountron in St. Louis speeded up the ordering process by having waitresses at its counter write them into a teleautograph machine, which repeated them instantly in the kitchen. They also wrote one of four colors on the order, so when the food was put on the conveyor belt in the kitchen it could be covered with a plate of that color. An electronic eye at the waitress's station would then sense the color and halt it.[63] A California inventor developed the drive-in "Motormat," where a customer would drive up to a window-high bin, mounted on rails, containing glasses of water, menu, pencil, and pad. He or she would then fill out the order, push a button, and send the bin scooting back to the kitchen, which lay at the center of the circular structure. While the order was being prepared, the bin would be sent back with the bill. After the bin was returned with payment, the food and change would be sent back down the rails, with no need to tip a waitress.[64] In the early 1950s the Insta-Burger-King chain—later to become simply Burger King—built a franchise operation in twenty-nine states around contraptions that cooked hamburgers on revolving trays by radiant heat in one minute and then automatically dropped them into sauce and joined them with buns from another revolving tray.[65] The Burger Chef chain bested them all by sending burgers on conveyor belts through infrared broilers at the rate of eight hundred an hour.[66] When Ray Kroc went to San Bernadino, California, to investigate why the McDonald brothers' hamburger outlet was so successful, he saw instantly that their system—not their food—was the key.[67]

Although speed was important, a major object of these systems was to eliminate any skill involved in food preparation. The industry journals of the 1950s, reflecting as they did restaurant owners' obsession with eliminating skilled labor, read like transcripts of meetings of nineteenth-century mill owners plotting their next move against the dwindling number of workers whose skills gave them a shred of bargaining power. "No skilled hands are needed to make hamburgers quickly," said an impressed industry journal of the Burger Chef operation. They were fed into the conveyor "by unskilled operators who then simply watch the finished product come out the other end."[68] In both instances the deskilling of industry resulted in more mass-produced sameness for consumers, who seemed to be largely content with the results. By 1960, however, the industry was already developing the new systems that would drastically change eating out in America and make the decade of the 1950s, with all its aura of efficiency and "automation," seem like the last hurrah of the industry's mom-and-pop era. But these would arise in tandem with other factors that were beginning to poke holes in what Gael Greene so aptly called "the Velveeta cocoon."[69]

CHAPTER 9

Cracks in the Façade: 1958–1965

Not until the mid-1960s did doubts that America led the world in virtually everything—and that all Americans would eventually share in its abundance—begin to seep into the national consciousness. Yet many of the events that would help shatter these illusions were already in motion by the late 1950s: Soviet successes in missile development, a growing number of nations unwilling to side with the United States in a bifurcated world, the vortex in Southeast Asia, the civil rights movement and decaying inner cities at home. While events on the food front may not have had the same global importance as those threatening nuclear holocaust, they too helped make a mockery of *Time-Life-Fortune* publisher Henry Luce's confident 1941 prediction that this would certainly be the "American Century."

The seeds of skepticism about the system were first planted at home—literally. By end of the 1950s, the first baby boomers were reaching adolescence, hard on the heels of a wave of concern over teenage delinquency and other signs that parents were losing control over their children.[1] Anxiety over the young was matched by growing doubts that the glittering new appliances and attractively packaged processed foods could really deliver on their promises of freedom. Nixon's boast to Khrushchev that modern American technology was making the housewife's life easier began to ring hollow in millions of exhausted housewives' ears. Any time saved by convenience foods and labor-saving appliances seemed to be gobbled up elsewhere. Studies showed hardly any diminution in the time devoted to housework since 1930. One even indicated that the amount of time middle-class urban housewives spent on food preparation actually increased slightly from 1931 to 1965.[2] Processors and appliance makers seemed to inadvertently admit this by portraying cooking less as a labor of love and expression of creativity and more as part of "kitchen drudgery." Deeper questions about the nature of middle-class home life itself fed the malaise. Sociologists wrote of suburbs inhabited by lonely, unfulfilled women whose main concern was gaining their neighbors' approval. Intellectuals denounced the baleful effects of television and the debased, commercialized "mass culture" that enveloped the middle class, stifling creativity. They bemoaned the rise of "the organization man," "mass man," and the "lonely crowd," whose values were derived from those around them rather than the promptings of their inner selves. The popular James Dean movie *Rebel Without a Cause* told of middle-class teenagers driven to delinquency by the sterility of their relationships with their cold, conformist, neglectful parents. While some disputed these notions of conformity and boredom, and most suburbanites were only vaguely aware of the critique, concerns over whether housework and child-rearing were worth the effort still seemed to seep into many middle-class households, creating the unease that Betty Friedan labeled "the problem that has no name."[3]

One of the offshoots of women's disillusionment with their role in the kitchen was the first significant step toward the sharing of responsibility for cooking with men: the barbecuing boom of the late 1950s and early 1960s. Heavily promoted by beef and other food producers, the success of the campaign was reflected in everything from a surge in lawn furniture sales to a booming market for hot dogs.[4] One of the weekend barbecue's main selling points was that, presided over by Dad, it provided welcome relief from the drudgery of cooking for Mom. "Mom gets damned tired of preparing and planning three meals a day," said one enthusiast. "Cooking, to her, is no longer an adventure. It's a chore, and she's sick of it."[5] The fact that the barbecue often replaced the traditional Sunday afternoon sit-down family dinner—the most elaborate one of the week—made it doubly welcome on that score.

Yet the traditional division of gender roles in the household remained secure. The kitchen still rested safely in Mother's hands, for the smoky ritual took place outside its confines. It was also understood that only the most easily prepared foods would be served, mainly hot dogs and hamburgers, accompanied by a range of purchased foods such as canned beans, cole slaw, potato salad, macaroni salad, ketchup, and relishes—nothing that could threaten the female monopoly over *real* cooking. To assuage any fears that Dad might really be serious about moving into the kitchen, he was encouraged to wear large aprons with macho slogans, which were the opposite of her frilly ones, and wield oversize utensils, which were clearly inappropriate for the kitchen. The idea that cooking over an open fire was itself a particularly masculine pursuit also helped, reflecting American images of cowboys on the range or fishermen in the woods rather than the fact that it is women's daily task throughout much of the world. "When a barbecue goes into operation, it automatically becomes a masculine project," said *Esquire's Handbook for Hosts*. "After all, outdoor cooking is a man's job. A woman presiding over a barbecue grill looks as incongruous as a man engaged in doing a trifle of lacy tatting [a kind of lace-making]."[6]

Of course, doubts about the value of housework were not entirely new. Even in 1952, at the height of veneration for what Friedan called the "Happy Housewife, Heroine," Marjorie Husted, a.k.a. Betty Crocker, warned advertisers that women were "uncertain, anxious, and insecure" about their role as "homemakers." They said "Oh, I'm just a housewife" apologetically, she said, warning that "homemaking no longer carries prestige." Husted told the advertisers that they had to correct that "misapprehension," but by the later 1950s they were being accused of exploiting this anomie.[7] In *The Hidden Persuaders*, his 1957 best-selling examination of modern marketing techniques, Vance Packard painted a disturbing picture of anxious suburban housewives being deftly exploited by expert psychologists, who helped sell products by appealing to the housewife's deep insecurities and unconscious instincts. He said that supermarkets arranged their colorfully packaged products to induce trancelike states in shoppers—studies showed their eye-blink rate slowed almost to that of people under hypnosis—mesmerizing them into picking things they did not need off the shelves.[8]

Suspicion of the food industries was also fueled by renewed questioning of the healthfulness of food additives and processing. Since the passage of the Food and Drug Act of 1938, the government had been generally passive in the face of the plethora of new chemicals being sprayed on or added to foods. The Eisenhower-era Food and Drug Administration was bent more on cooperating with industry than confronting it. Although grounded in part in the agency's inability to do much about the profusion of new food additives, this also fit in with the climate of the times, which admired rather than questioned the nation's productive achievements. Other public agencies were equally supportive, as were most politicians. For more than five years after the special hearings he headed ended in 1952, Congressman James Delaney's proposals that the Food and Drug Act be amended to make food processors prove that their chemical additives were safe for humans were, in his words, "completely ignored." In 1956 charges leveled by four doctors that a drug used to hormonize chickens and cattle left residues of potential carcinogens in their meat caused some stir, but the allegations were disputed by other experts and faded.[9] Then, in 1958, the bill's prospects soared when one of the top researchers at the National Cancer Institute reported that a number of chemicals long used in food might cause cancer in humans. The story received wide press coverage, ranging from the *New York Times* to the racy *Police Gazette*, which ran a sensational series on the charges. However, Delaney thought it was the intervention of the aging Hollywood star Gloria Swanson that was crucial in finally getting the bill through Congress. The still-glamorous star, a health-foodist disciple of Gayelord Hauser, buttonholed crucial congressmen's wives, persuading them to convince their husbands to support the measure.[10] Now, not only would processors have to satisfy the FDA that new additives had been tested and proven safe for human consumption, they were given two years to establish that those already in use were not "toxic" to humans—what the act called "Generally Recognized As Safe." While most processors acknowledged that this kind of government certification of food safety would ultimately be to their benefit, they were still upset by a clause in this "Delaney Amendment" that forbade the use of any substance deemed to have caused cancer in animals.

Their most effective argument was one that was becoming the mainstay of tobacco's defenders: The human body was different from a rat's, and what was bad for one was not necessarily bad for the other.[11]

The doubts about additives sowed by the 1958 cancer scare were soon followed by heightened fears of pesticides. Fortuitously, the new amendments had defined any chemicals used in growing foods that left residues in them as "additives," and therefore subject to FDA regulation. Sure enough, just before Thanksgiving of 1959, FDA inspectors found traces of a carcinogenic weed killer used in cranberry bogs on part of that year's cranberry crop. The affected cranberries were whipped off the market, and experts were trotted out to assure the public that a person would have to eat fifteen thousand pounds of contaminated cranberries a day to suffer any harm. The FDA head, George P. Larrick, reassured the public that the cranberries that remained on the market—the bulk of that year's crop—were perfectly safe. The nation's food was much safer than that sold fifty or one hundred years earlier, he declared. But Americans are notoriously ahistorical, and most of that year's holiday turkeys were eaten ungraced by their traditional accompaniment.[12]

To make matters worse, within weeks their confidence in fowl in general was shaken with the announcement that the main chemical used in hormonizing poultry had caused cancer in other animals.[13] This was followed by the revelation that one of the most common food colorings also caused cancer in rats. As if this were not cause enough for concern, it soon emerged that the FDA was powerless to ban it because lobbyists had quietly managed to have food colorings exempted from the act; this was hastily but belatedly rectified. The summer of 1961 brought another major blow to confidence in food producing and processing technologies, the publication of Rachel Carson's book *Silent Spring*. Its message was reflected in the chilling title: Chemical sprays and additives not only threatened to annihilate much of the animal, insect, and plant life on earth but could eventually lead to the virtual extinction of human beings as well. As a result of all this, the dormant consumer movement sprang back to life, flooding the FDA with demands that processed foods' labels carry lists of all their ingredients.[14]

Like the worries over sprays and additives, concern over malnutrition among the poor had also practically disappeared amidst the hoopla over the "best-fed nation on Earth." During and after the war, academic nutritionists had strengthened their ties to food producers and drifted further away from concern over the poor. The processor-supported Nutrition Foundation came to play an important role in funding and disseminating the results of nutrition research. Its journal, *Nutrition Reviews*, with its useful summaries of the latest nutrition research, reflected the steady drift among nutritional scientists toward indentification with the world-view of the large processors—one that was hardly concerned with the poor. Financial links between the two were not only taken for granted, they were encouraged. Cornell University's experience was typical of this process. It was only after it had been assured of food processors' financial support that its governing council had allowed the creation of the nation's first School of Nutrition in 1941. Freezer manufacturers and utility companies provided additional funding during the war, and after the war the Grange League, a giant New York farmers' marketing cooperative, gave it a huge grant to establish a biochemistry department that would help the co-op expand into food processing.[15]

Their links with the food industries had encouraged nutritionists to drift away from lingering concerns with malnutrition and go along with a benign view of the nation's nutritional status, which ignored inequalities due to income. In a 1955 review of the literature on malnutrition a government nutrition expert wrote that "the evidence supports the conclusion that the nation as a whole is fairly well fed." Any nutritional deficiencies were general ones, not confined to the poor.[16] The 1927 and 1935 editions of Lydia Roberts's standard text, *Nutrition Work with Children*, had at least included poverty as one cause—indeed, the single most important one—of malnutrition. The 1954 revision dropped it completely from consideration.[17]

By 1960, however, important people were coming to think that malnutrition was indeed linked with poverty. During his campaign that year in the Democratic presidential primary in West Virginia, John F. Kennedy was shocked by the emaciated appearance of unemployed Appalachian coal miners and their families. In the opening statement in his first television debate with Richard Nixon, he raised the old specter of breadlines knee-deep in wheat: "I am not satisfied," he said, "with nine billion dollars' worth of food rotting in storage while millions go hungry."[18] Later, he rattled off the charge that one in seven Americans was hungry (a figure that seems to have come from a 1948 Gallup poll, cited by Eleanor Roosevelt in her newspaper column, in which that proportion of Americans admitted to having gone hungry that year).[19] After he moved into the White House, he let it be known that he found the grim picture of widespread deprivation painted in Michael Harrington's *The Other America* to be truly shocking.[20]

Meanwhile, in 1961, another harbinger of the future arrived: the first cholesterol scare. For years Americans had been warned that their diets were deficient in essential nutrients. Then they learned that their foods might contain dangerous chemicals. Now came the first indications that their normal diet might be lethal. The most common everyday foods, milk, butter, cream, cheese, and beef—a good-sized chunk of the "Basic Four" charts—contained a substance that was clogging their arteries and causing their hearts to fail. Reports linking high levels of cholesterol in the bloodstream to atherosclerosis and heart disease had been appearing since 1950 but had been dismissed as inconclusive by the nutritional establishment and aroused little public concern.[21] As was so often the case, commercial considerations played a large role in overriding this complacency. The discovery in 1959 that eating polyunsaturated fats might lower serum cholesterol levels set off an explosion of health claims by the manufacturers of vegetable oils and margarine, whose huge advertising budgets ensured that few could escape the news. "The rush to get aboard the polyunsaturated bandwagon has become a stampede," said *Food Processing* in December 1961. "Each week new products containing higher and higher amounts of polyunsaturated oils are finding space on shelves. . . . Complete lines of salad dressings, mayonnaise, margarine, even bread and rolls are advising customers by means of labels, shelf talkers and ads that they contain polyunsaturated oils."[22] Conversely, consumers began to shun dairy products, provoking alarm in the dairy industry, which frantically tried to mount reassurance campaigns.[23] By 1962 fully 22 percent of American families claimed to have changed their diets as a result of the scare.[24]

Meanwhile, interest in dieting for weight loss had been revived. Despite the popularity of large-bosomed female sex symbols such as Marilyn Monroe and Jayne Mansfield, the very thin ideal had never disappeared, particularly among the most stylish segment of the population. In 1946 the British writer George Orwell, writing of the women pictured in *Vogue* magazine, remarked that "nearly all these women are immensely elongated. A thin-boned, ancient Egypt type of face seems to predominate: the narrow hips are general, and slender, non-prehensile hands like those of a lizard are quite universal."[25] But while dieting to lose weight never quite went out of style, it was also not particularly high on the middle-class agenda. Women's magazines ran weight-losing adivce, but the weight-loss diets shared space with ones that could cure muscle pain, migraine headaches, gout, or "Heberden's nodes."[26] (More serious medical journals reported on diets that promised to cure multiple sclerosis or heart disease—a milk diet—or raise IQs.)[27]

This reflected a perception that the audience for weight-loss advice was not all that large. Almost half of a cross-section of Americans interviewed by Gallup in 1953 said they were satisfied with their present weight. Only 35 percent wanted to lose weight, and 14 percent actually wanted to put on pounds. The proportion of women who wanted to lose weight (42 percent) was smaller than the combined total of those who wanted to remain the same (40 percent) or put on weight (12 percent). Warnings by prominent nutritionists such as Russell Wilder and Frederick Stare that overweight was now the nation's most serious health hazard caused little stir, even though they were supported by impressive-looking new statistics from the Metropolitan Life Insurance Company.[28] Surveys of Americans' attitudes toward weight and health in the late 1950s and early 1960s concluded that they did believe that overweight shortened life—but not by very much. "Most Americans think it shortens life by maybe five minutes," said one of the researchers.[29]

However, around 1958 Pandora's dieting box seemed to open again, pried open not by health concerns but by fashion-conscious women. Suzy Parker, the most celebrated model of the late 1950s, was decidedly tall and slim, particularly in comparison with Monroe, Mansfield, and company. According to Helen Woodward, a women's magazine executive, publications such as *Glamour*, *Seventeen*, and *Mademoiselle* suddenly discovered that reducing diets were "sure-fire circulation builders." "Next to fashions," she wrote in 1960, "the most important subject in all these magazines has been reducing. Each new diet . . . is treated as a miraculous new discovery. Then, in a little while there is a totally different diet, again a miracle."[30] Sales of saccharin rose, and those of sodium cyclamate, a recently approved chemical that did not have saccharin's bitter aftertaste, rose even faster, more than tripling from 1959 to 1961.[31] Dieting received a further shot in the arm in 1959, when the Mead and Johnson's chemists put the finishing touches on Metrecal, the first of what the trade called "metered calorie products." When mixed with milk, these adult versions of baby formula provided about the same balance of protein, carbohydrate, and fat as breast milk. Each glass of the glutinous drink was remarkably filling—enough to substitute, it was hoped, for a full meal—yet contained but a few hundred calories.[32] By late 1961 Metrecal and similar products were chalking up close to $350 million in annual sales.[33] In the same two-year period since Metrecal's introduction, supermarket sales of other "low-cal" products almost doubled.[34]

A wave of diet cookbooks began to wash into bookstores. In 1957 Avis De Voto, an editor for Alfred Knopf, wrote Julia Child that she was utterly depressed by the manuscripts for diet cookbooks that were appearing on her desk in increasing numbers. The only word to describe one she had just received, she said, was "gruesome." There was "not a single honest recipe in the whole book—everything is bastardized and quite nasty. Tiny amounts of meat . . . are extended with gravies and sauces made with corn starch and then further extended by the addition of canned vegetables. . . . Desserts, of which there is a fat section, are incredible—sweetened with saccharine and topped with imitation whipped cream! Fantastic! And I do believe a lot of people in this country eat just like that, stuffing themselves with faked materials in the fond belief that by substituting a chemical for God's good food they can keep themselves slim while still eating hot breads and desserts and GUNK."[35]

Books such as these, as well as the millions of people willing to substitute ersatz formulas for real food, did indeed say little for prevailing culinary standards, but some bright lights were still visible on the horizon. Although, as previously noted, fine restaurant cooking remained in its Depression doldrums through the 1950s, beneath the surface the foundation for revival was being laid. In October 1941, over one hundred years after Lorenzo Delmonico had initially introduced Americans to haute cuisine in his Manhattan restaurant, Henri Soulé took the lead in doing it again. The dapper Soulé, an appropriately portly man with slicked-back, receding hair, had been brought over to direct the restaurant at the French pavilion at the 1939 New York World's Fair. His high standards and well-trained *equipe* helped make it "the hottest ticket in town," recalled one New Yorker, a place whose food "was of a greater delicacy than we had in New York at the time."[36] When the fair closed he persuaded ten of the French staff to help set up Le Pavillon on Manhattan's East Fifty-fifth Street, across from the St. Regis Hotel and just around the corner from expensive Fifth Avenue shops such as Cartier's, Tiffany's, and Bergdorf's. It was elegantly furnished, bedecked with a profusion of fresh cut flowers (each table had a centerpiece of long-stemmed red roses), and boasted gleaming silver, sparkling china, and one staff member employed to do nothing but polish the Baccarat glassware. After navigating a few rough shoals, it became a favorite of the Vanderbilts, Astors, Whitneys, Rockefellers, Burdens, and others of the social elite who still regarded a taste for haute cuisine as a mark of distinction. Although privately contemptuous of many of their habits—such as swilling cocktails before dinner and smoking between courses—Soulé carefully cultivated their self-image as people of sophisticated tastes and worked hard at developing a reputation for being unimpressed by mere wealth. For him, it was said, only good taste, manners, and breeding counted.[37] The ascendancy of quality over quantity was symbolized by the restraint of the *buffet froid* that greeted guests as they entered one of the three dining rooms. "Instead of the overwhelming profusion of dishes so often encountered on an hors-d'ouevre wagon," Joseph Wechsberg reported to *New Yorker* readers in 1953, "here are only a few carefully chosen delicacies—salmon from Nova Scotia, caviar from Russia, *foie gras aux truffes* from France, and *terrine de canard*, *boeuf à la mode en gelée*, and *langue givrée* from the kitchen of Pavillon."[38]

During the 1950s the clientele for this kind of food and treatment expanded rapidly. The United States had emerged from the war led by a political and financial elite who spoke self-confidently of assuming the mantle of world leadership. Wall Street lawyers and financiers, corporate executives, and government officials assisting in the reconstruction of Europe—all trod a well-worn path to the Hotel Crillon in Paris. In the course of establishing amiable relationships with their European counterparts, many fanned out from there to avail themselves of the renewed marvels of French haute cuisine, which sophisticates still regarded (*pace* Duncan Hines) as the world's finest. Then, in 1949 and 1950, tourism to Europe began picking up smartly, as the wealthy took advantage of newly refurbished luxury ocean liners and a dollar that ruled the waves to revive the tradition of the elegant European grand tour, partaking of its finest culinary offerings.[39]

Thus, while some recalcitrants in the old upper class still clung grimly to conspicuous underconsumption, an elite clientele for French restaurant food was again growing, particularly in New York City. Some of the old Park Avenue elite remained loyal to Voisin, one of the few bright culinary lights of the 1930s, but the most celebrated members of the elite gravitated to Pavillon, where such powerful men as Bernard Baruch, Walter Bedell Smith, and J. Edgar Hoover dined alongside cultural celebrities such as Sol Hurok, Cole Porter, and Salvador Dali who were known for their refined tastes.[40] By the mid-1950s people trained by Soulé were spinning off and opening their own places. Like Soulé, they usually sought to serve the kind of haute cuisine that had taken root in finer restaurants in France before the war. While somewhat lighter and less elaborate than that of the turn-of-the-century *belle époque*, the food was still based on Escoffier. The *sauce espagnole* and *béchamel* simmered on the corner of the stove all week long, and the rat-tat-tat of truffles being chopped reverberated through the kitchens. To most Americans, whose finest restaurants were still shrines to shrimp cocktail, steak, and baked potatoes, it was all eye-poppingly sophisticated.

Along with the sophisticated food came large dollops of that traditional garniture of haute cuisine, snobbery. One area of the new restaurants was always conspicuously reserved for the best clients (at Pavillon it was called "the Sanctuary"), and no restaurant was complete without a "Siberia." Julia Child thought restaurateurs like Soulé actually enjoyed treating customers snobbishly, and they almost certainly did.[41] Craig Claiborne's obituary of Soulé, whose own origins in provincial France were distinctly humble, remarked on "his characteristic disdain for those who would climb social ladders via his red carpet."[42] "I am reluctant to have acquired a reputation as a snobbish restaurant," Soulé once sighed with a straight face to the fawning food writer Lucius Beebe, "but I have to protect my valued patrons against fourflushers who would like occasionally to make a big flash or consummate a business deal, but who have no proper knowledge of how to eat and drink."[43] He thus helped nurture a generation of restaurant employees whose specialty was the snub: the back permanently turned on the portly midwestern businessmen in checked jackets waiting patiently at the bar; the barely disguised sneer with which one reached for a patently unfashionable tie for the irreverent who dared show up without the requisite "proper dress"; the impatient glances over the shoulder while explaining the intricacies of French menus to the young couple celebrating an anniversary.

Of course, the new generation of French restauranteurs were by no means the originators of restaurant snobbery. The expectation of special consideration from the *patrón* or his minions has always been an attraction whenever and wherever restaurants have existed. It was an important part of the success of nineteenth-century America's greatest restaurant, Delmonico's in New York City. To a Delmonico alumnus, Oscar Tschirky, the famed turn-of-the-century maître d' at the Waldorf, belongs the dubious distinction of having been the first to use a plush rope behind which waiting guests would have to pass inspection by a man who had contributed absolutely nothing to the culinary output of the room to which they hoped to gain admittance.[44] It was this patina of exclusivity that had made the Stork Club in New York City the most famous restaurant of the 1940s, a mantle it maintained through the 1950s. Its proprietor, a barely literate migrant from Oklahoma named Sherman Billingsley, had perfected his skill at knowing who to turn away at the door while running a succession of speakeasies during the Dry Decade. Even gaining admittance did not mean escaping humiliation—it was he who was reputed to have invented the term "Siberia."[45]

The three Kreindler brothers, who owned and ran New York City's "21," known in 1952 as "the most profitable and expensive restaurant in the world," had also perfected their lucrative form of pseudo-snobbery by running speakeasies. They built a reputation as "New York's haughtiest eatery" by refusing to seat anyone without a reservation. In true speakeasy style, they had two muscular men guarding the door, ostentatiously telling nonregulars without reservations that there were no spaces available—even at the bar. (This was all part of an elaborate charade, for anyone could get a reservation in the cavernous place by ducking around the corner to a telephone.) Once inside, patrons would be subjected to a system of scrutiny, greeting, grading, and segregation that would have made court life at Versailles seem egalitarian. The maître d'hôtel was an expert at "dressing the house": seating the prominent people where they would like to be seen and banishing the less so to a hierarchy of wilderness locations, all the while making it look crowded. The food, prepared at the rate of one thousand meals a day, could not have reached any culinary heights. Patrons paid astronomical prices for the usual selection of steaks, chops, shellfish, and sandwiches—speakeasy food.[46]

The new French restaurants functioned in a rather different sphere. They appealed less to those who wanted to rub shoulders with show business celebrities or to show off the immensity of their fortunes than to people whose status derived in part, at least, from the sophistication of their consumption habits. The new maîtres d'hôtel guarded the gates, not just of the tables that denoted status, but also of the mysteries of French food. The most renowned regulars at Le Pavillon were the most sophisticated consumers of them all, the Duke and Duchess of Windsor, who devoted their lives to being symbols of impeccable taste. (Even when it was known that he was in Paris or on the Riviera, the duke's table would be held for him until 1:30 p.m. for lunch, and only then would it be released.)[47] That is one reason the French restaurant revival caused hardly a ripple among mainstream middle-class Americans: They seemed more concerned with the quantity than the quality of their material possessions. Bigger houses, more powerful automobiles, larger-screened television sets, giant freezers—these were the stuff that middle-class dreams were made of in the 1950s. Moreover, the central obsession of the middle class, raising children, was hardly conducive to making fine distinctions at the dinner table. True, in the late 1950s there were reports that "gourmet food" was sweeping the country—General Foods even set up a Gourmet Foods Division—but it was really just a boomlet, benefiting mainly vendors of "specialty foods" such as pickled mushrooms, canned wild boar, and chocolate-covered grasshoppers.[48] Even in Manhattan, epicenter of the French revival, there had been little fallout from the elite. The aforementioned survey of what people there were eating at home turned up hardly a trace of the "gourmet" or French trends.[49]

However, in early 1961 the realization that French cuisine had reoccupied its place at the pinnacle of status burst into mass consciousness. In the midst of the monumental job of staffing his new administration, John F. Kennedy let it be known that one of his more important appointments would be that of a new French chef in the White House. Well-traveled and sophisticated, Kennedy had been one of the glitterati who had never needed to concern himself about being seated beside the swinging kitchen doors. Emulating his bon vivant father, who was reputed to have been an original investor in Le Pavillon, he had developed an early appreciation for fine French restaurant cooking.[50] He had endured the rigors of campaigning for the presidency in America's culinary wastelands thanks in part to the vichyssoise and chicken in champagne sauce prepared for his campaign plane by Fred Decré, the Soulé alumnus who had opened La Caravelle.[51] One of the first things Kennedy did after moving into the White House was to have the small kitchen in the family quarters transformed into a professional kitchen, with two stainless steel hotel-style ranges and a huge commercial-sized stainless steel refrigerator. When, after over two months of transatlantic negotiations and rumormongering, he finally selected another New York City French chef, René Verdon, for the White House post, Verdon's name almost instantly became a household word, and White House menus were reported breathlessly in the press.[52]

Soon the East Side of New York was virtually awash in elegant French restaurants.[53] French restaurants even finally took root in Los Angeles, although none could compare with New York's finest. Lesser cities began to sprout, if not French restaurants in the Pavillon or Caravelle mold, at least "continental" restaurants with names of a kind Calvin Trillin satirized as "La Casa de la Maison House, Continental Cuisine." There, maître d's of often distinctly un-French origin assiduously honed their "zisses" and "zats" on patrons intent on reproducing the glitter of the White House by ordering escargots, while fire departments remained on high alert in fear of the orgy of flambéing that threatened to engulf the nation. Even the hidebound Senate responded to the new winds blowing from the White House, hiring a French maître d'hôtel to help revamp its menu.[54]

However, as the association of French cooking and elite society reached this new apogee, forces that would broaden its base and bring it down a notch were already at work. First came the 1960s explosion of middle-class travel to Europe, which followed the introduction of jets on transatlantic runs in 1959. To many of these first-generation overseas travelers, Europe meant mainly England, where—Duncan Hines notwithstanding—the food normally struck Americans as simply horrid, and France, where quite the reverse was the case. Whereas the older, better-heeled tourists of the 1950s had gravitated to haute cuisine sanctuaries such as Maxim's and La Tour d'Argent, the new tourists, inspired by visions of Hemingway and Fitzgerald, discovered the rest of the Left Bank, the provinces, and the food of modest restaurants and bistros: chicken that tasted like chicken, simple seafood dishes of startling freshness, omelets, and *pommes frites*, which seemed of a completely different genus from "french fries." The idea that French "bourgeois" cooking—unadorned by the elaborate sauces, complex service, and inherent snobbery of *la grande cuisine*—was superior to anything America had to offer gained currency. "You can't get a bad meal in France" became part of the new traveler's credo.

Leveling influences were also at work at home. One of the more unlikely forces in this direction was Craig Claiborne, who became a food writer and restaurant reviewer for the *New York Times* in 1958. There were a few excellent writers on food in the 1940s and 1950s—most notably A. J. Liebling and M. F. K. Fisher—but they did not review restaurants. Duncan Hines's operation, for all its faults, stopped reviewing restaurants in the 1956, as Procter and Gamble, which bought it in 1948, redirected it toward food marketing. The American Automobile Association guides, to which travelers were now forced to turn, were equally concerned with cleanliness, more impressed by ambience, and less concerned with food quality than Hines. Otherwise, restaurant reviewing was mainly in the hands of a few self-proclaimed "gastronomes" such as Lucius Beebe, who spent much of his time pandering to pretentious establishments like Trader Vic's. They were often so ignorant of basic cookery techniques that after the meal they would invite chefs out front to explain what they had done.[55] Others were part-timers working for newspapers' women's pages, where it was acceptable to receive "freebies," ranging from refigerators and trips to Europe down to canned food and free meals, from the subjects of articles.

Claiborne, on the other hand, had been trained in *haute cuisine* at the *Ecole Hôtelière* in Lausanne and refused to go along with the corrupting old system, which allowed expensive restaurants to surround themselves with fawning writers. Moreover, as Betty Fussell has pointed out, his rating system, which measured food quality in relation to price, had an inherent democratizing bias: "It put one of the most pretentious spots in town, the Forum of the Twelve Caesars, on equal one-star footing with a Chinatown joint like the Joy Luck Coffee Shop."[56] Although Claiborne remained enamored with French haute cuisine, he also extolled French bourgeois and rustic food as well as, most importantly, many other foreign cuisines and the southern cooking of his childhood.[57] The *Times* food pages and the resulting cookbooks were soon replete with recipes for dishes from an immensely varied repertoire. Inevitably, French *grande cuisine* came to be perceived as merely one of a number of exquisite ways of eating.

A more important force in democratizing the image of French food was another unlikely source: Julia Child, a New England–born, California-raised Smith College grad who could hardly cook a thing until after the war, when she and her husband, a United States Information Agency graphic artist, were posted to Paris. Her television show, "The French Chef," produced in Boston and carried by fledgling educational television stations across the nation, was not the first to give instruction in French cuisine. (Dionne Lucas, for one, had a French cooking show in New York City in the 1950s.) But none had come close to evoking the same response. It catapulted her to virtual stardom, making her, along with James Beard (who had pioneered television cooking shows in 1946) perhaps the best-known personality in the American food business. To a certain extent, it was a case of serendipitous timing: Her first cookbook was published in 1961, and the TV show premiered in 1962, just as the travel revolution and White House glitter were spurring middle-class interest in French food. But her popularity was also based on the fact that she was demystifying French cooking. Dionne Lucas, with her proper British accent, made much of her Cordon Bleu links and techniques, giving the impression that French cuisine was complex cooking for sophisticates. Although Child's techniques often reflected her classical training, her folksy manner was the exact opposite of America's image of "the French chef." Instead of a meticulous man with a thin moustache and a foreign accent performing magic tricks in the kitchen, here was a six-foot-tall woman who huffed and puffed as she hefted large joints of meat, dropped things, and encountered enough near-disasters to allow plenty of instruction in how to compensate for them. She repeatedly argued that French cooking was much simpler than everyone thought, and the first volume of the cookbook she coauthored—which had sold over one and a quarter million copies by 1974—was essentially a demonstration of this assertion, based as it was on the repetition of certain basic techniques with different ingredients.[58] "What I was trying to do," she said, "was to break down the snob appeal. There was the great mystery about it, and you didn't tell people what was going on. What I tried to do was to demystify it."[59] Claiborne tried to do the same in his articles and cookbooks, making a special point of disabusing the public of the old notion that there were arcane things called "chef's secrets."[60]

The idea that French cooking could be accessible to all was soon echoed by other cookbook authors and food writers. By mid-decade middlebrow magazines such as *Life*, *Look*, and *McCall's* were featuring French food and recipes (the latter even ran an article on how to make *millefeuille* pastry!), and French dishes were becoming standard features at urban middle-class dinner parties.[61] By then the culinary gridlock of the 1950s had been broken, and American eating habits were again in flux, buffeted on all sides by conflicting ideas and demands.

CHAPTER 10

The Politics of Hunger

To the cynical, it seemed to be merely clever politics: Senators Robert Kennedy of New York and Joseph Clark of Pennsylvania—one the presumptive heir of his fallen brother's presidential standard, the other the chairman of the Senate's antipoverty subcommittee—would follow up the committee's one-day hearings in Jackson, Mississippi, in April 1967 with a dramatic visit to the state's cotton-producing northwestern delta. There, thousands of black farm workers and sharecroppers who had been thrown off the land by the relentless push of mechanization, chemical farming, and government subsidies for crop reduction were reported to be living in conditions of near-starvation.[1] It would provide a newsworthy opportunity to demonstrate sympathy for the downtrodden, particularly for Kennedy, who was transforming himself into a tribune of the "underclass." The prospect of pictures of the handsome young white man, with his shock of sandy hair and flashing smile, amidst the blackness of the delta would surely attract the media; and so it did.

Of course, as is normal in these situations, the area had been selected in advance for maximum squalor, and the politicians had been briefed on what to expect. But no sooner had they plunged into the first of the rundown encampments than it became clear that one thing had been left to chance, the emotional reaction of the sensitive senator from New York. Kennedy seemed at a loss for a fitting reaction when actually confronted with swollen-bellied children with running sores and listless adults who told of living on only one meal a day. In a dark, windowless shack that smelled of mildew and urine, he reached down to the dirt floor and picked up a small boy from whose swollen belly dangled an uncut umbilical cord. Placing him on the filthy bed, Kennedy began tickling the child's stomach, attempting to evoke a giggle; yet the boy remained in a trancelike state. Tears running down his face, Kennedy turned to reporter Nick Kotz and said, "My God, I didn't know this kind of thing existed. How can a country like this allow this?" Vowing to do something about it, he returned to Washington, where he told the agriculture secretary, Orville Freeman, and the full committee of the "extreme hunger" he had seen. Several times during his committee testimony he became so disconcerted that he broke into nervous laughter.[2] Nevertheless, the determination of the glamorous heir to the Kennedy political mantle to carry through on the vow made in the shanties helped, albeit briefly, to make hunger in America something it had not even been in the 1930s: a politically sexy issue.

Hunger had actually returned to the limelight some years earlier, thanks to a synergistic relationship between swelling farm surpluses and renewed awareness of desperate straits abroad. Concern over foreign hunger had slumbered after 1948, when the European crisis had diminished. The term "population explosion" was as yet unknown; famine seemed to be an occasional result of natural disaster and crop failure rather than the future condition of much of the world. Faith that the technological wizardry of the American Century would soon conquer want reigned supreme. "Startling developments" in food production would soon make famines a thing of the past, a food industry leader confidently predicted in 1953. Mechanization in agriculture and food processing, along with such advances as synthetic foods and hydroponic gardening, would do the trick.[3]

Startling developments did indeed unfold over the next ten years, but they served mainly to increase America's farm surpluses. When swelling populations in the Third World put growing pressure on resources, calls arose to use them to alleviate hunger overseas, but, as in the Depression decade, the powerful farm lobby stood guard against anything that might undermine its markets and prices. In 1954, when the government began a modest Food for Freedom program to sell surplus foods to poor nations, it sold them for unconvertible local currencies in a way that ensured that sales in dollars were not threatened. As the program's title indicates, its main justification was political: to prevent Communists from taking advantage of these countries' temporary difficulties. Few took the prospect of worldwide starvation seriously, particularly as America's farm surpluses continued to increase. Distributing the world's food properly seemed the main problem, not producing it. In June 1963 Secretary of Agriculture Orville Freeman, who had inherited a grain stockpile that had risen from 12.4 million tons in 1948 to 115 million tons when he took office in 1961, assured the World Food Congress that science and technology had "opened the door to a potential abundance for all." Food aid projects were ushering in an era when wars caused by the pressure of population against food supplies would be a thing of the past. President Kennedy told the same Congress that "we have the ability, we have the means, and we have the capacity to eliminate hunger from the face of the earth in our lifetime. We need only the will."[4] In 1966 optimism on this score was fueled by a new Food for Peace program. Unlike the surplus-driven Food for Freedom program, which often left recipients awash in foods they did not like, want, or even know how to cook, it gave poor governments American dollars to buy American farm products at world prices, helping to support farm prices across the board.[5]

Optimists also took heart from advances in plant hybridization, fertilization, and pest control that seemed to be bringing a "Green Revolution" of increased harvests to the Third World. American scientists added to these high hopes by developing economical and nutritious new foods for the poor. Aid agencies in Latin American put great store in Incaparina, a relatively cheap Metrecal-like powder made of cottonseed, corn, and sorghum, which, when mixed with water, sugar, and flavorings, was comparable to milk in nutritional quality. In Indonesia UNICEF distributed Saridele, a soy bean extract that added protein to infants' diets. Pilot plants were set up in Chile to produce protein-rich fish meal flour for the poor out of "trash fish."[6] When an enthusiastic United States delegate at a United Nations meeting gave fellow delegates chocolate chip cookies made from fish flour, *Chemical Week* proudly proclaimed, "Fish Protein Joins the War on Hunger."[7] Many eyebrows were raised at a Michigan dairy farmers' meeting in April 1968 when delegates were told that the cold glasses of milk they had just drunk were made from palm oil, corn syrup, and seaweed extract. That month, the Atlantica Foundation announced that its Ecological Research Station had concluded that termites could provide an excellent source of nutrition in the protein-short underdeveloped world, providing an exceptionally high protein-per-acre yield.[8]

Large multinational corporations saw a happy combination of good works and profit in the new gospel according to the food technologists. Scientists labored in oil company labs to produce single-cell proteins that fed on the paraffins in crude oil; large millers developed ham, beef, bacon, and chicken substitutes from oil-seed; chemical companies created new fertilizers; and drug companies worked on new herbicides, fungicides, and food preservatives, as well as new feed additives, antibiotics, and drugs to control the breeding of animals.[9] International Telephone and Telegraph developed ASTROFOOD, a highly sweetened, nutrient-filled cupcake that eventually found its way into domestic school lunch programs. "U.S. Industry Set to Feed the Poor," said the *New York Times* headline in 1967 when General Mills and Phillips Petroleum formed a jointly owned company to develop and market new products for the hungry world.[10]

Not everyone was convinced that the problem could be solved in the nation's laboratories. At the very 1963 congress where Freeman had waxed so optimistic about technology, the British historian Arnold Toynbee had sounded a dissonant note—one that would become much louder as the decade drew on. "Science cannot increase our food supply ad infinitum," he warned. What was needed was for "the planet's hundreds of millions of wives and husbands to voluntarily decide to regulate the number of human births" and "political cooperation on a world-wide scale" to help equalize food supplies.[11] Pessimism was also fostered by the growing realization that most of the Green Revolution's advances came in grain production. Protein supplies seemed to lag, fueling fears of massive, worldwide protein deficiencies. Moreover, since the new techniques made farming more capital-intensive, they seemed to be encouraging the expulsion of poor peasants from the land, adding to the millions who could not afford an adequate diet.

But while concerns over the population explosion and the pressure it was putting on food supplies rose among the well-informed, international hunger soon lost much of its appeal to politicians. In the first place, as Toynbee had noted, it was obviously connected with the birth control issue, a political bombshell. Second, there were not, as yet, any spectacular famines: no equivalents to the Ethiopian famines that galvanized the West—albeit briefly—in the 1980s. Indeed, in 1967 and 1968, after two years of generally good harvests, the international food situation seemed so good that U Thant, the secretary general of the United Nations, warned against overconfidence that the race between food and population could be won.[12] Yet even his own organization failed to heed him. By 1970 the Green Revolution seemed to be so successful that the UN's Food and Agriculture Organization confidently estimated that the earth's agricultural potential was great enough to support 157 billion people.[13] That year, N. E. Borlaug, the American agronomist credited with being the "Father of the Green Revolution," was awarded the Nobel Peace Prize.

The growing complacency about the food situation abroad was counterbalanced, however, by increased concern at home. Since 1963, when President Lyndon Johnson had initiated his War on Poverty, there had been a steady stream of reports about deprivation in America, accentuated by an increasing tempo of inner-city riots. But the emphasis had been on jobs and housing, not on hunger, and the focus had been on urban ghettos, not the countryside. Indeed, in January 1966, almost two years before Kennedy toured the Delta, Representative Samuel Resnick, another congressman from New York, had also visited Mississippi and, shocked by what he had seen, had returned to Washington to demand that something be done about "the desperation point of starving Negroes." But Resnick's name did not carry quite the same weight as Kennedy's, and nothing happened.[14] Kennedy could not be so easily ignored, and Freeman ordered an immediate investigation into the Mississippi situation.

However, President Johnson, a wily politician if there ever was one, was skeptical of the growing demands for emergency food shipments to Mississippi and an easing of food stamp requirements. He feared, suspected, and indeed despised Robert Kennedy and inevitably saw the senator's presidential aspirations lurking behind them.[15] Secretary Freeman, meanwhile, felt hamstrung by congressional conservatives who could eviscerate his department's budget. They would brook no "coddling" of the rural black labor supply and argued that any increase in money to buy food would not go for food but for nonessentials such as television sets.[16] But the pressure to do something about rural hunger began to rise. Hard on the heels of the Kennedy-Clark visit, a team of doctors and nurses sent to Mississippi by the Field Foundation returned with graphic descriptions of hunger and near-starvation among poor blacks there.[17] "Wherever we went and wherever we looked," they reported, "we saw children for whom hunger is a daily fact of life and sickness, in many forms, an inevitability. The children we saw were more than malnourished. They were hungry, weak, apathetic. Their lives are being shortened. . . . They are suffering from hunger and disease, and directly or indirectly, they are dying from them—which is exactly what 'starvation' means."[18] Kennedy arranged for them to present these findings to Clark's committee, a response they contrasted favorably to the "runaround" they received from Freeman.[19]

Other critics sensed the administration's vulnerability on the issue and began lashing it over hunger. In April 1968 the Citizens' Crusade Against Poverty, headed by liberal labor leader Walter Reuther, reported that its investigation revealed that chronic hunger and severe malnutrition existed in all parts of the United States, particularly in the South and Southwest. Using a special formula, it classified 256 areas as "hunger counties."[20] Black activists took up the cause. The comedian Dick Gregory vowed to make an around-the-world trip to "try and beg money" from other nations to buy food to feed hungry black people in the South. The boxer Cassius Clay (later Muhammed Ali) promised to devote the proceeds of a special bout to feed poor blacks.[21] The Southern Christian Leadership Conference (SCLC), which, under Martin Luther King's leadership, had led much of the drive for civil rights in the South, altered its course to focus more on alleviating black poverty. During its Poor Peoples' March on Washington in May 1968, the recently assassinated King's successor, the Reverend Ralph Abernathy, led a foray by demonstrators from their "Resurrection City" on the Mall to the Department of Agriculture on Capitol Hill. There they confronted a discomfited Secretary Freeman with demands that surplus foods be given to the poor and that food stamps be made cheaper and more accessible.[22]

Initially, the crusaders against hunger concentrated on the situation in the rural South, but since the 1930s there had been a dramatic shift in where most of the nation's poor lived, from the rural South to the urban North.[23] Investigators, activists, and politicians soon discovered malnutrition and hunger there and lambasted the government for doling out millions in aid to large farmers while the urban poor went hungry.[24]

As was now so often the case, television played the decisive role in arousing the public. On May 21, 1968, the night before the Abernathy-led march on the USDA, a CBS News documentary called "Hunger in America" almost single-handedly turned hunger into a national political issue. The opening segment was devastating: a wasted baby onscreen with narrator Charles Kuralt's offscreen voice saying, "This baby is dying of starvation. He was an American. Now he is dead. . . . America is the richest country in the world, the richest country in history."[25] Juxtaposing contrasting images with striking effect, the program then showed well-fed tourists savoring the delights of the San Antonio World's Fair and some of the more than one hundred thousand Mexican-Americans in the city who, Kuralt said, were "hungry all the time."[26] The sumptuous homes of celebrities in Loudon County, Virginia—beautiful horse country only twenty-five miles from Washington—were contrasted with shacks in the same county where thousands of white tenant farmers and their undernourished children were said to scrape by with no help from federal food programs. Vivid descriptions of hunger's mental and physical toll accompanied pictures of underweight and sick Navajo children in Arizona and black children in rural Alabama, evoking comparisons to the worst of the "underdeveloped" world. Yet, the program concluded, not only was the existing USDA surplus food program inadequate, it had actually turned back over $400 million in unspent food aid funds to the treasury and intended to give another $227 million back. Recently the USDA had even—shades of the 1930s—overseen the slaughter and burial of fourteen thousand hogs.[27]

Not everyone was impressed. Congressman James Whitten of Mississippi, a powerful opponent of food aid, had J. Edgar Hoover assign him several FBI agents to investigate the program's supposed distortions. (They eventually uncovered evidence that two infants shown dying of malnutrition-related ailments had likely died of other causes and that some other visuals did not correspond to what the narrator was describing.)[28] The chairman of the House Agriculture Committee had his staff survey public health officers in the 256 "hunger counties" Kuralt had cited and announced that they had not found a single case of starvation.[29] Secretary Freeman, a good liberal who resented being portrayed as hard-hearted, was particularly perturbed. After all, his department had revived food stamps with a pilot program in 1961, and he had spent considerable political capital shepherding a permanent program through Congress in 1964. It was more flexible than that of 1939–1943, allowing the poor to spend bonus stamps on any foods they wished, and was undoubtedly better than the existing surplus foods program, set up in the 1950s, which made available only those foods that happened to be in surplus.[30] However, as in the 1939–1943 program, enough stamps for a month's supply of food still had to be bought all at once, and many of the poor could not muster that kind of money. Consequently, in almost every place that switched from surplus foods to food stamps the participation rate dropped dramatically. (Indeed, it was exactly this switch that had caused the destitution Kennedy and the Field Foundation doctors had witnessed in the Mississippi Delta.)[31] Critics thus argued that the poorest of the poor were being bypassed by the program and demanded that stamps be provided to them either free or at only a nominal cost.[32] But President Johnson, facing mounting deficits brought on by the Vietnam War and his Great Society programs, was adamant about not spending billions more on this, particularly since the demands had such a clear Kennedy stamp on them. On at least twelve separate occasions, Nick Kotz reported, Johnson turned down recommendations from his cabinet or aides for increased food aid. When he discovered that Freeman was quietly trying to divert some of his department's money in this direction, he flew into a rage.[33]

The day after CBS's "Hunger in America" aired, Freeman acknowledged to a House committee that "time and again when the poor cried for a full loaf of bread they were forced to settle for half," but this was because "public support to fund anti-hunger campaigns was weak or non-existent." For this he blamed "the middle class that lacks motivation and is complacent."[34] Alas for him, even if this were true then, quite the reverse was soon the case. Liberal "advocacy groups" such as the National Council on Hunger and Malnutrition and the Citizen's Advocate Center emerged to flay him and his department. After Kennedy was assassinated in early June, Senator George McGovern, a Democrat from North Dakota, took up his campaign against hunger—and administration policy—in the Senate. Republicans such as Jacob Javits of New York and Charles Percy of Illinois supported his successful demand that the Senate create a special Select Committee on Nutrition and Human Needs to investigate the problem. Richard Nixon, the Republican nominee for the presidency, who had also been alarmed by "Hunger in America," demanded more food aid. He appointed an antihunger activist, Robert Choate, as an advisor and called for "total reform" of the food aid system, including easier access to food stamps. His rival, Hubert Humphrey, was forced into making similar promises.[35] Soon after taking office, Nixon committed himself to the "elimination of hunger in America for all time."[36]

The political trajectory of the hunger issue could not have been more dramatic. In less than two years, the loose coalition of antipoverty groups, activists, and politicians called the "hunger lobby" was on the verge of achieving what amounted to a political miracle: wresting control of food policy from the farm lobby. One of their central ideas—that the problems of hunger and malnutrition should be separated from that of agricultural surpluses—was becoming government policy, and the government was committed to massive intervention, albeit of an indeterminate kind, to meet their demands.[37]

Yet how widespread were hunger and malnutrition? Of the two, only malnutrition could possibly be measured, but all attempts to do so were flawed. Despite the repeated assertions that malnutrition was rampant, experts were little closer to commonly accepted definitions of what it was than they had been in 1940. The most popular approach of the earlier era—to calculate how much income was required to purchase diets supplying the "recommended" amounts of certain vitamins and minerals—remained the basis of most estimates. Indeed, in 1964 statisticians at the Social Security Administration took the USDA's calculation of how much a family would have to spend to obtain a diet that met the NRC's Recommended Daily Allowances of nutrients, multiplied it by three, and declared the result to be the "poverty line."[38] Any family living on an income below this level was assumed to be undernourished. Yet the old problems with this approach remained unresolved. It treated the RDAs as minimum "requirements," rather than what they were intended to be: "minimum no risk standards," which were set at far above average requirements in order to ensure that no one, not even those who required much more of the specific nutrients than average, would be deficient.[39] (The RDAs "are set in excess of practically everyone in the population," a Harvard nutritionist acknowledged. "Nearly everyone in the population can consume less than the standard yet be adequately nourished."[40]) Moreover, experts were little closer to agreeing on what the standards should be than they had been in 1940. Indeed, in some cases they were farther apart.[41]

However, as in the prewar years, the public was fed the statistics unadorned by qualifications. Many experts based their estimates on a Field Foundation report, "Hunger, U.S.A.," which used the USDA family budget method to conclude that ten million Americans were undernourished. Margaret Mead, testifying before the Senate nutrition committee, added, without additional proof, that many of these ten million were "on the verge of starvation."[42] Dr. Jean Mayer, professor of nutrition at Harvard, put the number of malnourished poor people at from ten to twenty million; Dr. Michael Latham, a Cornell nutritionist, said "probably far more than 10 million." In March 1969 the USDA estimated that five to ten million Americans were "suffering from severe hunger and malnutrition," while other government officials called hunger "a daily fact of life" for at least six to nine million.[43] The black activist Jesse Jackson used the "poverty line" method to number the poor at forty million, all of whom, he said, fit the government definition of malnourished, including ten million children "with bloated bellies and brain damage."[44]

Other kinds of statistics were equally questionable, yet lent credibility to the outcries. In January 1969 the first results of a national survey of the extent of malnutrition, which Congress had ordered in late 1967, came out. It was based on physical examinations of twelve thousand people, mainly poor blacks and Latinos, in Texas and Louisiana, a majority of whom were children. The study's director, Dr. Arnold Schaefer, estimated that 16 to 17 percent of those examined represented "real risks" who needed medical attention. Malnutrition was just as bad in the United States as in recently surveyed countries in Central America, he said.[45] The results of examinations covering fifty-eight thousand more people, released in April, filled in more of what he called "an ugly picture of hidden hunger in America." Ninety-two percent of Head Start children examined were said to be more deficient in vitamin A than children already blinded by the deficiency. Anemia among them was so severe, he told the National Association of Science Writers, "that if it were your child or mine his condition would require immediate therapy." Moreover, 10 to 15 percent of the children examined had "retarded growth levels," representing "a high risk in retardation of mental and physical performance."[46]

But, although it sometimes used biochemical tests to measure levels of iron, hemoglobin, and some vitamins in the bloodstream to supplement the traditional physical examinations and age/height/weight scales, the National Nutritional Survey was still grounded in the shifting sands of disagreement over definitions of malnutrition. Even its own examiners used different methods and criteria. Midway through it, the Public Health Service, under whose auspices it was conducted, changed the levels of nutrients to be considered "low." Reduced to a Ten-State Survey, it was abandoned in 1971, just as doctors who had measured the nutrients in the bloodstreams of two hundred New York City children were admitting that they had no idea what any deficiencies meant. "We don't know how to define malnutrition," said the head of the team. "Nobody can say this is normal and that is not normal and this is good and that is bad. At the moment it's a very imprecise science."[47] Three years later, a government study said there was a critical need for methods of determining nutritional status. "The present array of biochemical and clinical measures is unsatisfactory for a world which must base its food assistance programs on facts rather than demands."[48] Some months earlier, after interviewing scientists and reviewing much of the literature, Jane Brody concluded: "Only a relative handful of scientists know much of anything about human nutrition, and most of them readily concede that their knowledge is far from complete."[49]

Critics charged that the Nixon administration shelved the Ten-State Survey and banished Dr. Schaeffer to the Pan American Health Organization because he was thought to be in league with Senator McGovern and the hunger lobby.[50] But even if this were true, it was not a great loss to their cause. When all was said and done, it was not statistics about malnutrition that moved the public, it was descriptions of hungry Americans in the land of plenty. Indeed, as one reporter later noted, "the idea that some Americans go hungry stirs more political passion than the evidence that many go without adequate income, health care, or housing."[51] The media thus focused on hunger, rather than malnutrition, as the new breed of investigative reporters scoured the country in search of scandalous living conditions. "Of course the kids are hungry!" a Mississippi mother told a *Washington Post* reporter, as she lifted her skinny five-year-old daughter's blouse to expose the skin lesions on her chest.[52] In February 1969 *New York Times* star reporter Homer Bigart wrote a five-part series on hunger that took readers on a national tour of miserable slums, shanties, and grim diets. They read of silent black children in shacks in Beaufort County, South Carolina, suffering from kwashiorkor, scurvy, and anemia. The article on migrant workers in Florida told of families subsisting on little more than grits, beans, rice, and fatback, while in the Mississippi Delta he still found emaciated children in families that lived on nothing but pinto beans and bread. In the Southwest he found Mexicans and Indians who were "stoical victims" of hunger. Many of the Navajo children were said to be permanently stunted and brain-damaged by malnutrition. In Appalachian shanties, where the Kennedys had first become sensitized to the problem in 1960, Bigart visited silicosis-stricken miners and heard again of how federal food aid rarely lasted beyond the third week of the month, forcing people onto a diet of bread and gravy.[53] Even *Esquire* and *Good Housekeeping* ran long articles on hunger, the latter featuring "America's Hungry Families": "the hidden, the invisible, the forgotten—in a land of plenty."[54]

As the Senate nutrition committee toured the country during the first half of 1969, holding hearings in South Carolina, Florida, California, New York City, and Illinois and touring the slums of Washington, its visits were often preceded by media exposés of local hunger. Just before it met in impoverished East St. Louis, Illinois, the *St. Louis Globe-Democrat* ran a series beginning with a word-portrait of a seventy-six-year-old wisp of a man living in a junk-filled garage "surviving on a diet of pessimism and two anemic meals a day."[55] In April the *Chicago Sun-Times* ran a week-long series on hunger in that city, describing a listless infant who never uttered a sound while its brothers and sisters tried to still their hunger by eating lead paint chips off the rotting tenement walls. It told of a pregnant mother who ate laundry starch and hungry children in ghetto schools whose heads flopped down on their desks halfway through the morning. A social worker said the elderly poor suffered "a slow death from malnutrition that goes on daily."[56]

Hunger lobbyists were careful to paint hunger as a multiracial problem. Robert Choate met Bigart in Appalachia and took pains to point out that only three million of the estimated twelve million rural poor were black.[57] Yet the matter inevitably became enmeshed in the racial politics of the day, particularly as black activists sensed that they had an issue that might help them mobilize white support for their broader causes. Testifying before the McGovern committee in May 1969, the Reverend Ralph Abernathy now called the previous year's March on Washington a demand for an end to hunger in America, implying, quite incorrectly, that it had been the SCLC's central concern all along.[58] At that very moment, his rising young rival, the Reverend Jesse Jackson, who headed the SCLC's recently organized Operation Breadbasket, was leading close to three thousand followers chanting "I Am Somebody" on a "hunger march" on the Illinois state capitol in Springfield in a successful effort to pressure state legislators into rescinding proposed welfare cuts.[59] Later, testifying before the McGovern committee in East St. Louis, Jackson took pains to depict hunger as a problem affecting whites as well as blacks. "Hunger knows no color line," he said. "Even though percentagewise there are more black children going hungry than white children, in terms of absolute numbers more white children than black go to bed hungry every night in Illinois and throughout America." He could understand, he said, how racism could "allow white men to rationalize the starving of black men . . . but what is the rationale for white men starving their own people?"[60] For the gun-toting Black Panthers of Oakland, California, and Chicago, whose black uniforms and dark sunglasses were intended to instill pride in blacks and fear in whites, this was a nonquestion. They began dishing out free breakfasts to ghetto children, greeting them with the salutation "power to the people." "Everybody knows hunger exists," said one of their leaders. "We're the only ones doing something about it."[61]

Jackson and the other activists often said that opposition to food aid stemmed largely from an unwillingness to appropriate funds to help blacks. Some even charged that it was part of a deliberate campaign to drive blacks from the South.[62] But others saw less mean-spirited motives behind the opposition. Its main source, explained Representative Thomas Foley of Washington, the ranking supporter of aid on the House Agriculture Committee, was "the residue of the Puritan ethic attitude. . . . What it essentially comes down to is viewing food as a spur to work." "People do not work for nothing," the crusty committee chairman, Robert Poage of Texas, told a hunger lobbyist. "Bob Poage . . . is not going to help some deadbeat who is sitting down at the pool hall waiting for his wife and kids to go out and see what the neighbors have brought in."[63] But this principled stand in favor of self-sufficiency was undermined by Poage's and his allies' continuing support for subsidies to large well-off farmers in many of the states—particularly their own—where hunger was said to be most prevalent. In February 1969 the full Senate delivered them an extraordinary rebuff by rejecting a severe cut in the McGovern committee budget proposed by the Southern-conservative-dominated Rules Committee. The last-minute conversion of the "moderate" Senator Ernest Hollings of South Carolina helped. He toured poverty-stricken rural areas of his state with reporters in tow and then confessed that as governor he had "supported the public policy of covering up the problem of hunger" in the interest of attracting new industry to the state. "I know the need for jobs," he said, "but what I am talking about is downright hunger. The people I saw couldn't possibly work."[64]

Even had it wished to, the new Nixon administration could not turn its back on the clamor for more food aid, and it clearly did not wish to. In his first address to USDA employees, the new secretary of agriculture, Clifford Hardin, spoke not of farm policies but of how CBS's "Hunger in America" had shown "that in this rich land there is hunger."[65] The question now was not whether more food aid was needed; it was the form it would take. Most of the hunger lobbyists wanted an expanded food stamp program as the core of any solution. Estimated to cost upwards of four billion dollars per year, it would be supplemented by improvements in school lunch and school milk programs to redirect them toward the poor. They also called for special programs for pregnant women, infants, and preschool children. In May 1969 Nixon sent a special message on hunger to Congress proposing that the food stamp program be expanded by making the stamps cheaper and forcing all states to participate in it. His intention, he proclaimed, was "to put an end to hunger in America for all time." But while welcoming the new attitude in the White House, some in the hunger lobby criticized the extent of the liberalization (it would cost only about one billion more) and the timetable (it would not come into full effect until 1971).[66]

Later that year Nick Kotz infuriated Nixon by revealing that in a meeting preceding the message Nixon had said, "Use all the rhetoric you need, as long as it doesn't cost money." But it is likely that Nixon was merely making one of his feeble jokes.[67] As Kotz well knew, the president was really of two minds on the food stamp issue, for he was much taken with the suggestion of his social policy advisor, Daniel Patrick Moynihan, that the entire social welfare system be revamped and replaced by a Family Assistance Program based on a guaranteed minimum income. If this was put in place, programs such as Aid for Dependent Children, which discouraged stable family life, and—so it seemed—food stamps would be replaced by cash payments to families. The poor would then be free to decide what proportion of their income should be spent on food, shelter, and other needs. The apparent threat to the food stamp program prompted outraged protests from the hunger lobby. For many poor people, said John Kramer, executive secretary of the National Council on Hunger and Malnutrition, the Family Assistance Program would be a "Family Deprivation Program."[68]

But by then Nixon had astutely recruited the chairman of Kramer's organization, the Harvard nutrition professor Jean Mayer, as an advisor on hunger and nutrition. Mayer, who had done research on the inadequate diets of poor black migrant laborers and urban ghetto dwellers, came to the post with impressive credentials as a hunger lobbyist.[69] As one of the first witnesses called by the newly created McGovern committee in December 1968, he had lambasted the USDA, calling it indifferent to the poor and interested only in aid programs that would rid it of surpluses. He had then joined Abernathy and Reuther in forming the National Council on Hunger and Malnutrition to act as a watchdog on government antihunger programs, with himself as chairman.[70] The French-born Mayer sported fourteen decorations for wartime service in the Free French Army and French Resistance. Alternately charming and blunt, he could stand up to and face down the most persistent of critics, a faculty he would soon need.

Mayer said that Moynihan's proposals would not affect the food stamp program, but the hunger lobby and the majority of the McGovern committee remained unconvinced.[71] Nixon's formal message to Congress proposing welfare reform failed to clear the air, leaving a pall of confusion over the administration's intentions with regard to food stamps.[72] To help clarify matters, Nixon had asked Mayer to organize a conference—modeled loosely on the 1940 nutrition conference—to forge a consensus on how to deal with hunger and malnutrition and also give direction to other national food and nutrition programs. Moynihan had particularly high hopes for it. It might be "a landmark in American social history," he wrote the president: "a national gathering simultaneously concerned with the remnant of pre-industrial problems, such as hunger and malnutrition, the onset of post-industrial problems, such as overeating, and the industrial era problems such as the toxic effects of prepared foods."[73] But despite the organizers' commitment to ruminating over the latter two, it was the first question, hunger, that held center stage.

Mayer's war experience came in handy, for he soon found himself attacked from all sides—besieged, he told the press, by "radical" and "reactionary" groups.[74] He worked diligently to have several hundred putative representatives of the poor among the three thousand delegates, but there were not enough to satisfy the hunger lobby. Food producers also demanded more representation. Hunger lobbyists charged that the conference was being funded with money earmarked for food aid.[75] Meyer was accused of favoring Republicans over Democrats and big business over consumers. His stubborn streak bred complaints that he ran roughshod over critics. John Kramer, whom he himself had chosen as executive director of the

Council on Hunger and Malnutrition, called him "a classic example of a man swallowed by an institution. . . . I call him Dr. Mayer and Mr. Hyde. He used to be a gadfly in the previous Administration. Now that he's on the inside he tries to gag dissent. . . . He's an arrogant authoritarian."[76] Just as the conference was to begin, while hunger lobbyists denounced it for addressing the concerns of industry at the expense of the poor, its most prominent corporate participant, Donald Kendall, a strong Nixon supporter who was the head of Pepsi-Cola, which owned Frito-Lay, quit in protest against disparaging remarks Mayer had made about snack foods: Mayer had told *Life* magazine that fried worms were more nutritious than most of them.[77]

Mayer responded to the hunger lobbyists' charges by affirming that hunger and malnutrition were indeed the conference's top priorities, but that they could not be separated from broader issues concerning the food and nutrition of the country as a whole. They were not impressed. "The various panels have given us 600 pages of recommendations," said Kramer. "They can be junked. Only one thing matters—immediate relief for the 25 to 30 million poor people who are malnourished."[78] The conference soon degenerated into a free-for-all among the organized interest groups, each voicing its own demands while ignoring the others. "The whole tone of this carnival is a deaf one," said a disillusioned New Yorker.[79] A shell-shocked Kansas dietitian seemed bewildered to discover that in the world of nutrition there were so many "vested interest groups" who "wanted things done only their way."[80] Nevertheless, the hunger lobbyists managed to make their mark: Conference leaders drafting its final recommendations replaced the guaranteed income with emergency action on hunger as the nation's top priority. The next day, when Nixon pledged to the conference that he would "put an end to hunger and malnutrition due to poverty in America," the guaranteed income remained at the top of his program for doing so, but he also went along with demands for an expanded food stamp program.[81]

The guaranteed income plan was soon stalled and eventually torpedoed on Capitol Hill, in part because the hunger lobby failed to support it. But, as Nixon had vowed, food aid expanded rapidly. By 1972 the number of people receiving food stamps had quadrupled, and the school lunch program had been broadened and redirected to aim much more at the poor. Indeed, as Nixon, hounded from office, boarded his presidential helicopter for the last time on the White House lawn on August 9, 1973, even his severest critics were conceding that he had overseen an expansion in food stamps and school lunches that seemed to be significantly improving the diets of the poor. In late 1972 Kramer, noting that the number of people receiving food stamps had risen from 1.8 million in 1967 to 11.8 million, acknowledged that his administration, "far more than was true of its predecessor, had a willingness to move forward."[82] Moreover, during Nixon's presidency one of the worst of the New Deal legacies—that tying food aid to farm surpluses—had been shattered, as the Department of Health, Education, and Welfare (HEW) and other government agencies broke the USDA's hammerlock on food aid programs.[83] By 1977 the positive results of the food assistance programs seemed apparent. A nationwide food consumption survey among the poor indicated that those receiving food stamps and/or the special supplements for women, infants, and children consumed significantly more nutrients than similar nonparticipating households. Likewise, children who took part in the school breakfast and/or lunch programs had higher nutrient intakes than those who did not.[84]

Nevertheless, throughout much of the decade hunger lobbyists demanding even more food aid continued to claim that hunger and malnutrition stalked the land.[85] As the number of food stamp recipients rose, so did estimates of those in need. In 1972 the Citizen's Committee claimed that half of the nation's twenty-six million poor—eleven million of whom did not receive food stamps—were still hungry.[86] The startling rise in food prices in 1973–1974 led to warnings that inflation was undermining the value of the stamps and creating many more people in need. A 1974 study by hunger lobbyists claimed that, although seventeen million were now collecting food stamps, a further seventeen to twenty-two million who were eligible were not.[87] Stories circulated of poor people eating canned cat food instead of expensive tuna; it was said that one-third of the dog food sold in slum stores was consumed by humans. The *Nation* characterized the government attitude as "Let 'Em Eat Alpo."[88]

But despite the alarming stories and numbers, hunger passed from public consciousness almost as quickly as it had risen. Within a year of Nixon's departure, his successor, Gerald Ford, sensing the falloff in public concern, began a drive to cut back on what he called the "scandal-ridden" food stamp program. Secretary of Agriculture Earl Butz said that the American people were "fed up" with it. Others called it "a haven for welfare chiselers."[89] Now, instead of heartrending stories of hunger amidst plenty, the media told of the millions of dollars wasted on food aid. Food stamps were "a program that has run amok," said *Reader's Digest*.[90] *Time* found a well-off twenty-three-year-old student at Stanford University living on food stamps, prompting charges that it was "a federal scholarship for affluent students."[91] The hunger lobbyists, after years of pressing for more generous programs, were now on the defensive. Only the wiliest efforts of the McGovern committee and other congressional friends—not a concerned public—headed off the savage cuts.[92]

A number of factors had helped turn hunger in America into one of those issues that, as the historian David Potter once noted, galvanize Americans in one decade and leave them cold in the next.[93] The inflation in food prices in the early 1970s redirected middle-class attention away from the cost of the shopping baskets of the poor and toward their own. Some of the remaining concern for the underprivileged was diverted to new food crises abroad. Drought and famine ravaged sub-Saharan Africa; fears of world overpopulation outstripping its food production capacity reached a fever peak. Climatologists warned that a pronounced global cooling had begun—a new Ice Age, which would severely curtail world food production.[94] The recession of late 1974 and 1975 led to a sharp increase in the number of people eligible for food stamps and a surge of mean-spiritedness among those who were not. Many of the newly unemployed were white-collar workers, and supermarket checkout lines often became charged with tension as middle-class people saw other middle-class-looking people using food stamps, sometimes for foods they felt they themselves could not afford. Congress was deluged with "hamburgers and steak" letters: complaints that the writer could only afford hamburger while those with food stamps were buying steaks.[95]

Concern over domestic hunger had also been dissipated by its links with the race question. Despite the efforts of the hunger lobbyists, it had become closely associated with other efforts to help poor urban blacks. As middle-class whites became disillusioned with or indifferent to black demands, it became a symbol of government handouts to undeserving blacks. That poor black women weighed considerably more than their white counterparts also undermined defenders of food aid, forcing them to argue to an uncomprehending public that obesity was often a sign of poor nutrition.[96] "It's not how much they eat but what they eat," Kuralt had said on "Hunger in America." "Fat people can be malnourished."[97] The caption under a *New York Times* photograph of an enormous black woman sitting with her three chubby-looking children read: "Though she is so fat that she cannot rise unaided, Aline Johnson's very obesity is a sign of malnutrition, according to food experts."[98] Yet the experts also condemned the poor for spending inordinate amounts on expensive snack foods, helping to foster images of obese welfare mothers pushing supermarket carts loaded with crispy snack foods, soft drinks, sweets, and sugared breakfast cereals to be paid for with food stamps.[99] Not only did this make it difficult to think of the poor as hungry, it also put any blame for their inadequate nutrition back on their shoulders: on their ignorance or, more commonly, their inability to resist temptation and postpone gratification. The "culture of poverty" in which people live only for the present was not easily understood by "the future-oriented middle class," wrote a Texas economist. "A good diet is something that is carried out over time, and the time concept is just not congenial to the culture of poverty. . . . An increase in income in this culture is put into the home, clothes, transistor radios, movies, or the purchase of lottery tickets." Any additional government spending on food stamps or income maintenance would just be wasted.[100]

Middle-class disdain for those bereft of the Protestant virtues was reinforced by traditional notions of manhood, especially that a real man is the family "breadwinner." If the idea of hunger in the land of plenty exerted a powerful moral pull in the late 1960s, its countervailing image—that men who did not strive to "bring home the bacon" were parasites who deserved no help—made a comeback in the 1970s. Critics of food aid often focused effectively on the small proportion of recipients who were able-bodied men. "In the Book that most all of us accept it says somewhere 'By the sweat of thy brow shalt thou eat bread,' " said Congressman Poage in justifying his opposition to free food for the men hanging around pool halls. "Some poor widow and her kids" were a different story.[101] Thanks in large part to feelings such as these, by the late 1970s the food stamp program had plummeted to an all-time low in public esteem. The large majority of Americans now thought that far too much was being spent on it.[102]

But hunger and malnutrition had also been driven offstage by the same thing that diverted middle-class concern away from malnutrition among the poor in the late 1930s: fears over its own food supply. Few of the hunger lobbyists who criticized Mayer for arranging sessions on the purity and nutritional value of the American diet at the 1969 White House Conference could have known that these issues would soon drive theirs into the background; but this is exactly what happened. In 1971 the federal government replaced the Ten-State Nutritional Survey, which had concentrated on the poor, with a survey of the nutritional status of "the average American."[103] Even the McGovern committee, set up mainly to deal with hunger and malnutrition, began to concern itself much more with the food of the rest of the nation than with that of the poor.[104] In this, it was following rather than leading public opinion. Middle-class concerns over the safety and nutritive value of the food supply, which had been virtually snuffed out in postwar America, were snowballing. As the snowball gained size and momentum, the millions of people whose low incomes pushed them below the nutritional margins—however defined—seemed to disappear. The plight of the poor was again of little concern to a middle class becoming increasingly fretful about its own diet.

CHAPTER 11

Nutritional Terrorism

The antipesticide campaign of the 1930s had been mainly directed against lead arsenate, a particularly dangerous-sounding substance.[1] The Great Cranberry Scare of 1959 had been over a suspicious weed-killer—aminotriazole—hardly anyone had heard of. But DDT, the most-damned pesticide of the 1960s, had actually been a public favorite. It had first been synthesized in 1874, but only in 1939 was it discovered to be an effective pesticide. The U.S. Army, facing the prospect of fighting in malarial swamps and typhus-infested grasslands throughout the world, spurred its development. In 1942 it tested it on human volunteers, who were sprayed with large amounts and even drank it with no ill effects. It was then immediately shipped to the Pacific and other insect-infested areas, where it was hailed as one of the most important weapons in winning the war. After the war, questions about its safety were routinely dismissed by those impressed by the boon it was for the world's farmers. In 1948 Dr. Paul Müller, the Swiss who first discovered its pesticidal properties, received the Nobel Prize for what was called a major contribution to the betterment of humanity. Attempts by the Delaney committee to question its use in 1950 were met by a stone wall of scientific and agricultural opposition.[2] In 1951 the National Research Council's Food Protection Committee said there was no cause for concern about residues of DDT and other pesticides in food. On the contrary, it said, the country would not be able to feed its people were it not for such pesticides.[3] In 1954, when the chief of toxicology at the Public Health Service's Communicable Disease Center made public a study indicating that traces of DDT were present in every meal consumed by the average American, he was quick to add that the amounts were not dangerous, for they were far below what human volunteers had consumed without apparent injury.[4]

The study noted that DDT had a particular propensity to collect in fats, but the toxicologist mentioned only the fat in foods, not in humans. Eight years later, Rachel Carson's revelation—first in a series of articles in the *New Yorker* in June 1962, and then in her best-selling book *Silent Spring*—that DDT also accumulated in human fat caused a sensation. Carson said that the new sprays, such as DDT and malathion, and the new families of organic chemical fertilizers seeped into foods and groundwater and were ingested in a multitude of ways, including through mother's milk. Once in the body, they were not easily flushed out but built up relentlessly in its fatty tissue.[5] For the first time in history, she warned, every human being "is now subject to contact with dangerous chemicals from the moment of conception until death." The pesticides, fungicides, and chemical fertilizers that had revolutionized post–World War II food production were actually "elixirs of death." Pesticidal poisoning, she concluded, had gone "beyond the dreams of the Borgias."[6]

The *New Yorker* was swamped with mail, 99 percent of it favorable. Newspaper editorialists expressed their distress, members of Congress read parts of it into the record, and President Kennedy told a press conference that he would set up a special committee to study pesticides. But farmers and chemical producers reacted like Carthaginians watching Romans unloading salt to plow into their fields. The chemical companies mounted a major counterattack. Carson's methods, evidence, expertise, and conclusions were assailed in *Time*, *Saturday Evening Post*, *Reader's Digest*, and other popular magazines. The substances Carson condemned were the very ones that had helped make Americans the best-fed people on earth, said her critics.[7] The Nutrition Foundation, whose president attacked the book as unscientific, joined the Manufacturing Chemists' Association to rush out a "Fact Kit" of negative reviews of the book and defenses of pesticides. "Silent Spring Is Now Noisy Summer," commented the *New York Times*.[8] Some years later, Dr. N. E. Borlaug, the "Father of the Green Revolution," called Carson's book "the genesis of the . . . vicious, hysterical propaganda campaign against agricultural chemicals." The kind of environmentalists it spawned, he said, were "scientific halfwits, callous to human suffering," whose chief interest lay in creating an environment that would give them personal pleasure.[9]

Carson's book did not have the immediate impact on American eating that might be expected. In part this was because most of its frightening scenarios dealt with wildlife, not humans, and ecological concerns were still in their infancy. It was also damaged by the wall of industry-inspired flak. By the time the president's committee issued its report in August 1963, serious holes seemed to have been shot in it. Although the committee's report did urge more testing of agricultural chemicals and increased safeguards in their use, the agribusinesses managed to dilute the outcome into a call by Congress for more research on chemicals in agriculture—to be done by the farmers' friends in the USDA. The department's report, issued a leisurely four years later, conceded that the use of chemicals as fertilizers, disease killers, defoliants, desiccants, and growth regulators had indeed shot up in the past fifteen years. Nevertheless, it concluded, there was no need for concern: All chemicals were used "under federal controls designed to keep foods free of unsafe, high-level chemical residues."[10]

Yet the very success of the USDA, the chemical companies, and the agroindustries in heading off bans on DDT and other suspect chemicals soon backfired, for it helped feed a rebirth of concern over the safety and nutritive value of the nation's food supply. The revelations in 1967 that many of the nation's freshwater fish, particularly in the Great Lakes, had ingested such high levels of mercury as to be deadly to humans brought home the fact that, as Carson had warned, much of the effluence of industrial and agricultural production did find its way into the food supply, and some of it, at least, could be dangerous. By 1969, a survey indicated, almost 60 percent of Americans thought that agricultural chemicals, even if used carefully, represented a danger to their health.[11]

Enterprises producing and selling "organic" foods thought to be free of these chemicals were among the first to benefit from the new fears. The term had been coined by J. I. Rodale, a well-off New York City publisher of health food books and magazines who in 1940 had begun to experiment with chemical-free farming on an abandoned farm in Emmaus, Pennsylvania, applying ideas derived from an Englishman, Sir Albert Howard.[12] Born Jerome Irving Cohen on New York City's Lower East Side, the diminutive ex-auditor for the Internal Revenue Service was quite unlike most of the others in the huckster-filled health food business. For one thing, he was a talented writer and made his money on his books and magazines, not on lectures or the products he sold; for another, he sincerely believed in what his publications advocated. For over twenty years, Rodale and his assistants dug away industriously at the huge compost heaps on their farm and turned out a monthly report on their activities and ideas. The magazine, *Organic Gardening and Farming*, elicited little interest outside health food circles. The bulk of the Rodale Press income continued to come from publications reflecting more conventional faddist material—the usual reports of the healthful or curative properties of new and old vitamins, foods, and regimens. By the mid-1960s, however, interest in chemical-free foods was perking up. Yet despite the millions it spent on research, the USDA had no information to dispense on chemical-free farming. Rodale's was practically the only game in town when it came to discovering how to do it. *Organic Gardening and Farming*'s circulation climbed rapidly, soaring from 60,000 in 1958 to 650,000 in 1970.[13]

As the number of farms claiming to produce organic foods increased, enterprising middlemen picked up the slack between producers and consumers. "A farmer won't stop spraying poisons on his crop because you hand him a copy of 'Silent Spring'—*but because you promise to buy that unsprayed crop*," explained the Ecological Food Society, which sold "DDT-less apples" by mail from its midtown Manhattan office. "Twenty states (so far) have banned their fish, poultry, and game because of mercury poisoning that has killed entire families," it warned. "Doctors now suggest infants should not drink their mother's milk, because the DDT content of mother's milk in America is now *four times higher than the permissible 'safety' level*. . . . The paraffin-wax coating applied (for 'visual appeal') to 70% of all fruits, vegetables, and produce in this country is a known cancer producer, which cannot be washed off or cooked out." "Organic" food, on the other hand, had "NOT been sprayed, stimulated, bleached, colored, fortified, emulsified and processed to within an inch of its life (and yours)."[14] A New York City science teacher, interviewed while shopping in an organic food store, said she ate as much organic food as possible "to dilute the poisons" in the regular food supply.[15]

Rodale was carefully vague on the contentious issue of whether foods grown with modern chemical fertilizers were more deficient in nutrition than those produced without them. He knew that agricultural scientists could find no difference in nutritive value in, say, individual kernels of corn from plants fertilized either "organically" or "chemically," because all nutrients enter plants from the soil in the same inorganic form. Yet he did claim there was "a relationship between a chemicalized soil and the increasing amount of human degenerative disease," citing the many readers who wrote about how their health had improved since they began growing vegetables organically as proof. The composer Richard Wagner was right, he liked to say, in telling man to "make his life a mirror of nature and free himself from thraldom to artificial counterfeits."[16] Gentle evasions such as these helped make the dubious idea that the use of chemical fertilizers had led to decline in the nutritive value of the nation's food a basic tenet in health food circles, from where they soon seeped outward into the population at large. By 1969 polls indicated that a majority of Americans believed that foods grown with "natural" fertilizers were more nutritious than those raised with chemical ones.[17]

The food industry's defenses of chemical fertilizers and additives paralleled the agroindustries' arguments for pesticides. "Without lawful chemical additives, the food industry could not begin to feed even the population of New York City," warned the research director of General Foods. "Our food scientists agree," said a 1965 article in *McCall's*, "that were it not for . . . chemical additives . . . we would literally know famine."[18] But these arguments failed to still the growing doubts about the healthfulness of the food supply. Soon the doors were opened to the usual horde of health food advocates and fad diet promoters impatiently waiting for the moments when they could again hold center stage. Now, the enormous expansion of the media—particularly television and the weekly newsmagazines—gave them much more national exposure than the last time they had held the spotlight, during the 1930s. Yet, since so many of them had cut their marketing teeth in the 1930s, perhaps it was inevitable that they would help revive the old concern of that era: that processing itself robbed food of its nutrients. When tied to the critique of modern agriculture, it made up a potent double-barreled indictment of all that happened to American food from the seedling stage to the supermarket.

The tall, handsome author and lecturer Gayelord Hauser, one of the health food superstars of the 1930s, was among the first to benefit from the renewed concerns. Again he enthralled women with the names of the elegant and high-born ladies he had persuaded that the nutritionally deficient foods grown with modern fertilizers made you age and die before your time: the Duchess of Windsor, Greta Garbo, Paulette Goddard, Gloria Swanson, "the lovely Barbara Hutton," Fred Astaire's mother. All seemed to be sending their servants out to buy Hauser's Five Wonder Foods—brewer's yeast, powdered skim milk, yogurt, wheat germ, and molasses—and trying his One-Day Hollywood Liquid Diet.[19]

But Hauser's posturing soon took a back seat to the more scientifically based theories of the decade's best-selling health food advocate, Adelle Davis. Unlike Hauser, whose credentials were as dubious as most in that often-sleazy business, she had formal training in nutrition, with credentials from undergraduate courses at Purdue, Wisconsin, and Berkeley, as well as some graduate work in biochemistry from the University of Southern California. She had also practiced as a dietitian in New York City's Bellevue Hospital and in a Greenwich Village settlement house, where she had wrestled with the Herculean task of giving dietary advice to poor Italian-Americans. Her books, unlike those of Hauser and other faddists, seemed well supported by numerous references to medical and scientific studies. Her best-selling work, *Let's Eat Right to Keep Fit*, listed 2,402 of them.[20] She also differed from Hauser in being a firm believer in vitamin supplements; her nickname at Purdue had been "Vitamin Davis." But she agreed with Rodale and Hauser on the most important things: The modern American diet was the cause of most of the ill health in America, and modern food production techniques were the root of the problem. "Thousands of adults and millions of children in our country have never once had a mouthful of genuinely wholesome food," she wrote, "not one sip of delicious medically certified raw milk or one bite of delightful freshly stoneground, 100 percent whole-grain bread or cereal or of unbelievably good organically grown fruits and vegetables." Vitamin supplements were necessary, said Davis, because wholesome food, "grown on naturally mineralized, naturally composted soil," was no longer available. Once harvested, the debilitated foods were further enfeebled by the food-processing industry, "which allows nutrients to be lost during processing in order to increase profits."[21]

Although her critique was a radical one, a large part of Davis's success (her book sales totaled over ten million copies) derived from her unwillingness to stray too far from current nutritional ideas. Her regimen was not a strict one revolving around a small number of foods; instead it was based on the familiar idea of a balanced diet. Her assertions that drinking a quart of milk a day prevented cancer did not seem out of line, even if the milk was to be unpasteurized; after all, for years Americans had been told that milk was a miracle food. Nor did her high regard for eggs run counter to what Americans had been taught; the eggs just had to be fertilized. She even had a benign view of meat, particularly beef. Her nostrums for the new fears of processing and additives were an artful blend of the old and new: Return to the "natural," she said, but supplement it with vitamin pills (she took six such pills after every meal) and other special concoctions. Her "Pep-up" drink, guaranteed to cure a variety of illnesses, exemplified this old/new melange: egg yolks, oil, lecithin, calcium salts, magnesium oxide, yogurt, granular kelp, milk, yeast, wheat germ, and soy flour.[22]

By 1966 Davis was a popular guest on Johnny Carson, Dick Cavett, and other radio and television talk shows; she was quoted in the daily press and the mass circulation magazines. Unlike the older generation of charismatic blowhards such as Hauser, "Adelle," as her followers called her, seemed to exude down-to-earth, practical common sense, supported by rock-solid science. That members of the medical and nutritional establishments called her work "hogwash," "garbage," and "potentially dangerous" (the panel on deception and misinformation at the 1969 White House Conference called her probably the most damaging single source of false nutrition information in the land) was drowned in a sea of impressive-looking footnotes citing medical and scientific authorities.[23] One of Davis's British admirers noted that this scientific grounding allowed "the advocates of healthy eating [to move] from Greenwich Village to Fifth Avenue. Instead of citing dubious nineteenth-century naturopaths, they quoted doctors and the latest biochemical research. Instead of addressing themselves to a chosen few, they found an audience among the thoughtful, the intelligent, the influential and among doctors themselves."[24]

Organic foods, which had been accumulating their own scientific backing, also basked in this new respectability. A 1970 *New York Times* article on converts to the organic featured two professors, a science teacher, an ex–advertising executive, the Broadway musical star Gwen Verdon (who had praying mantises prowling the organic garden of her Central Park West penthouse apartment), and the actress Jane Fonda.[25] *Newsweek* noted that "the organic food community now includes large numbers of environmental activists, housewives with tired blood and sophisticated gourmets who are out to break the additive habit." Among the California devotees were the crusty film mogul Jack Warner, the aging radio comedian Edgar Bergen, and Pat Boone, the clean-cut singer who embodied the "all-American" ideal. There were already three-hundred-odd health food stores and twenty-two organic restaurants in southern California alone. No wonder an *Organic Gardening and Farming* writer called it "a food shopper's paradise."[26]

Meanwhile, the nation's health food stores had been transformed from dusty, cluttered, cramped places where the proverbial little old ladies in tennis shoes bought powders and potions into minisupermarkets with wide aisles, bright lighting, and fancy shelving. Their owners were no longer local weirdos; they were usually the most conventional of middle-class business people. (However, Mrs. Mary Hatch, who at age sixty-five left her job as a mortuary organist to open one in San Ramon, California, was probably more akin to the old school. She said she did it because "I saw so many dead young people when I worked in the mortuary—so many who would have lived if they had realized that you are what you eat.")[27] Their clientele was now decidedly upscale, for health food usage increased with increasing income and education.[28]

But health food regimens were usually too complex, restrictive, or expensive for most Americans. Instead, they responded to the new food concerns as they had thirty-odd years earlier—with a wave of vitamin-mania. By 1969 over 50 percent of Americans were regularly taking vitamin pills or other dietary supplements.[29] Many went further and followed new "megavitamin" theories, which claimed that massive doses of certain vitamins could cure everything from cancer to schizophrenia to alcoholism. Vitamin E, for example, was not only the "sex vitamin"—enhancing male sexual performance, preventing impotence and sterility, and changing the sex of babies in the womb—but massive doses of it were also credited with preventing cancer, heart disease, ulcers, hair loss, and skin problems. It was even said to alleviate the effects of air pollution and to slow the aging process.[30]

Megavitamin therapy, or "supernutrition," had a particular appeal to practitioners of "alternative" medicine forced to circumvent the "regular" MDs' monopoly on prescribing medicines. Some such therapies had eminently respectable backing. Norman Cousins, editor of the staid literary journal *Saturday Review*, wrote of a young female graduate of a "fashionable college" whose schizophrenia, caused by "pellagra of the brain and nerve cells," had been cured by massive doses of niacin. The revered scientist-missionary Albert Schweitzer, a Nobel Peace Prize winner, wrote a testimonial for Dr. Max Gerson, who claimed that he could cure cancer with twelve glasses of orange, "green," carrot, and liver juice a day along with injections of vitamin B_{12} and coffee enemas. In 1966 megavitamins gained their most effective advocate when another Nobel Prize winner, Linus Pauling, who won the prize for chemistry in 1954, began a vigorous campaign on behalf of megavitamin therapy—particularly vitamin C—as the cure for a number of ailments, including the common cold. Adelle Davis became convinced that vitamin C cured not only colds but also anxiety and "every form of injury."[31]

But the Food and Drug Administration had continued to operate within the old nutritional paradigm, the one that had gelled during that fateful time in 1941 when the NRC decided that only enough nutrients would be added to wheat flour to bring it back up to premilling levels. The government had thenceforth sided with the medical establishment as it battled to keep the public from turning to vitamins—rather than doctors—to solve their health problems. In early 1942 the AMA had reaffirmed this stand, severely condemning the "indiscriminate administration of vitamins to workers in industry," saying it would do nothing to raise productivity and could well cause harm.[32] This continued to be the official line in the postwar period, epitomized by the head of pediatrics at Stanford University Hospital who told the 1954 home economics convention that vitamins should only be prescribed by doctors, not by parents. "Vitamin intoxication" was more prevalent than vitamin deficiency, he warned. Children intoxicated with vitamin A, for example, felt listless and fatigued, had dry hair and skin, suffered pain in their extremities, and—perhaps most disturbing of all to mothers—lost their appetites.[33] In 1960 the president of the AMA denounced "vitamania," lamenting that "ingenious advertising and misleading claims have helped cram Americans full of vitamins which they don't need." This was a "tragedy," he said, for "masked organic disease may be the cause of the symptoms that are being treated erroneously with vitamins."[34] The FDA and AMA stance received solid backing from the food industries. Food processors applauded an FDA pamphlet that denounced "the myth of overprocessing" and vigorously supported the government contention that all essential nutrients could and should be gained through a "balanced diet" of the foods American farmers and industry made readily available.[35]

But the half-billion dollars a year Americans were spending on vitamins by the early 1960s meant that there were now forces with economic muscle that could challenge the official line. The most potent of these were the vitamin producers themselves. A small number of large pharmaceutical corporations such as Merck & Co., Squibb, Pfizer, and Hoffmann–La Roche, Inc., had emerged on top as their industry underwent the same process of corporate concentration as the food industries. Together they formed the National Vitamin Foundation, which sought to lessen their dependence on dreary health food stores and corner pharmacies and break into the mass market. To do this, they thought, they would have to rid vitamin pills of their faddist connotation, something that could be done by enlisting the best of respectable science on their side. They therefore hired the well-known nutritional scientist Robert Goodhart as executive director and provided him with the funds to rally scientific wisdom to support their pills and potions.[36]

The defenders of the conventional paradigm rose to the challenge. When the Vitamin Foundation announced it was hiring a public relations firm to educate the public on the value of supplemental vitamins, the Nutrition Foundation mobilized a countercampaign to remind the public that there was an abundance of food available from which all Americans could select a balanced diet. "Last year Americans were robbed of $500,000,000 through food fads, extreme diets and cure-alls," said its executive director. "We hope to prevent some of this loss." "This is not a scientific battle," Goodhart acknowledged, "but an economic one."[37]

It was then that the FDA took what seems in retrospect to have been a major step down a self-destructive path. Convinced, quite correctly, that a new Golden Age of Food Faddism was dawning, it put its prestige, and even some of its money, on the side of the physicians and food processors. In October 1961 it joined with the AMA to organize a National Congress on Medical Quackery at which six hundred leaders in public health and nutrition gathered to hear Secretary of Health, Education, and Welfare Abraham Ribicoff and other government and AMA officials promise a full-scale assault on "quackery," which was deemed to cost the nation over a billion dollars a year. The most widespread and expensive form of quackery, said George Larrick, commissioner of the FDA, was "the promotion of vitamin products, special dietary foods, and food supplements." Dr.

Leonard Larson, the new president of the AMA—obviously perturbed by the respectable experts gathering in the Vitamin Foundation stables—said the problem was that the public thought "quackery went out with river boats, sideburns and the snake oil hawker. . . . Quackery of today is commercial, it is almost respectable, it is cosmopolitan, it is modern."[38]

In the years that followed, the FDA kept up the struggle to reassure Americans about the quality of their food supply. To many in the agency it seemed to be a continuation of the same good fight against greedy charlatans and hucksters that had inspired its revered founder, Dr. Harvey Wiley. In 1966 the FDA, backed by the NRC's Food and Nutrition Board, announced that it intended to curb the sale of megavitamins and require that all vitamins carry labels saying:

> Vitamins and minerals are supplied in abundant amounts by the food we eat. The Food and Nutrition Board of the National Research Council recommends that dietary needs be satisfied by foods. Except for persons with special needs, there is no scientific basis for recommending routine use of dietary supplements.

In May 1967 it issued a fact sheet labeling the claims of vitamin supplement advocates and organic food proponents as "nutritional nonsense." It was "not so" that "our soil has lost its vitamins and minerals." The fact was that "if plants will grow at all, they will have the vitamins and minerals that you need." Other "facts": Chemical fertilizers were not poisoning the soil; organic fertilizers were not safer than chemical ones and did not produce healthier crops; the FDA protected consumers from any danger from pesticide residue on crops; there was nothing to fear from additives. The FDA commissioner called stories about depletion of the soil and loss of food values because of processing techniques "a lot of nonsense." "Frankly," he said, "it's time we faced facts about our American diet. Our soil is naturally rich and the envy of every other nation. Our ability to grow, pack, ship and sell food is a modern marvel because the natural value of the food is not lost in the process. In fact, the reverse is true: foods can get better in the process."[39]

But the majority of Americans were unconvinced. A 1969 survey done for the FDA indicated that 75 percent of them thought that many foods lost much of their nutritive value "because they are shipped and stored so long" and that 60 percent agreed that "much of our food has been so processed and refined that it has lost its value for health."[40] Rather unwisely, the agency now flew right in the face of these beliefs, and the vitaminmania they underlay. It proposed to prohibit the sale of vitamin pills containing more than 150 percent of the vitamin's RDA without a doctor's prescription. This meant, for example, that those who wished to take as much vitamin E as its advocates said was necessary to prevent cancer would have to take thirty-three pills a day. The political uproar should have been predictable. The large drug companies had plenty of the influence economic power brings in the corridors of Congress, while the health food vendors—much of whose income now derived from vitamin supplements—could rally "people power" to their side. Thousands of their patrons, many of them still the proverbial little old ladies in tennis shoes, wrote and wired their representatives to protest the proposed outrage. Their indignation was not difficult to understand, for to them it seemed that the government was trying to prevent them from making up for the nutritional deficiencies its own negligence allowed and encouraged. Members of Congress heeded the howls, forcing lengthy hearings to be held on the FDA's proposed regulations.

The result was two years of hearings, thirty-two thousand pages of testimony, and, ironically, media exposure for a raft of vitamin advocates who—because the Federal Trade Commission prohibited making health claims in food advertising—would otherwise have remained cloistered in the back of health food stores, churning out mimeographed advice to a few hundred converts.[41] Given the nature of nutritional science, which would require Nazi-like experiments on thousands of human beings to prove anything conclusive about most vitamin deficiencies, the megavitamin advocates could not provide demonstrable scientific proof of the efficacy of their various nostrums. The usual anecdotes constituted the bulk of their arguments. But neither, for the same reason, could the FDA back up its warnings to the public, who in any event remained confused about what vitamins actually were. (The 1969 FDA survey disclosed that over 75 percent of Americans still believed that extra vitamins gave them more "pep" and energy.)[42]

The hearings also helped publicize the vitamin advocates' warnings that malnutrition, particularly in the form of vitamin deficiencies, was undermining the nation, rich as well as poor. As a result, in 1969 the government was forced to allow fortification on a wide scale. Nutrients could now be added to foods of any kind—whatever their preprocessing nutrient levels—if it was thought that the diets of a significant number of people were deficient in them.[43] But the resulting flood of fortified breakfast foods and beverages such as Hi-C hardly whetted the public's appetite for vitamins. In 1973 members of Congress reported having received more mail about the proposed restrictions on vitamin pills than about Watergate. Congress not only turned down the FDA's proposals but followed this up with the so-called Vitamin Amendments to the Food and Drug Act severely limiting its power over vitamin sales.[44]

But it was a more traditional crusade, a clean meat campaign, that led to the rebirth of the kind of consumer food campaign Schlink and Consumers Research had tried to mobilize in the 1930s. In 1967 U.S. Department of Agriculture inspectors tipped off another young consumer advocate, Ralph Nader (already well known for his auto safety crusade), that for four years their department had been suppressing reports of filthy conditions in meatpacking plants that were not subject to federal regulation because they did not ship products across state lines. The gaunt, ascetic Nader, who had turned his back on the easier career paths his Harvard law degree might have opened to him, was driven by one overriding compulsion: the idea, according to his biographer, "that men are biologically obsolete, that their senses can no longer protect them from harm, that almost everything Americans eat or drink or breathe has been corrupted by the shadowy order of corporate poisoners." He not only refused to eat anything containing additives but also abjured any foods that were ground, stuffed, or processed.[45]

Working with Nick Kotz, the young Washington correspondent for the *Des Moines Register* and *Minneapolis Tribune* whose reporting on hunger later enraged Nixon, he began digging for evidence to support a bill to bring these plants—and their ground, stuffed, and processed foods—under federal supervision.[46] In July 1967 Nader published the first of his indictments in the *New Republic*. There he echoed Rachel Carson's theme: that the very same new chemicals and processes in which the postwar food and agriculture industries took such pride had upped the ante in food dangers, making Dr. Wiley's crusades look relatively easy:

> It took some doing to cover up meats from tubercular cows, lump-jawed steers, and scabby pigs in the old days. Now the wonders of chemistry and quick freezing techniques provide the cosmetics for camouflaging the products and deceiving the eyes, nostrils and taste buds of the consumer. It takes specialists to detect the deception. What is more, these chemicals themselves introduce new and complicated hazards unheard of sixty years ago.[47]

Meanwhile, Kotz had been handed more damaging material. He wrote a series of articles on contaminated meat, which earned him flattering comparisons to Upton Sinclair, whose novel *The Jungle* had led to the passage of the original federal meat inspection law, and a Pulitzer Prize. As was so often the case, however, it was television that played the crucial role. After the Public Broadcast Laboratory, predecessor of the Public Broadcast System, aired three programs on the topic in November, the three major television networks picked up the story. They repeated charges such as that meat packers relied on the "4D" animals—dead, dying, diseased, and disabled—for processed meats. Realizing that public confidence in all of their products was being shaken, the large meat packers, some of whom owned intrastate packing companies, abandoned their opposition to the bill, allowing it to pass easily.[48]

The quick victory amazed the reformers. Rarely had a bill shot through Congress in less than six months, particularly in the face of stiff opposition in the key committees responsible for it. It showed, some Naderites said, that American politics was not, as current intellectual fashion had it, based on irrational appeals; the public had been given the facts, and its representatives had been forced to respond.[49] Yet in fact the critics soon began to use food issues for quite the opposite reason—their emotional punch. After all, the news that hot dogs contained rat hairs and that breast milk harbored DDT hardly provoked cool reasoning. The activists hoped that a public emotionally aroused by issues such as these would support their wider campaigns to protect the nation's health and welfare from the effects of business greed. Indeed, Nader was rather rueful about the packers' quick cave-in. They had thereby managed, he said, to put the lid back on the Pandora's box of other issues—chemical adulteration of meat, microbiological contamination, misuse of hormones and antibiotics, pesticide residues, ingredient standards—that would have been useful in rallying the public to his larger crusade to curb corporate abuses.[50]

But the food industries were hardly more successful than Pandora. After confidently announcing to a skeptical audience of newspaper editors in 1968 that food would become the top news story of the coming year, Nader recruited enthusiastic young law and college students to ferret through the food industries' dirty linen—combing the obscure reports of government agencies, analyzing lists of ingredients, asking chemists and biologists about their additives, assembling statistics to document the continuing concentration of economic power in fewer hands. By mid-1969 he was ready with enough evidence to begin the massive assault along the broad food-processing and agribusiness front that he relished.

Appearing before the Senate's new Select Committee on Nutrition and Human Needs, he charged that the food industry was dominated by immensely powerful oligopolies who cared only about selling their products, not about their nutritive value. Their "manipulative strategies" bilked the consumer; "the silent violence of their harmful food products" caused "erosion of the bodily processes, shortening of life or sudden death." Geneticists feared that "the river of chemicals that all of us use and breathe" might be causing genetic defects among the newborn. Yet the "skilled salesmen" of the food industries used "applied social science" to shape consumer preferences in order "to maximize sales and minimize costs no matter what the nutritional, toxic, carcinogenic, or mutagenic impact may be on humans or their progeny." Again, Nader cited the beloved frankfurter as an example of corporate greed and irresponsibility. They should really be called "fatfurters," he said. They were loaded with so much cholesterol that they were "among America's deadliest missiles." The oligopolies had also ruined the hamburger, which was often so adulterated as to be worthy of the name "shamburger." Nor did baby foods escape nefarious corporate influence. They were flavored with salt, which would adversely affect a child with a hereditary susceptibility to hypertension, and sugar, which was "nutritionally poor, carcinogenic, and possibly atherogenic [i.e., caused clogging of the arteries]." Why? Not to please babies, said Nader, but to please the mothers who tasted the food. The same went for modified starches, which added smoothness and bulk to the baby foods, and monosodium glutamate (MSG), whose flavor-enhancing properties were unappreciated by babies but which seemed to pose dangers to their health. Strong congressional action was needed to educate the public and combat this "multimillion dollar fraud" with its "massive assaults on human health such as fat content, unnecessary salt content, untested chemical additives." Congress must pass strict food labeling laws, greatly increase funding for research on chemical additives, strengthen the FDA, force the USDA to take the interests of consumers into account, and alert the nation to the dangers it faced.[51]

The raft of criticism of its supposed failure to safeguard the food supply certainly put the government on the defensive, but one of the greatest blows to confidence in it was delivered from within. On October 7, 1969, HEW secretary Robert Finch startled the nation with the announcement that he was banning the use of the artificial sweetener cyclamate—the food chemists' wunderkind of the 1960s—in food and beverages.

Saccharin, which had ruled the artificial sweetener roost since 1879, had always been hampered by its bitter aftertaste, which was particularly apparent in beverages. Cyclamate, a noncaloric sweetener without an aftertaste, had been discovered by accident in 1937 by a University of Illinois scientist. The FDA approved its commercial use in 1951, and it was marketed by Abbott Laboratories as Sucaryl. Sales had climbed steadily during that decade, as it found its way into canned fruits, chewing gum, and even toothpastes. In the early 1960s, as weight-consciousness again began to sweep the nation, it became the key ingredient in a diet soda craze. By the time it was banned, foods and beverages containing it were to be found in an estimated 75 percent of American homes.[52] In 1965 the FDA had summarily discounted studies apparently showing that cyclamate caused cancer in test animals. When, in late 1968, other studies indicated it caused chromosome damage, the agency warned only against consuming too much of it, stating its suggested limits in terms almost incomprehensible to laypeople.[53] Then, on October 1, 1969, Dr. Jacqueline Verrett, an FDA scientist, appeared on the NBC Evening News to describe the horrible malformations that had occurred among chicks born of eggs injected with cyclamates. Hard on the heels of that shocker came an admission from Abbott Laboratories that 8 of 240 rats injected with high levels of cyclamates had developed cancerous tumors in their bladders. It seemed that under the 1958 Delaney Amendment to the Food and Drug Act an immediate ban was mandatory, whether or not any risk to humans had been demonstrated.[54]

Realizing that the government was vulnerable to charges that it had allowed carcinogenic foods onto the mass market, Finch took the lead in damage control. He announced the ban himself to assure the public that concern for their health was paramount at the highest levels of government. At the same time, however, he implied that the ban was not really necessary—it was unfortunately "required" under the Delaney Amendment—and he gave the industry some months to implement it, during which time their warehouses could be emptied of the drinks by diet soda devotees stocking up for the future.[55] Donald Kendall, the head of Pepsi-Cola, told President Nixon that the beverage industry appreciated the way Finch had handled the difficult situation.[56] But the FDA and the rest of the food industry could hardly have been so impressed, for public confidence in the regulatory process was now shaken to the core and the entire food industry on the defensive. Shortly thereafter, when a study indicated that infant rats injected with large doses of MSG developed brain damage, panicky baby food manufacturers—who only days before had produced scientific defenses of the substance which the commissioner of the FDA found to be utterly convincing—fell all over themselves in a rush to eliminate it from their products.[57]

Not everyone was pleased by the food industry's sudden cave-in. Even the head of the laboratory that had recommended that cyclamates be banned said it was "absurd" to draw conclusions about MSG's safety from these studies. The food processors' "panic," he said, was causing "a great deal of unnecessary alarm."[58] Dr. Arthur Schramm, the head of the National Academy of Science's Industry Liaison Committee, lashed out at the media, "particularly TV," which he said had publicized experimental data on cyclamates and MSG "in such a manner as to dispose a large majority of the lay public to draw dire conclusions." This had helped create "an atmosphere . . . of economic terrorism."[59] Ex–FDA commissioner Herbert Ley agreed. He blamed the government's "rather impetuous actions on cyclamate" on the "probably self-serving efforts . . . of some scientists who were relatively low in the organization and who felt no esprit . . . to bring this issue to the eager attention of the public as a topic for national debate on evening TV." But even President Nixon, whose suspicion of the media verged on paranoia, succumbed to this kind of "terrorism" by ordering a special twenty-million-dollar study of all the additives on the FDA's Generally Recognized As Safe (GRAS) list, including salt and sugar.[60]

His instincts proved correct, for the media blitz continued, with revelations about mercury in fish, botulism in pizzas, pesticides in turkeys, arsenic in chickens, antibiotics in cheese, hormones in meat, salmonella in soup, and DDT in practically everything. "Any housewife buying groceries for her family these days must be groggy from trying to figure out what not to put on her shopping list," said the *New York Times* in January 1971.[61] Four months later more than two hundred thousand chickens had to be destroyed after eating feed contaminated with a potent carcinogen, one of the polychorinated biphenyls (PCBs), but sixty thousand of their eggs had already been sold in Washington, D.C., markets. FDA commissioner Charles Edwards was hardly contrite. Instead he attacked the media, saying public confusion had been "compounded by a few alarmists seeking headlines aided and abetted by unbalanced reporting."[62] But large food producers could ill afford such bravado. A number of them now cringed before the "terrorists" and put up only the feeblest of defenses against demands that they stop using the chemicals DES, TCE, and certain food colorings before succumbing completely.[63]

At first glance, it would appear uncharacteristic that the media should play such an important role in terrorizing such large advertisers, particularly in light of the 1930s, when they rallied to industry's side in the fight to disembowel the Tugwell Bill. It was not that the stories themselves were simply too sensational to ignore: Schlink's 1930s' charges were at least equally shocking, yet they had trouble finding space in magazines with a larger readership than the *Nation*. But the media were no longer as dependent on food advertising as they had been in the 1930s, when food-processing had been one of the few growth industries. National food advertising now played only a minor role in the daily press and was not a major factor in the weekly newsmagazines, which had blossomed since the 1930s. Most food processors concentrated their media dollars in the still-docile women's magazines and in radio and television.[64] While the latter's entertainment programs were notoriously vulnerable to advertisers' pressure, their more prestigious news divisions were less susceptible.[65]

But perhaps the most important difference was that the new assaults had such impressive-looking scientific backing that they were almost impossible to ignore. Schlink and Kallett had no scientific credentials to speak of and could be dismissed as little more than muckraking journalists. The new warnings came from scientific and medical people with the most respectable credentials. The people who first linked "Chinese Restaurant Syndrome"—headaches some people experienced following Chinese meals—with MSG were four scientists at New York's highly respected Einstein College Hospital.[66] The man whose experiments indicated that it might cause birth defects was a research scientist at Washington University in St. Louis, one of the nation's top research establishments.[67] Michael Jacobson, the Schlink admirer who joined the food crusade in 1970, had a Ph.D. in microbiology from Massachussetts Institute of Technology. His two cofounders of the Center for Science in the Public Interest also had impressive scientific credentials.[68]

All of this reflected the enormous changes in the world of health science research since 1940. Before World War II, prominent chemists such as Sherman at Columbia and McCollum at Johns Hopkins were among a small minority who could work practically full-time on nutrition research.[69] Most other nutrition research was produced more or less on contract from food producers, as at Wisconsin, or consisted of relatively amateurish papers written by doctors about things they observed in five, ten, or perhaps fifteen of their patients. Wilder and colleagues had conducted their Mayo Clinic experiments on thiamin almost as a sideline while they continued the clinical work for which they drew their salaries and fees. Although the government finally began to provide large-scale funding for basic medical research during World War II, most of this went to research on antibiotics and other fields remote from nutrition. Vitamin research, the core of nutritional science in the 1940s, tailed off. Indeed, the dearth of exciting discoveries even encouraged the idea that there was little more left to find out. In 1961, according to a government study, nutritional science "appeared to be a completed discipline." By the late 1960s, however, that idea had become laughable, for the field had shattered into a multidisciplinary activity involving close to fifteen hundred full-time scientists calling themselves biochemists, physiologists, microbiologists, endocrinologists, plant and animal geneticists, and food technologists, as well as the more traditional physicians, dentists, chemists, and dietitians.[70]

A number of things had combined to transform the field. The Sputnik scare of the late 1950s had spurred federal government aid to scientific research, including research with medical implications. The coming of age of the baby boomers had stimulated a great wave of university expansion, helping university hospital complexes to develop as impressive centers of research. They were staffed by full-time faculty whose careers depended on what they published rather than on what they did in their clinics. Until then the USDA—with its mandate to help agribusiness and food producers—had been the major source of government funding for nutritional research. Now government funds were coming from agencies such as the National Institutes of Health (funder of the MSG research), the United States Public Health Service, and other government agencies that were not beholden to those interests. By 1973 the NIH alone had a budget of $1.5 billion, much of which was devoted to probing the relationship between diet and disease.[71]

Government largesse was supplemented by funds from the growing number of charitable foundations devoted to combating a single dread disease or ailment. Emulating the immensely successful March of Dimes, which had enlisted President Roosevelt himself in its campaign against polio, they hired professional fund-raisers who mounted slick campaigns to raise money for research into their maladies, often for work that differed from established paradigms. The avalanche of publicity that greeted the climax of the March of Dimes campaign, Jonas Salk's 1955 discovery of a vaccine against polio, spurred hopes for research breakthroughs to conquer other dreaded diseases. Not only did it prompt a great increase in congressional appropriations for the National Institutes of Health, each of which was devoted to finding cures for one type of ailment, it also boosted the fortunes of the new single-affliction foundations.[72] Meanwhile, of course, the pharmaceutical companies, always major financers of medical research, were pursuing their own agendas in nutrition research, while food-producing interests began to play the game more seriously as well.

The story of cholesterol-consciousness provides an example of what could result from the competitive, high-stakes atmosphere that developed. The issue received a burst of publicity in 1961, when medical scientists in the American Heart Association reported that cutting down consumption of saturated fats was "a possible means of preventing atherosclerosis and decreasing the risk of heart attacks and strokes." They were quickly denounced by the National Dairy Council, whose own experts warned that the recommended changes "could be dangerous" for the public at large, and the processor-supported Nutrition Foundation, which advised the public "not to be unduly alarmed by scareheads about fat and cholesterol." Other powers in the nutritional science establishment reinforced the dairy interests. The AMA's Council on Foods and Nutrition called the Heart Association's recommendations "premature," while the FDA cautioned against dietary change, noting that "a causal relationship between blood cholesterol levels and artery diseases has not been established."[73] Secretary of Agriculture Orville Freeman denounced those who suggested any kind of a link, declared calcium deficiencies to be a major health problem, and urged Americans to eat more butter and milk.[74]

But the well-heeled Heart Association was not beholden to any of them and was therefore able to persist in fighting what, by mid-decade, still looked like a lonely battle. Their 1961 statement had caused a spurt of cholesterol-phobia that had spurred margarine sales, but they were unable to make headway with their demand that foods be labeled with their saturated and unsaturated fat content. In 1965 an FDA official confidently predicted that the agency would continue to have no trouble resisting the demand. It faced a wall of opposition from the powerful dairy industry, the meat industry, and other affected food companies, he reported. Makers of corn and other edible oils were noncommittal because they were uncertain how it would affect their products. "The only real support for the proposal are a few doctors, three or four drug companies, and three medical societies," he said. The AMA's opposition "seems to leave a stand-off, medically, allowing the FDA to resist the demand."[75] Meanwhile, the National Dairy Council feverishly expanded its research activities, recruiting a professional staff of 320 who were spending fourteen million dollars a year on "educational/scientific" activities by the end of the decade.[76] But the edible oils producers threw in their lot with the Heart Association and started to support research into the beneficial effects of unsaturated fats on cholesterol levels.[77] More important, research funds from other government agencies began to undermine the FDA and USDA. Among the doctors supporting the Heart Association was the director of a major study on the detrimental effects of saturated fats funded by the U.S. Public Health Service, an agency free of Agriculture Department (and dairy industry) control.[78] By the early 1970s, the writing was on the wall—or at least would soon be on the labels—and cholesterol-phobia was well on its way to receiving the government's benediction. The new world of nutritional research funding had encouraged the established wisdom to be challenged in a fashion that would have been inconceivable in prewar America.

This was but one front in a number of scientific jungle wars, where barely camouflaged interest groups stalked each other, first in scientific journals, then in the mass media. Vitamin producers funded research into the benefits of their products. (Hoffmann–La Roche, the largest wholesaler of vitamin C, bankrolled the Linus Pauling Institute to the tune of a hundred thousand dollars a year.)[79] Manufacturers of noncaloric sweeteners supported research into the dangers of sugar, while sugar producers, in turn, financed more than half the research into cyclamate in the five years before it was banned and then supported studies that led to saccharin being banned as a suspected carcinogen.[80] Food processors revamped the Nutrition Foundation, directing it to produce specific scientific defenses to rebut charges against their products. In November 1971 its genial, public-relations-oriented director was replaced by one of the nation's most established nutritional scientists, William Darby of Vanderbilt University, chairman of the NRC's Committee on Food Protection and member of the NRC's powerful

Food and Nutrition Board. It would now abandon its funding of "research in new, uncharted, speculative areas," it announced, and would instead concentrate on research that would "focus objective scientific attention on current issues in food safety and nutrition."[81]

One reason for the Nutrition Foundation's new posture was that critics of the food industries were organizing into pressure groups that publicized the growing number of accusations against the country's food supply and diet. In 1971 Jacobson helped found the Center for Science in the Public Interest, directing much of its attention to food processors' depredations and scams. The "Raiders" on Nader's Health Research Group delved into similar matters. Nader graduate James Turner, compiler of the disturbing book *The Chemical Feast*, formed Consumer Action for Improved Foods and Drugs. Other advocacy groups, such as Action for Children's Television, produced studies condemning such things as food advertisers and breakfast cereals.[82]

By the beginning of the new decade, then, attacking the food industries was becoming a mini-industry in its own right. Researchers with career interests in questioning the healthfulness of many of their products were proliferating; organized groups were disseminating their findings to the media, who were more than willing to feature them. These found a receptive audience among a middle class that was rapidly losing faith in both the food industries and government. Some of the reasons for this were far removed from scientific studies of the food supply and diet. Rather, they were connected with the startling social and political upheavals of the late 1960s and early 1970s.

CHAPTER 12

The Politics of Food

Strange as it may sound, some of the public's disenchantment with the food supply can be traced to the break-up of the American Communist party in the 1950s. In 1959 and 1960, small bands of young radicals disenchanted with its ways pronounced Communism irrelevant to America and began calling themselves the New Left. Students for a Democratic Society, founded in 1963, became its best-known organization, but "the movement" was much more than SDS. By the late 1960s, thanks mainly to growing sympathy for the black civil rights movement and swelling opposition to the war in Vietnam, the New Left had mushroomed into a large, informal network. While never a serious threat to the political establishment, it did have considerable influence in the nation's elite campuses and therefore inevitably left a mark on the media and government.

Ironically, it was American liberals—whom New Leftists often pilloried as creatures of the corporate state—who responded most positively to crucial parts of the New Left message. That people whom the radicals regularly denounced as "sellouts" should agree with many of their criticisms is not all that surprising: The essence of the New Left critique of modern American society harked back to one of the original appeals of modern liberalism—the attack on the large corporation as an undemocratic, socially irresponsible force corrupting American society. While liberals differed on the solutions to the problem and refused to go along with nationalization, "community control," disarmament, or the other vague leftist nostrums, after two decades in which they had been lulled into more benign views of Big Business, the attacks on "corporatism" had the ring of "Tenting Tonight on the Old Camp Ground."

Initially, New Leftists derided the liberal Ralph Nader because he believed that government regulation could rein in the amoral corporations—failing to see that government *was* the corporations, they said—but his critique of the nefarious practices of Big Business differed little from theirs. Only belatedly did the New Left realize that, in alerting the nation to the corporations' role in ruining their food, Nader was doing much more to turn Americans against corporate America than any of their carefully argued analyses of Big Business's imperialistic, antidemocratic, or poverty-creating activities. He, and the growing environmental movement (hitherto dismissed by leftists as elitist "wilderness freaks"), showed that corporate America could be blamed for much of the pollution of the nation's air, water, and, ultimately, food. Yet not until late 1969 did activists in places like Berkeley, Ann Arbor, and Madison begin to seriously venture into the politics of the environment and food.

Once involved, however, their self-righteous zeal was soon manifest. Demonstrations expressing outrage against polluters became almost as common as protests against the war in Vietnam and—after 1970, when American involvement in the war began to wind down—more so. Along with these came denunciations of many of the same corporations for poisoning and/or denutrifying food. In 1971, the same year in which the first national Environment Day was held, many of the same groups—influenced by Nader, the New Left, or both—also participated in the first Food Day, much of which was devoted to denouncing food processors. An article by Judith Van Allen in the glossy New Left magazine *Ramparts* reflected the common ground that the leftists and Naderites came to share. The American diet was getting worse, she said, because the oligopolistic food industry concentrated on selling sweets and junk food with little or no nutritional value. It pumped them full of dubious additives, pretended that they saved time, and poured millions into creating markets for them. "The food industry's propaganda runs a close second to its products in nausea level, and has just about as much to do with consumer 'demand' or consumer need," she said. "We buy what they choose to sell, and we pay the costs of their telling us it's what we want. . . . It controls what we eat now and what we'll eat tomorrow."[1]

Nader, meanwhile, was moving leftward, losing his confidence that tough laws alone could protect Americans. Increasingly, like the leftists, he saw government as the creature of the very corporate interests the laws were meant to regulate. In 1971, commenting on the measures he helped have enacted, including two pure food laws, a disillusioned Nader said, "I have no pride of authorship . . . associated with five bills! They are frauds! . . . They're written much better than they're enforced or administered."[2] Yet he remained a popular figure among the public. In 1971 Americans ranked him seventh among their most admired men—an honor tarnished only slightly by the fact that Vice-President Spiro Agnew, later forced to resign over corruption charges, nosed him out for the sixth spot.[3]

By then mainstream book publishers had begun to turn out New Leftish denunciations of the food industry's depredations on American health. Simon and Schuster published Beatrice Hunter's *Consumer Beware! Your Food and What's Been Done to It* (1970) and Jacqueline Verrett's *Eating May Be Hazardous to Your Health* (1974); Grossman did James Turner's *The Chemical Feast* (1970), the report of a Nader Summer Project; Holt, Rinehart and Winston undertook Judith Van Allen and Gene Marine's *Food Pollution: The Violation of Our Inner Ecology* (1971), and Random House published *The Grubbag: An Underground Cookbook* (1971) by Ita Jones, food writer for the

New Leftist Liberation News Service.[4] In 1969, when she first started writing *Diet for a Small Planet* (1971), Frances Lappé thought that she would be lucky to find a small Berkeley publisher. Yet so rapidly did interest in the topic mushroom that she had no problem enlisting Ballantine Books, a large publisher of mass market paperbacks, which eventually sold over two million copies of the book.[5]

Like many leftists', Lappé's interest in food problems had begun with a concern over hunger and malnutrition in the Third World. However, she came to the conclusion that American eating habits were largely to blame for this. The livestock raised to satisfy Americans' lust for meat, she concluded, represented a monumental waste of protein, which could be going to feed the malnourished in the Third World. Beef-eating was particularly wasteful, for cattle were extremely inefficient protein producers, consuming twenty-one pounds of vegetable protein to produce one pound of protein from their meat. Although she did not suggest Americans abandon meat completely, Lappé herself became a vegetarian, something she said was based "more on my feelings than my rationality." It also reflected her New Leftist fear of corporate manipulation. "When I went to a supermarket," she wrote, "I felt at the mercy of our advertising culture. My tastes were manipulated. And food, instead of being my most direct link with a nurturing earth, had become mere merchandising by which I fulfilled my role as a 'good' consumer." As a result, she felt compelled to eliminate the middleman—and, as it were, middle animals—and eat as "low on the food chain" as possible.[6]

By 1973 the organized New Left had virtually disappeared. SDS had self-destructed some years earlier, imploding like a dying sun and releasing a gaseous cloud of bizarre radicalisms. However, although they drifted away from "the Movement," ex–New Leftists had moved into influential positions in academia and the media, as well as in Congress and state governments. But while many of their old political reflexes remained, they were dulled by dillusionment and failure. Many moved away from the search for a better world through far-reaching social and political change and instead turned inward, seeking salvation through personal betterment and changing lifestyles. Concern over food and diet naturally came to the fore. Some took up food preparation itself. In 1972 Berkeley radical Alice Waters opened Chez Panisse, a restaurant whose emphasis on fresh, local, additive-free ingredients made it one of the pioneers in creating a new style of American cooking. The fact that her efforts helped change the eating habits of the rich, not the poor, could stand as a metaphor for the ultimate fate of the whole New Left critique of America's food.

Meanwhile, another early tendency of the New Left, a longing for "community," had set down roots among the younger generations. For the most part, the teenagers of the 1950s and early 1960s had subscribed to the prevailing cult of the family. They thought that "nothing, except the family, deserves their wholehearted allegiance," the social psychologist David Riesman observed in 1959, and aspired to nothing more than "suburban domesticity and a quiet niche."[7] By the mid-1960s, however, large fissures were appearing in this family-centered consensus. Young people were breaking out of conventional family life and seeking other kinds of security or salvation, often in communities of "dropouts" or "hippies" who claimed to be searching for new forms of social organization that would lead to fuller, more satisfying lives.

Many of these ideals were derived from the New Left, particularly the hope of creating "parallel structures" and "loving communities."[8] However, the young people of what was labeled the "counterculture" did not share the New Left interest in theory and politics. Instead, like their nineteenth-century Romantic predecessors, they looked to the heart rather than the mind for inspiration. Like the New Leftists, they condemned industrial America's pressures for conformity, its constraints on behavior, and its apparently destructive effects on health and human relationships, but they thought that these evils could be escaped by returning to the simple rules marked out by Nature, particularly on what to eat. This would not be easy, they thought, for the message from within could barely emerge through the thick crust of poor food habits cultivated by a decadent civilization. "The very fact that this column is being written," said the nutrition columnist for a Santa Cruz, California, underground newspaper, "shows how far we have departed from an intuitive harmony with natural law. We should *know* what we require for radiant vitality, for the body has its own intelligence. However, the conditioned mind, the cultural overlay of cokes and french fries . . . have perverted this infallible knowledge that is within each individual."[9]

Like many others, San Francisco's Diggers, one of the best publicized of the new communities, blamed private property for all that was wrong. They therefore distributed free food (some donated, some stolen) every day in a neighborhood park. "It was always the dream of the white man to live in a natural state, and that's what we got to do," one of them explained. "The Polynesians, the black man, every race has done it but the white man, except for seventeenth-century England."[10] After 1968—when the atmosphere in the two main meccas, San Francisco's Haight-Ashbury and New York City's East Village, began to sour—many counterculturites joined an exodus to farms and communes where they hoped to grow their own food and live simple lives in tune with Nature's rhythms.

One of the oldest of the ideas that reemerged was that diet was reflected in personality. Counterculturites regarded English versions of Feuerbach's famous aphorism "*Man ist was er isst*" (Man is what he eats) as particularly "deep."[11] Certain foods were to be avoided because they produced antisocial or otherwise undesirable behavior. High on most lists, of course, was meat. "When carrion is consumed, people are really greedy," sniffed the leader of a northern California commune.[12] "In every religion I know of," said a Bay Area vegetarian, "people who aim at maximum spiritual growth are cautioned against eating too much meat. It has been suggested that the present frenzied preoccupation with sex (not love) and violence (not strength) is connected with the unprecedented quantity of meat in our diet."[13]

Yet nagging questions persisted. Was Hitler not a vegetarian? (He was not a *real* vegetarian, but somehow the idea persisted.) How did one account for the well-known tribal peoples who, while living the admirable natural life, relished meat? Did not gentle Eskimos, for example, consume enormous amounts of it? "There is no *bad* natural food," said one of the more Talmudic attempts to answer questions such as these. Nature provides different kinds of indigenous foods for different climates. It was only when foods that Nature provides in abundance for humans in one climate were taken out of that context and given to those in another that trouble began:

> While it is very much in accordance with Nature to eat meat in an arctic climate, it violates the natural order to eat meat in a temperate or tropical climate. When a person ignores this order . . . he becomes narrow-minded, materialistic, aggressive, and preoccupied with gold, possessions and machines that kill. . . . If you doubt [this] simply eat nothing but meat for a month and observe what happens to your mentality. A friend tried such an experiment and at the end of three months he had degenerated into an animal. All he could think of was sex and violence.[14]

Others shunned meat for the traditional moral reason, a refusal to extinguish lives for food. But most of those who turned to vegetarianism did so for health reasons. Nature seemed to say that foods low on the food chain were the healthiest ones, and, as Lappé had noted, livestock, which put grains and grasses through a complex process of transformation, did not qualify as such. Of course, health and moral concerns often complemented one another. Being a vegetarian and eating nothing but "natural" foods was "a moral as well as a physical commitment for the rest of my life," explained a California college student.[15]

Vegetarianism also has deep religious/philosophical roots in America. The idea that eating meat stimulated carnality and aggression, for example, was popularized by the early nineteenth-century Protestant reformer Sylvester Graham. But these American roots were too intertwined with those of repressive Puritanism or faith in modern science to inspire the new converts. Instead, they often turned to Asian philosophies for inspiration. "The only nutrition rules we disregard are modern ones," explained the twenty-one-year-old cook for a Boston commune.[16] The most popular of the counterculture eating regimes, the Zen or macrobiotic diet, made its first inroads in the nation's campuses and bohemias in the mid-1960s. The bohemians of the the late 1950s—the "Beat Generation"—had stirred avant-garde interest in Oriental religions, but their food tastes had remained decidedly mainstream. The heroes of one of the Beats' sacred texts, Jack Kerouac's novel *On the Road*, sustained themselves on their cross-country hegira on apple pie à la mode ("because it's nutritious, man").

Much of the culture of the apostles of Zen, on the other hand, centered on an extremely complex system of food preparation, perhaps not unconnected to the fact that their leading light, Georges Ohsawa, lived in Paris. Ohsawa's great innovation was to apply to food the Buddhist idea that enlightenment came from bringing the two life forces, Yin (expansive) and Yang (contractive) into balance. Yin foods were those, like fruit and vegetables, which grew in the summer, while Yang ones were winter foods such as meat, fish, eggs, and caviar. Vitamin C was Yin, while vitamins A, D, and K were Yang. Most alcoholic drinks were Yin, but Scotch whiskey, for some inscrutable reason, was declared Yang. To complicate matters further, the human, an animal, is Yang, so Yin food had to be "Yangized" (through heat, pressure, and/or salt) in order to be eaten. The initial aim, for the entry level diet, was to achieve a balance of five parts Yin to one part Yang.[17]

Although it did permit small amounts of fish and chicken, this regimen was mainly vegetarian, emphasizing whole grains, pulses, and rice dishes. At this stage it posed no danger to health and even offered a number of recipes attractive to non-Zen types. (Ohsawa's Udon noodles in clam sauce or buckwheat noodles gratin, with a sauce of cauliflower, onions, soy sauce, and bonita, could pass muster on many a demanding table.)[18] Some macrobiotic restaurants even managed to attract custom from among the nonbelievers. (In the 1980s it was still necessary to reserve two weeks in advance for a weekend dinner in the one in San Francisco's Fort Mason.) However, following the macrobiotic road meant progressively upping the ante—or rather, reducing the intake—through ten stages, each more restrictive than the last. At the final stage, the diet consisted of nothing but brown rice and no more than eight ounces of fluid a day. (Horace Fletcher, the "Great Masticator" of the early twentieth century, would have been pleased to note that every bite of food was also to be chewed from 50 to 150 times.)[19] At this level, which proponents claimed led to the highest stage of enlightenment—a "natural high," according to reformed druggies—severe protein deficiency could cause kwashiorkor, scurvy, and kidney failure. In one often-cited case, it was said to have caused a death, although this was never confirmed. In any event, few adherents ever reached this exalted, and unhealthy, stage.[20] Nevertheless, an AMA committee condemned it as unhealthy and dangerous, "an extreme example of a general trend towards organic and natural foods."[21] It was reports in 1970 that a number of young people in Berkeley had developed kwashiorkor as a result of the macrobiotic diet that spurred Francis Lappé to finish *Diet for a Small Planet*.[22]

One reason the macrobiotic, organic, and other such diets gained popularity among the young was that they touched a very American nerve: an obsession with food and filth. This may sound incongruous, for the hippies, who were often their most visible apostles, were not known for their concern over bodily cleanliness. But a constant theme in counterculture thinking about food was the necessity to purge oneself of the dirty things modern eating put into one's systems. "The first step in halting pollution is to stop consuming the devitalized, plastic, pseudo foods which turn men into walking cesspools," said an advertisement for a California "natural foods store."[23] We have to "empty ourselves of the garbage so that nature can function unimpeded," said a Santa Cruz nutritionist. "The debris that has to be eliminated is both mental and physical, for the two are inseparable."[24] Barron Bingham, founder of the Back to Eden organic food store in Hollywood, described by *Seventeen* magazine as "a fruitarian with long hair, messianic eyes and a Viking body," told a reporter, "A few years ago in Mexico I realized that my body-temple was unclean. From then on I decided not to eat anything but the purest, most delicate, prettiest fruits." (He also planned a chain of drive-in vegetarian restaurants.)[25] A refugee from seven years in a macrobiotic group said they "hate their bodies. They see them as sick and full of toxins that must be purged."[26]

There was a paradox in all this: While cleanliness in American food culture is usually associated with whiteness, the New Left and counterculture went in quite the opposite direction. As Warren Belasco has noted, they warned against "eating white. . . . Whiteness meant Wonder Bread, White Tower, Cool Whip, Minute Rice, instant mashed potatoes, peeled apples, White Tornadoes, white coats, white collar, whitewash, White House, white racism. Brown meant whole wheat bread, unhulled rice, turbinado sugar, wildflower honey, unsulfured molasses, soy sauce, peasant yams, 'black is beautiful.' "[27] Granola, a mixture of various brown-hued foods, seemed ideal. When students at Yale University hosted a conference of Black Panther supporters in the spring of 1969, instead of the usual hot dogs and Cokes they provided a granola-type recipe of oats, dates, sunflower seeds, peanuts, prunes, raisins, and cornflakes.[28] Third World peasant foods were also highly regarded. Liberation News Service's food writer, Ita Jones, recommended Latin American and Asian recipes for New Left political reasons—as a way to express solidarity with their national liberation movements.[29] However, to counterculturites their good taste, healthfulness, and economy represented an affirmation that the cumulative wisdom of the world's poor brown, yellow, and black folk cultures was superior to that of the rich, white, industrial world.

The New Left and counterculture also came together at a number of other points on the food front, sometimes literally. Stores selling macrobiotic, organic, and natural foods often ended up merging with New Leftists' food cooperatives set up to bypass supermarkets and other capitalist middlemen. "Are you tired of eating crap?" said a co-op organizer in Oshkosh, Wisconsin. "At high prices too? Are you tired of reading lists of chemicals and preservatives on every packaged food you buy? Are you tired of buying old meat, low quality produce and rotten fruit? If you know what I mean, you must do your shopping in a supermarket, right?" Although they often aspired to provide "real" foods in bulk at reasonable prices to the poor as well as to the middle class, it was the college-educated who tended to be the backbone of the successful ones. The Free People's Store, a typical one in Rochester, New York, grew out of the Friendly Vegetable, a vegetarian co-op that operated out of people's homes, and the leftist Genesee Co-op. By late 1971 it had moved into a large old fire station, where it offered "food you won't find in a supermarket," mainly "organically grown" foods sold in bulk out of large bins. Customers were reminded that by scooping their brown rice into paper bags and buying their oils "on tap" they were freeing themselves from corporate processing, packaging, and manipulation.[30]

Predictably, given the deep hostility to the idea that food should be profitable to anyone but small farmers, economic success proved to be more difficult to explain than failure. The Buddhist Fred Rohe's New Age Foods, which expanded from the Haight to three other Bay Area locations and even mulled over franchising, was often condemned for having gone capitalist.[31] The purists who opened the Willamette [Washington] People's Grocery Store in January 1969 derided the thriving co-op in Berkeley for "having turned into a hip supermarket." Apparently, Berkeley's transgression was that it sold some packaged foods. The People's Grocery Store would sell only unpackaged items like huge wedges of cheese, barrels of beans, and "real peanut butter in bulk."[32]

"Real peanut butter" was one of the cult foods of the counterculture: In its "natural" chunky form, this protein-rich fruit of the tropical earth symbolized Nature's simple wisdom. "Genuine, old-fashioned, unhomogenized peanut butter," said the counterculture theorist Charles Reich, was "the very symbol of the world that has enjoyed technology and transcended it." However, when the nuts were ground until smooth and mixed with salt, sugar, and hydrolized vegetable oil, it epitomized all that was debased about America. The greatest abomination, of course, was premixed peanut butter and jelly.[33] (Twenty years later, "real peanut butter" devotees would be shaken by reports that it was the natural kind, which could contain deadly aflatoxins, that was potentially dangerous.)

By 1970 "back to Nature" had come to mean "back to the farm," as thousands of young urbanites sought out scrub land in passed-over rural arcadias to set up communes. Despite J. I. Rodale's striking resemblance to Leon Trotsky, social and political radicalism was not his cup of tea, and *Organic Gardening* was never completely comfortable with its long-haired new readers. New periodicals thus supplemented his in giving these latter-day pioneers practical advice on how to grow and cook their natural food. *Mother Earth News*, the most successful of them, mixed self-described "old-timey" recipes with suggestions for macrobiotic and organic foods. It mailed out its own root beer base and yogurt culture and was particularly partial to whole grains and home-baked bread. To help make bread properly it sold hand-cranked grain mills of a nineteenth-century design, cast-iron stoves and ovens, and sourdough starter. Significantly, it made little effort to cater to vegetarian sensibilities; readers were told how to cure beef and venison and "slaughter hogs the way it was done in '49."[34] This reflected the surprising fact that many communes were not vegetarian. One investigator found only half of those he surveyed to be so, and those were mainly on the West Coast. Those eastern communes that went easy on meat did so mainly for economic reasons.[35]

The new interest in home production of food was a reflection of the forces domesticating the counterculture. By 1971 many young people had already paid terrible physical and psychological prices for excessive drug use; others, forced to face economic reality, were trying to scrape together existences in more settled fashions. The days of aimless drifting from "crash pad" to "crash pad" while living off of Ritz crackers, Cheesies, and Oreos were fading. Now even the most political of the underground papers began running recipes for whole grain breads, vegetable soups, bean stews, and oatmeal cookies. Some papers concentrated on recipes for foods that were distributed by government surplus food programs, such as corn meal and cheese.[36] Alice Waters contributed French-inspired recipes using local natural foods to the *Berkeley Barb*. In Washington a number of hippies set up a catering service called Mother Nature on the Run, which offered legume soups, homemade macrobiotic breads, vegetarian stuffed eggplant, and fruit mixed with homemade yogurt and organic honey to "people who want to impress their friends with something different." (While it managed to elicit some interest among the city's young professionals, its biggest catering job was, predictably, an Environmental Action gathering.)[37]

On the left, only the new women's liberation press resisted the burgeoning interest in food and cooking, in large part because it still reflected traditional ideas about the division of labor. Indeed, from a feminist perspective, the counterculture was riddled with male chauvinism. It fought for free expression of sexuality by attacking restrictions on pornography. Underground papers were often festooned with ads for topless bars, sex movies, and peep shows. The *Berkeley Barb* regularly featured pseudo-reverential photos of women with large, bare breasts. The essays and poetry in the *East Village Other* seemed to extol fellatio almost as much as they did drugs. Counterculture domestic arrangements also often left much to be desired from a feminist viewpoint. Only women did the cooking among Zen adherents, something that was justified by declaring them to be more Yin than men.[38] "We have come full circle and are doing the things our mothers did," admitted the cook for a Zen commune in Boston.[39] The Paradox, a popular counterculture restaurant in New York's East Village, advertised "real good food . . . cooked by real women," hardly auguring a great new era of revised sex roles.[40]

This inability to break free from traditional gender roles alienated the emerging women's movement—which was coming to play an important role in shaping attitudes toward home, family, and cooking—and contributed to the rapid marginalization of the counterculture. Equally important in limiting its impact on food habits, though, was one of its great initial attractions: dope. The passing parade of drugs of choice brought with it a kaleidoscope of attitudes toward food, none of which sat well with the ideal of eating "natural." Marijuana enhances the taste of sweet foods, and pot smokers were particularly enamored with ultramanufactured foods such as Cool 'n

Creamy—an artificial chocolate pudding with an extraordinarily long list of chemical ingredients—Oreo cookies, and Cool Whip (declared the most impressive new processed food of 1970 by the Grocery Manufacturers of America). LSD, which gained favor in the late 1960s, transported people into imaginary worlds remote from pedestrian items like food. ("Who thinks of eating when there's LSD?" said Paul McCartney, recalling how he "nearly perished" when he was a Beatle.)[41] As for the final drugs in the cycle, amphetamines, or "speed:" As a generation of 1960s dieters could have predicted, these appetite suppressants caused "freaks" to lose all interest in food and, in many cases, to waste away.

Intellectuals such as Charles Reich and Theodore Roszak, who tried to put a rational gloss on the "freaks' " ideas, tended to ignore the central reality, which is that drugs practically cut off communication with the "straight" world. Because it was virtually impossible to do any sustained reading or writing when under their influence, communication had to be oral and was usually unintelligible, particularly to those not under the same influence. How does one classify vegetarians who refused to eat meat because it emitted "bad vibrations?"[42] "I haven't read a whole lot in the past year," a Haight-Ashbury "head" with a philosophy B.A. from Berkeley told Nicholas Von Hoffman. "Reading is hard when you're taking dope for some reason, but man, I like to rap."[43] Limited reading meant that there was little to communicate outside of personal experience and anecdote. With few exceptions—such as *Rolling Stone*—the counterculture press was full of self-absorbed testaments that were barely comprehensible to the unconverted.[44] Yet, as we have seen, by the late 1960s other critics of American food habits were talking in sophisticated scientific, medical, and politico-economic terms. The counterculture simply could not keep up.

It was the more conventional critics, then—the New Leftists, Naderites, and liberals—who, translating the growing body of critical medical and scientific literature into lay terms, provided much of the ammunition for the major assault on American eating habits and the government, industry, and scientific establishments that seemed to perpetuate them. In January 1973, for example, Consumers Union, publishers of *Consumer Reports*, joined with the Environmental Defense Fund, the Consumers' Federation of America, and Nader's Center for the Study of Responsive Law to assail the FDA's refusal to ban sodium nitrate and nitrites, which were believed to cause cancer when mixed with other substances, from foods.[45] Using a standard New Left tactic, the Naderites and radicals ferreted out evidence of the food counterpart of the "military-industrial complex." Liberals had hitherto regarded FDA administrators as well-meaning but underfinanced handwringers who were ill equipped to ride herd on processing giants. Now they were said to be industry-suborned wimps who spent their time at the agency cozying up to the food processors. Like Pentagon officials, they would then depart through the "revolving door" to "deferred bribes"—lucrative jobs in the companies they had just been regulating.[46] They also attacked the USDA-funded complex of experiment stations and extension services in the nation's land grant universities, which for almost one hundred years had been lauded for their leading role in making scientific research accessible to farmers, as mere servants of the agrocorporations. The Agribusiness Accountability Project calculated in 1972 that of the 6,000 scientific person-years of research conducted at land grant colleges "a mere 289" went into the Agriculture Department's categories of "people-oriented research"—that is, programs to help the rural poor. The rest went to research on mechanization, which drove the poor off their farms, and genetic and chemical research to enhance productivity, which endangered the health of the nation. "Money," they concluded, "is the web of the tight relationship between agribusiness interests and their friends at the land grant colleges" who had "a long list of satisfied corporate customers."[47]

During the lengthy debate over vitamin supplements, researchers funded by vitamin producers had regularly peppered the nutritionist establishment with charges that they were in the pay of the food processors.[48] Now, a new breed of dissenting research scientists expanded on charges such as this, joining the leftists and Naderites in the denouncing the government-science-industry nexus. They charged that the food industry had bought off prominent scientists with lucrative research and consultancy contracts, creating an "unholy alliance—science and the food industry" and putting at their disposal a clique of "Hertz Rent-a-Scientists."[49] These suborned scientists, in turn, were said to control the National Academy of Science–National Research Council's Food and Nutrition Board and Food Protection Committee through a series of "interlocking directorates."[50] Dr. Jacqueline Verrett, the FDA researcher who blew the whistle on cyclamate on national TV, charged that many of the experts on the NAS–NRC advisory panels such as the one which had refused to ban cyclamate were financially beholden to the very industries or institutions whose products they were assessing.[51]

Dr. John Olney told the McGovern committee that the NAS–NRC committee that rejected his warnings that MSG caused brain damage had ignored competent scientists and solicited the views of people who were not qualified to judge his research. Those who disputed his findings were "almost exclusively from a certain element of the scientific community; a group of individuals who maintain close ties with the food and drug industries; individuals . . . who function as a team and swing into action whenever a food safety issue arises. Some members of the team specialize in generating made-to-order evidence, while others are asked—by FDA through NAS—to evaluate the evidence." They were, he charged, "individuals whose credentials are in science but whose loyalties are with industry."[52]

Later in the decade, in his book *Everything You Always Wanted to Know About Nutrition* (successor to his best-seller *Everything You Always Wanted to Know About Sex but Were Afraid to Ask*), David Reuben amplified these charges into an allegation that the NAS–NRC Food and Nutrition Board was "organized and owned by food manufacturers and vitamin sellers" (a statement that legal action forced him to admit was "a factual inaccuracy").[53] What a change these public charges exemplified! Before the late 1960s they may have been voiced, but it would have been in sotto voce grumblings over coffee in hospital cafeterias or drinks at faculty club bars. For respectable research scientists to publicly accuse other scientists of allowing financial considerations to influence their scientific judgments was simply unheard of.

The new, charged atmosphere helped instill grave doubts about the wisdom and even the probity of the leading lights of the nutritional science establishment—people whose advice had always been treated with reverence by government, industry, and the media. Most remarkable was the assault on the nation's most prominent nutritionist, Frederick Stare, founder and head of the department of nutrition at Harvard University's School of Public Health. An M.D. with a Ph.D. from the University of Wisconsin's biochemistry department—renowned for its vitamin research and lucrative ties with the dairy industry—he had been the first editor of the industry-supported Nutrition Foundation's scientific journal, *Nutrition Reviews*, founded in 1942. Over the years he played a major role in the recurring offensives against "cranks" and "quacks" and was one of the key organizers of the 1961 National Congress on Medical Quackery. He was also a vocal supporter of the FDA's assault on vitamin sellers. Handsome, fit, and bespectacled (he bore a vague resemblance to Clark Kent), he was a regular guest on television and radio interview shows. Twice a week, he turned out a syndicated column on food and health, which ran in 123 newspapers. There, and in other forums, he continued to warn that quacks and hucksters were trying to arouse false fears about food safety and nutrition and divert people from mainstream medicine into useless or even dangerous therapies. "I would like to dispel the notion that our foods are increasingly tampered with in undesirable, unhealthy ways," he wrote in *Life* magazine in 1970. "Nutritive qualities have actually increased in many of our foods." There was nothing wrong with the nation's food supply: A healthy diet was available to all who would learn to choose the right foods.[54]

Stare seemed unfazed by critics' charges that food industry grants had made him an industry apologist. He testified to the Senate against banning MSG and appeared as an expert witness on behalf of six major food trade organizations and processors at the FDA hearings on vitamins.[55] But over the next few years Stare's links with industry, the foundation of his successful career, came to be his Achilles heel. His testimony before Congress and the FDA on behalf of such enterprises as Kellogg's, Carnation, and the Sugar Association was easily tainted by their grants to his department.[56] When, during Senate hearings on dangerous fad diets, the faddist Dr. Robert Atkins was confronted with Stare's condemnation of his high-protein diet as "nonsense," the agile Atkins replied, "I think the Harvard School of Nutrition depends on outside funds and this could produce a bias."[57] In 1976 came the unkindest cut of all, a special report of the Center for Science in the Public Interest, labeled "Professors on the Take." It held Stare mainly responsible for Harvard's nutrition department being "riddled with corporate influence." He concealed these corporate connections, it said, while defending sugar and the other additives his clients used.[58] Stare responded by setting up an independent foundation called the American Council for Health and Nutrition "to investigate chemicals in our society." It pledged not to accept contributions from anyone with a commercial interest in the topics being investigated. But Jacobson's dogged researchers would not let go of the prey. They dug up evidence of contributions from such presumably nutritionally nefarious sources as Coca-Cola, the National Soft Drinks Association, and International Fragrances, Inc.[59]

As was often the case with the more zealous of the critics, "Professors on the Take" had cast a very wide net. It had even included Mayer, who was hardly uncritical of processors, in its indictment. Although he had joined with Ralph Nader and Francis Lappé in a teach-in during the previous year's Food Day, he had disassociated himself from some of their more extreme charges, which smacked of ritual corporation-bashing and left-wing conspiracy-theorizing.[60] The report accused him of concealing his membership on the boards of two chemical companies that produced "additives, fertilizers, pesticides," and even "synthetic fruit bits."[61] But Mayer had not really joined Stare and the other wholesale defenders of the food industries. Unlike Stare, he attacked snack foods and questioned the nutritional value of many processed foods.[62] Most important, he distanced himself from Stare on what was perhaps the most contentious nutritional issue of the day, the sugar question. Mayer was on the side of those who were now loudly denouncing American overconsumption of sugar, while Stare professed to be unconcerned about the matter.[63]

Why sugar should have become so reviled provides an interesting insight into the forces that merged to change American attitudes toward food in the late 1960s and early 1970s—in particular, how reputable science now buttressed the traditional forces of faddism and the new political activism. Hitherto, refined sugar's whiteness had often symbolized its purity and healthfulness. Indeed, in seventeenth-century Europe its very whiteness was regarded as proof of its superiority over other sweeteners. It was thought to be more "civilized," pure, and wholesome—"marvellously white" according to one of the era's top medical scientists. Later, Nature-besotted philosophers such as Rousseau and the French Encyclopedists associated it with lush, fragrant forests and noble savages.[64] In the United States, late-nineteenth-century innovations in refining made it affordable to the masses, who rapidly made it the country's favorite sweetener.

There had always been countercurrents of suspicion, however, which were fed by the post–World War I ascendancy of the Newer Nutrition. As a growing number of vitamins and minerals were found in a host of other foods, the fact that sugar was bereft of these saving graces became increasingly obvious. By the 1930s those who bemoaned undernutrition in America regularly singled out the nation's sweet tooth, and particularly its ever-rising consumption of refined white sugar, as a major culprit. It was said to cause obesity, diabetes, and tooth decay. But for the most part these concerns were confined to health food circles and the small consumers' movement. Among the general public they were more than counterbalanced by positive thoughts about sugar: that it tasted good and was an excellent source of quick energy. Wartime sugar hoarders were undeterred by any thoughts that they might be squirreling away a dangerous substance; the government told them it was a morale booster. This insousiance had prevailed into the 1960s, thanks in part to industry campaigns hailing it as "Nature's miracle food," the perfect source of "quick energy."[65] Indeed, two national Roper surveys found Americans more favorably disposed to sugar in 1967 than in 1945.[66]

However, by 1967 a wave of sugar-phobia was already underway. The substance's most prominent scientific critic was John Yudkin, a professor of nutrition at the University of London, England, who made a number of appearances in the United States. Concern over heart disease was growing, and Yudkin's claim that sugar was its major cause gained considerable attention. His statistics linking this century's rising levels of sugar consumption with parallel increases in deaths from heart ailments seemed, to many of the laity at least, convincing. Since this disputed the charges that saturated fats were the villain, it is not surprising that the egg industry funded at least one of his American tours.[67] He subsequently added diabetes, arthritis, cancer, mental illness, dental caries, and a dozen other ailments to the indictment.[68] The English edition of his book bore the chilling title *Pure, White, and Deadly*.[69] Others amplified the charges, blaming sugar for practically every conceivable ailment and evil, including coronary thrombosis, hypoglycemia, impotence, scurvy, ulcers, strep throat, hemorrhoids, hair loss, menstrual cramps, and varicose veins. Responsibility for high rates of drug addiction, alcoholism, highway accidents, and suicide was also laid at its door.[70] J. I. Rodale thought sugar "caused criminals." He warned that Coke drinkers would become sterile and told a reporter that he would live to be a hundred unless he was "run down by a sugar-crazed taxi driver."[71]

The growing suspicion of refined and processed foods helped turn sugar's whiteness into a cross to bear. In the new age of reverence for darker foods, brown sugar and honey simply *looked* healthier. Its pleasurable taste now worked against it. Its detractors tapped into the nation's puritanical streak by portraying sucrophiles as fallen souls who had sacrificed their health and morals to the pursuit of pleasure. Scientists who did not agree with him, said Yudkin, were blinded by their own cravings for sugar.[72] Many experts, such as Mayer, disagreed with Yudkin's contention that sugar itself caused atherosclerosis and found most of the other arguments far from convincing. However, they still condemned sugar because it displaced foods made up of more nutritious calories; its "empty calories" (a deliciously catchy phrase) thereby contributed to obesity and its partners, diabetes and heart disease.[73]

Nevertheless, as late as 1970 sugar still teetered at the edge. Most Americans seemed to regard it as a pleasing food whose good and bad points more or less balanced each other out.[74] A key event in pushing it into the abyss of foods to be dreaded was a concerted attack on the most popular dry cereals. It was led by Robert Choate, an engineer who had already made a name for himself in hunger lobby circles as Mayer's assistant at the 1969 White House Conference. A scion of an old New England Republican family, Choate had moved from Boston to Phoenix, where he had done well in business. Exposure to the poor living conditions of the area's Hispanics and blacks led him to rediscover his family's Republican reform heritage. The fact that the tall, bearded, loner bore an eerie resemblance to Abraham Lincoln did him no harm in Republican circles, and his position papers on hunger and nutrition for Nixon's presidential campaign helped him gain a post as a consultant to the White House and the Department of Health, Education, and Welfare. It was in the latter guise that he prepared the analysis of the nutritional content of breakfast cereals that caused a national sensation when presented to the Senate Commerce Committee on July 23, 1970.

Choate had calculated how much of each of nine nutrients there were in a serving of each of sixty dry cereals. He then determined the percentage of the RDA of each nutrient this would provide and turned this percentage into a score, from 0 to 100. The total scores, out of a possible 900 (called "optimal nutrition content"), were then tabulated and the cereals ranked from one to sixty—from Kellogg's Product 19 at the top, scoring 700 out of the possible 900, down to Nabisco Shredded Wheat, which, to many people's astonishment, scored only 10. Americans love rankings, and Choate's table naturally grabbed headlines across the nation. Few could resist consulting it to find where their breakfast favorite stood.[75] The result was widespread disillusionment, for many of the most popular cereals, such as Kellogg's Corn Flakes and Rice Krispies, ranked near the bottom.[76] In vain did cereal manufacturers' scientists point out that the chart left much to be desired, that it made little sense to rank all nutrients equal in importance, and so on. Choate's subsequent disclosures that laboratory experiments seemed to show that rats fed on cereal boxes had fared better than those fed the cereals themselves provoked even more apoplexy in the cereal industry.[77]

Choate's condemnation of the cereal companies was intimately linked with the assault on sugar, for he charged that they replaced nutrients with sugar in order to hook younger children on their products. Television commercials played an important role in this, he said. Kellogg's Sugar Frosted Flakes, for instance, fifty-eighth on his chart, had the third largest advertising budget. "Our children," he said "are being countereducated away from nutrition knowledge." They were "deliberately being sold the sponsor's less nutritious products" and "being programmed to demand sugar and sweetness in every food."[78]

This idea, that fondness for sugar was an acquired rather than an inherited trait, was a favorite of sucrophobes, who also attacked baby food producers for putting sugar in their products. Yudkin condemned scientists for standing by while food processors and producers "thrust sugar down innocent and uncomplaining throats."[79] Even Jean Mayer warned that for the young it was just as addictive as tobacco and alcohol.[80] A California doctor said it caused hyperactivity in children—Feingold's disease—providing many thousands of parents with a convenient explanation for why their offspring were so poorly behaved. Michael Jacobson made the sugary road to perdition sound like the supposed stages in drug addiction, with "soft" baby foods in the role of "soft drugs": "For a child whose taste buds were initiated in blueberry buckle, raspberry cobbler and other sweetened and salted baby foods," he wrote, "the step to artificially colored and flavored sugar-coated breakfast 'cereals' is a small and natural one."[81] William Dufty, author of the sensational best-seller *Sugar Blues*, called it "the white plague." His first taste of it, he said, led him down the "road to perdition," where, in true drug-addict fashion, he even stole from his mother to satisfy his craving.[82] The target thus widened from sugared cereals to all sugared foods for children. Choate attacked television commercials for candy-coated vitamin pills.[83] Parents' groups pressed the Federal Trade Commission to restrict commercials that urged children to eat sweet foods and force the networks to grant them free air time to warn parents against ads for sugared foods.[84]

The effects of these campaigns on public opinion were palpable. A 1975 survey showed a great increase over 1967 in the proportion of people who thought sugar harmful. It was now thought to be antithetical to nutrition. "They must take out all the nutrition when they put in all the sugar" was a typical housewife's comment. There was now particular concern about its effect on children. Instead of quick, useful energy, it was now thought to give them "an induced sense of energy, a high of sorts."[85] By then, although there was still no proof of any link between sugar and diabetes, let alone heart disease, that one existed had become accepted with little question in lay and academic circles.[86]

The attack on dry cereals also added fuel to the fire already burning under concerns over processed foods. The very things processors' advertising had boasted about—that they were "shot from guns" or made to "snap, crackle, and pop"—now became symbols of how they had denatured and denutrified the foods they were made of. Yet while crusaders like Nader, Choate, and Jacobson hoped that the exposés would translate into demands for strengthening the regulatory system, the public reacted in ways more consonant with the American individualist tradition, taking matters into their own hands. In one sense, they did it literally. Criticisms of "plastic" white bread led to a boom in home bread-baking. James Beard, author of a popular book on how to do it, wrote that the new home bakers aimed "to fill the bread with vitamins and *health*." Consequently, he noted, they had "a tendency to acquire as many different flours as possible and incorporate them all into a single loaf, without thought for texture, for crumb. . . . The coarser it is, the healthier, some people think."[87] Also, as we have seen, they began administering vitamins to themselves on an impressive scale and tried to seek out alternatives to the foods they thought were harmful.

As Claude Fischler has pointed out, modern humans had now negotiated an almost complete about-face from the food attitudes of their forebears. As omnivores, we have natural anxieties about nature's own foods, a skepticism that allowed our ancestors to discover which ones will kill or sicken us. Wary of Nature's perils, they sought to "civilize" foods by cooking, seasoning, and then processing. Now, however, peoples' fears were directed not at nature but at its opposite: Industry. The food-processing plant had come to embody "man's Promethean impudence," the place where he "challenges Creation's dark forces and devotes himself to tasks which at any moment threaten to unleash that modern image of the Impure: the artificial."[88]

Clearly, ideas such as these represented a grave threat to the bottom lines of those who owned the processing plants. How could cereal manufacturers, for example, continue to convince a now-wary public to buy grains that had been milled, boiled, mashed, rolled, extruded, roasted, dried, and even—yes, it was true—shot from guns? Yet the danger was soon averted, for the giant corporations—whose "hegemonic" forces the New Left decried in every other sphere of American life—proved just as adept at co-opting (another New Left term) their critics in this terrain as in the others.

CHAPTER 13

✗ ✗ ✗ ✗ ✗ ✗

Natural Foods and Negative Nutrition

In late 1969 *New York Times* reporter Sandra Blakeslee thought she saw the shape of things to come: "Some even feel," she reported, "that it may be time to take seriously what many members of the younger generation, especially the hippies, have been saying for quite some time: You Are What You Eat."[1] Before the next year was out, this was indeed happening. Calls for a return to natural foods resonated far from the hippie enclaves, striking sympathetic chords among the kind of thoughtful middle-class Americans who read the *Times*. By 1972 the "straight," conservative *National Review* was extolling "natural foods."[2]

The four leading cereal makers, who dominated 80 percent of that market, picked up the "natural" beat with breathtaking speed. This was hardly surprising, for their industry was a peculiar one in which the traditional leading brands, oriented toward no specific demographic group, had to be regularly supplemented with new ones aimed at children. This meant that manufacturers were used to being faced with a segment of the market that, as they progressed through childhood and adolescence, was constantly being replaced by new, younger consumers subject to different fads and whims. It was thus a market that even in normal times demanded new products—or at least new packaging—at a prodigious rate.[3] With the Choate charges ringing in their ears, product development specialists raced to their labs, where they revived the nostrum of 1940, fortification, and began injecting their nutritionally vapid products with nutrients their constituent parts had never contained.

Within a year even Choate admitted that forty-five of the sixty cereals he had analyzed had improved their nutritional content, some of them quite dramatically. Junky-sounding products such as Fruit Loops, Apple Jacks, Cocoa Krispies, Sugar Pops, Puffa Puffa Rice, and Sugar Frosted Flakes were now nutritionally respectable. Later, he acknowledged that most of the worst ones had now been reformulated. Even Corn Flakes, which Kellogg had advertised as "one of the few things around someone hasn't tried to change," was being bolstered with an array of additional nutrients.[4] Frederick Stare, who had denounced Choate's claims as "grossly misleading" and "absolutely meaningless" and had often argued against fortification, now helped Kellogg's fortify Special K with additional nutrients and boost its protein content.[5] General Mills, sensing the beginning of a protein race, whipped out Protein Plus, whose protein content topped that of Special K and its other competitors by a wide margin.[6] The Quaker Oats entry made no bones about its health-related origins—it was called Life.

Some of the most successful of the new cereals were versions of the best-selling health food of the late 1960s, granola. This sweetened mixture of roasted grains, nuts, and dried fruits, initially an object of media jokes and derision, had become the most popular item among the new, younger patrons of health food stores. (One bearing Adelle Davis's signature was one of the fastest-moving brands.) "People are hypnotized by its sweet, icky taste," reported a New York health food store owner.[7] In 1974 the Quaker Oats version of granola, Quaker 100% Natural Cereal, became the first new brand to break into the top five cereal sales leaders in a quarter of a century. General Mills countered with Nature Valley; Kellogg's came up with Country Valley; Colgate-Palmolive went in the opposite direction—to the mountains—for inspiration for Alpen; Pet made a feeble double entendre on health and its midwestern roots with Heartland.[8]

Still, many food industry executives refused to be panicked into changing course. They stuck by the old industry adage that nutrition does not sell food, that price, taste, convenience, and packaging are all more important.[9] *Fortune* magazine saw little future for natural foods. Consumer demand for convenience and the unrelenting drive toward national distribution of foods made the use of more preservatives and processing inevitable, it said. "The very existence and salability of many new foods is possible only because of chemicals that preserve, stabilize, leaven, thicken, emulsify, or contribute color, taste, or nutrients."[10] Convenience would remain the name of the game for the foreseeable future, the head of Corn Products said in April 1969.[11] *Food Processing* pointed out that 40 percent of marriages involved teen-aged brides "with relatively meager home making skills [who] comprise a huge market for frozen and canned foods—particularly frozen-prepared main courses, and entrees with 'built-in' butter-, cream- and mushroom sauces."[12] Ads for the Staley Manufacturing Company, makers of "STA-O-PAQUE" modified starch, featured one of the young brides processors banked on. Standing by a fireplace, dressed in slacks and playing a guitar, with candles glowing and two cocktails sitting on a coffee table beside her, this "new kind of mom" had prepared an "elegant dinner tonight: Pre-mixed cocktails and 'pop-in-the-oven' hors d'oeuvres . . . frozen asparagus, packaged hollandaise, frozen roast turkey with giblet gravy, heat 'n serve rolls . . . for dessert, frozen peach pie or canned pudding, with pre-whipped topping . . . instant coffee. And it's all delicious."[13]

In 1973 an article in industry-supported *Nutrition Reviews* had conceded that nutrition had been a minor consideration in most new product development.[14] This was understandable, for industry wisdom still had it that only one subgroup of the population was sufficiently preoccupied with health to buy products promoted on the basis of nutrition: older people "hoping to stay on the young side."[15] The poor certainly seemed uninterested. Hardly any of them bought the specially formulated high-protein flour Pillsbury developed for them in 1971. Skeptics could also take heart from well-publicized setbacks for the organic and natural foods advocates. The New York City revelations that organic foods were often faked and grossly overpriced were widely, and gleefully, publicized. J. I. Rodale did not, as he predicted, live to be one hundred. In 1971, at age seventy-one, his life was snuffed out—not, as he feared, by a sugar-crazed driver, but by a heart attack, while taping an interview for the Dick Cavett show. In 1974 it was confirmed that, despite her cancer-preventing diet, Adelle Davis had contracted that dread disease.[16] Traditionalists also noted that, when food prices skyrocketed in 1974, sales at recently opened supermarket health food sections dwindled.[17] A General Mills product development specialist put it succinctly, if rather bluntly: "You can't sell nutrition," he said. "Hell, all people want is Coke and potato chips."[18]

Industry leaders even bet that Americans would not mind if those potato chips were processed practically beyond recognition. Procter & Gamble mounted an enormously expensive assault on the $1.5 billion potato chip market with Pringles—self-described "newfangled" disks—which could hardly have been more highly processed. Because chips' short, six-week shelf-lives frustrated attempts to produce, distribute, and market them on a national basis, the chip market was dominated by local and regional brands. Pringles, on the other hand, were made from dehydrated potato mash mixed with mono- and di-glycerides and butylated hydroxyanisole, which made them well-nigh eternal. Perfectly round so that they could be stacked in containers shaped like tennis ball cans, they could be shipped anywhere from their automated Tennessee production plant with no fear of breakage. But neither P&G's marketing clout, which gained them crucial supermarket shelf space, nor its massive advertising budget (fifteen million dollars was spent on Pringles' glittering debut alone) could overcome suspicion of what competitors called this "fabricated" product. Within a year after their introduction in 1975, they had failed to capture even 10 percent of the market and were being labeled a "washout" and a "bomb" by industry experts, a sign that consumers preferred the natural to the artificial. (Others thought taste was Pringles' undoing, pointing out that the most common reaction among those who tried but rejected them was that they tasted "like cardboard" or, as a P and G executive later recalled, "more like a tennis ball than a potato chip.")[19]

By then many in the industry had already concluded that the "natural" wind was not a mere squall that would soon blow over. General Foods, which produced foods under four hundred different brand names, bemoaned what it called "eroding consumer confidence in the food industry" and mounted a campaign to explain to the public what was obvious to the industry: that foods could have the keeping qualities, convenience, flavor, and appearance they cherished only because of additives, processing, and dabs of artificial flavor and color.[20] But it had already joined the other conglomerates who were trimming their sails to catch the new breeze. Jones Dairy Farm offered sausages devoid of "unnatural" preservatives; Dannon boasted that its yogurt had no chemical additives; Borden's even test-marketed organic tomato juice.[21] For almost a hundred years the bread industry had been in a spiral of increasing concentration of ownership, larger production facilities, and greater use of additives that turned out breads of blinding whiteness, bland taste, and plastic texture. Now, its giants took note of growing companies such as Pepperidge Farms and Arnold's turning out "home-style" breads and rapidly gobbled them up, using them as platforms for expanding into healthier-looking and tastier lines. New high-fiber breads such as Fresh Horizons—loaded with enough indigestible cellulose to make the *New York Times* Worst Foods of the Year list in 1976—joined them in squeezing aside old favorites like Wonder Bread on the supermarket shelves.[22]

This nimble response to public concerns caught New Left and liberal critics quite off guard. They were enmeshed in the idea, popularized by the liberal economist John Kenneth Galbraith, that corporations created rather than responded to consumer demand. "The food industry . . . advertises those products that make the most money," explained a radical feminist health book. "Unfortunately, we, as well as our children, have succumbed to this advertising and have in many cases adapted our tastes to the foods best suited to mass production, rapid turnover and longer shelf-life. In other words, we have been conditioned to like the foods that give the food industry the most profit."[23] Mesmerized by this idea that corporate profits depended on creating artificial demand for overprocessed and denatured foods, they watched in disbelief as the processors grabbed the "natural" banner from their hands. Could Kellogg *really* be responding to its critics by producing natural, nutritious products? The answer seemed to be a definite no, but exactly why was never very clear.[24]

Yet, as the giants soon showed, the word *natural* could be a natural money-maker. There were also magnificent opportunities in the companion concepts of "less" and "light" in the growing market for sugar-reduced, saturated-fat-reduced, sodium-reduced, or simply calorie-reduced foods. Indeed, people seemed prepared to pay more for foods in which relatively expensive ingredients had been replaced by cheaper ones. Of course, it is ironic that the processors were responding in part, at least, to the radical critique. It is even more ironic, though, that the food industries' promotional budgets in turn helped stimulate demand for natural foods. Food companies had long since become the nation's major sources of nutrition information. They spent infinitely more on advertising than governments spent on teaching people about nutrition. Coca-Cola's advertising budget alone approximated total federal government spending on nutrition education.[25] Furthermore, much publicly funded nutrition education, particularly in the schools, was highly dependent on industry-produced instructional materials.[26] The fact, then, that by 1977 more than one-quarter of the food advertisements in women's magazines involved "natural" appeals—quite something when one realizes that the alternatives were such popular appeals as price (i.e., discount coupons), convenience, status, and beauty—indicates the extent to which the industry itself was now popularizing "natural."[27]

The effect of the natural trend on consumers was measurable. In 1975 the manager of the chemical division of Hoffmann–La Roche told processors that, whereas nutrition had come in fourth as a consideration among food purchasers it surveyed in 1970, by 1974 it had risen to second. (Price, which had assumed extraordinary importance due to inflation, was first.)[28] In 1977 market researchers told the food industry that the word *natural* was the most convincing sales claim on a food package. Almost 60 percent of consumers thought that the name connoted a food that was healthier, safer, and better for them. Among other words cited as convincing were *organic, no chemicals, pure, real*, and *no preservatives.*[29]

By then much of the industry had joined in a mad scramble to repackage and reformulate products to make them seem more natural. The Nielsen Company reported that from five to seven thousand new grocery items were being introduced each year, most of them reformulated and rebranded versions of old ones.[30] Shoppers now walked down supermarket aisles flanked not only by "100 percent natural" cereals but also by natural potato chips, beer, deodorant, and even dog food with "natural beef flavor."[31] What this really meant in terms of additive-free food was questionable. Tang, one of the more successful new artificial products of the time, advertised its "Natural Orange Flavor" even though it was made up of sugar, citric acid, maltodextrin, calcium phosphate, potassium citrate, artificial flavor, cellulose, xanthum gum, artificial color, and BHA. "Natural" meant that it also contained a small amount of flavoring derived from real oranges.[32]

While they may have been commercial successes, the reformulated and repackaged foods hardly seemed to dent mistrust of the food industries. "Several years of anti-processed food propaganda is having an impact on consumer buying patterns, especially on upper-income opinion leaders," warned an editor of *Food Engineering* in 1975. A consumer survey he helped conduct showed that "the fact that processed foods are believed potentially dangerous was very much on the minds of articulate survey respondents." They now wanted foods made with "natural ingredients" and "basic techniques."[33] FDA experts were chagrined to find that almost 60 percent of shoppers thought (mistakenly, said the experts) that foods prepared "from scratch" were more nutritious than the same food bought canned or frozen.[34] A professor of food science told food processors that one reason 81 percent of shoppers doubted that supermarket food was good for you was the "Woodward-Bernstein phenomena"—the media spreading "the kind of nonsense" that led to the downfall of President Nixon. But, he said, processors themselves were also to blame, for their own advertisements implied that natural food was a panacea.[35]

The reaction against processed foods was even felt at the earliest stage of human life, in infant feeding, as an almost century-old trend toward artificial infant feeding was decisively reversed. Since the early years of the century, mothers and doctors had been increasingly persuaded that formulas, which were scientifically blended to contain proper proportions of all the necessary nutrients, were nutritionally superior to many women's breast milk. By the late 1950s, the large majority of mothers put their infants on the bottle within a few weeks of birth. In the 1960s, however, the pendulum began to swing back, as middle-class mothers (and pediatricians) exposed to skepticism about experts' advice on other food and medical issues paid more heed to critics who advocated following Nature's lead. As a result, from 1970 to 1980 the proportion of mothers feeding their children "naturally" doubled, from 30 to 60 percent.[36]

A 1977 Harris poll indicated that food manufacturers ranked highest among the industries the public would like to see investigated or charged by government.[37] However, to many people the government itself seemed to be part of the problem. Only 40 percent of the consumers surveyed by the FDA in 1974 thought the government made sure food was "good and nutritious"; only 56 percent thought it tried to ensure honest food advertising; and almost all thought it should be doing more in both areas.[38]

The lack of confidence in government's watchdog role was understandable. FDA spokespersons still spent much of their time on reassurance campaigns, lauding processors for improving the taste, digestibility, and economy of food.[39] The agency seemed more concerned with banning what it called "irrationally overfortified products" than facilitating fortification. It also lagged in supporting compulsory nutrition labeling. It eventually allowed processors to claim their foods were high in polyunsaturated fats but inadvertently encouraged misleading claims by refusing to allow them to say how much of which kinds of fats they contained.[40] Ralph Nader pointed out in 1970 that supermarket shoppers found more detailed nutritional information on dog and cat foods than on processed human foods.[41] In July 1975 the FDA finally required that food labels show what percentage of the RDA of nutrients each serving provided, but this applied only to those products that made nutritional claims.[42] Semiofficial government bodies such as the NRC's Food and Nutrition Board seemed equally unconcerned, particularly with regard to anxiety about processing and denutrification. To nutrition educator Joan Gussow they seemed more worried about what was happening in the "health food" stores than in the "unhealth food stores."[43] Only in July 1974 did the FNB allow all flour to be fortified with more nutrients than those approved in 1940–1941, and this was mainly a belated response to distress over the nutrition of the poor.[44] The government response to concern about children's eating habits seemed half-baked, at best. A 1973 FTC ban on advertising candy-coated vitamins on children's TV apparently paved the way for a 25 percent increase in commercials for the kind of sugar-coated cereals, candies, and cookies Choate and his supporters abhorred.[45] When the FTC finally proposed to restrict these in 1977, it was rather easily defeated by the food industry lobbies in Congress.

The government's apparent reluctance to ban suspect additives also provided ammunition to its critics. Year after year the FDA resisted demands from Nader and others that it ban Red Dye No. 2. It stuck to this position in the face of a 1975 uproar caused by revelations in the *New York Times*—repeated in a CBS-TV program called "The American Way of Cancer"—that a number of its own scientists thought it was still questionable whether it was carcinogenic, mutogenic, or toxic.[46] The FDA and Department of Agriculture also refused to ban the suspected carcinogen DES (diethystilbestrol) from cattle feed, forcing the Senate to take the lead in mollifying a frightened public by having it banned.[47] When an expert panel set up by the Agriculture Department to investigate the safety of the nitrates and nitrites that preserve and color cured meats reported that they should be restricted and possibly banned because in the course of digestion they produced carcinogenic nitrosamines, the department, prodded by its secretary, Earl Butz, led a dogged and successful defense of the chemicals.[48] Yet even food technologists sympathetic to industry admitted that there were solid grounds for concern.[49] The Carter administration was hardly more successful in resisting pressure from the powerful cured meat industry, whose leaders warned the public that to ban nitrates would spread botulism and make their products "time bombs resulting in death or disease."[50]

On the other hand, when the FDA finally did ban a commonly used additive in 1977, all hell broke loose. Forced by the Delaney Amendment to act on Canadian studies showing that rats fed massive amounts of saccharin developed tumors, it declared that it would have to ban the substance. The artificial sweetener industry quickly took out full-page ads in newspapers across the nation alerting dieters to the impending danger and defending the product's safety. The "ridiculous" action was "just another example of the arbitrary nature of BIG GOVERNMENT," it warned. An irate citizenry immediately arose to head off this tyranny, swamping the agency and members of Congress with demands that the sweetener be left alone. Diabetic and saccharin-dependent women chartered a "Saccharin Special" train to take protesters to Washington. Pollsters reported that an overwhelming majority of the public opposed the ban. Congress temporized by declaring an eighteen-month moratorium on the ban while the National Academy of Sciences studied the question. The NAS knocked the ball back into the congressional court by reporting that saccharin was indeed a carcinogen, but a very weak one, which so far had been shown to affect only rats. Congress then extended the moratorium twice more, until 1983, by which time FDA approval of a generally preferable sweetener, aspartame, allowed the ban to quietly take effect.[51]

The response to the threatened saccharin ban exemplified a new anti-Washington, antiregulatory mood in the country. In a way, radical attacks such as Nader's and Jacobson's had backfired. Their critiques of government regulatory agencies had contributed to the rise of a kind of neoconservative populism into which they could not easily tap. The debate over the Vitamin Amendments severely restricting the FDA's ability to regulate vitamin sales, which the Senate passed by an overwhelming majority in April 1976, had helped crystalize this kind of sentiment. "What the FDA wants to do," said Senator William Proxmire of Wisconsin, who represented this often-contradictory kind of laissez-faire populism, "is strike the views of its stable of orthodox nutritionists into tablets and bring them down from Mount Sinai where they will be used to regulate the right of millions of Americans who believe they are getting a lousy diet to take vitamins and minerals. The real issue is whether the FDA is going to play God."[52] The election that November of Jimmy Carter, an outsider who capitalized on the anti-Washington sentiment, reinforced the idea that the public was fed up with government telling it what to do and what to eat. The new FDA chief promised that the agency would now help reverse the trend toward federal intervention and return to the philosophy of "caveat emptor." As a step toward that, he said, the Delaney Amendment should be modified to allow small amounts of carcinogens in food.[53]

Meanwhile, organic foods had shed much of their hippie associations and were edging into the mainstream. This was due as much to environmental concerns as to health considerations. Senator Alan Cranston of California, cosponsor of a 1972 bill for government certification of organic food, argued for it on environmental, not nutritional, grounds. Organic farming was superior to modern farming, he said, because it was based on "respect for the soil."[54] The Texas Department of Agriculture began conducting serious experiments with organic farming, presaging the day—not far off—when even the USDA would take it seriously. By 1978 the *New York Times* was portraying J. I. Rodale's son and heir, Robert, as a sober environmentalist rather than a health food nut.[55]

By then nutritional science was in the midst of a sea-change, veering away from a half-century or more of concentration on vitamins and additives toward a concern with lifestyle. Epidemiological studies that followed large numbers of people over a number of years, matching health records with such things as smoking, diet, and exercise, were now producing results. The correlation between smoking and health problems had been clear almost from the outset. That exercise also played a role seemed likely but less certain. The connections between diet and health were least clear of all. Nevertheless, a consensus of sorts was developing on two points: first, that there was a connection between obesity and higher mortality, and second, that high levels of blood cholesterol were linked to heart disease.

That obesity was thought to be a risk factor was hardly news. Since early in the century, when analyses of insurance company records seemed to demonstrate a clear connection, this had been a relatively commonplace idea. By the 1950s deaths from infectious diseases had plummeted, and those from degenerative diseases such as heart disease and cancer were therefore taking a proportionally greater toll, especially in men over forty-five. This prompted renewed warnings, particularly from the same insurance company circles, against overweight. Still, as we have seen, the public did not seem to take its dangers too seriously.[56] Toward the end of the decade, however, due in part to the efforts of the American Heart Association and high-profile nutritionists such as Jean Mayer and Frederick Stare, overweight came to be regarded as a killer ailment. Most important, it was now viewed as a national rather than an individual problem. In 1963 and again in 1968, the National Research Council revised the number of calories in its RDAs downward.[57]

During the 1970s epidemiological studies, such as the one begun with five thousand Framingham, Massachussetts, males in 1954, seemed to indicate that losing weight could significantly lower blood pressure.[58] But their most dramatic results linked high levels of cholesterol in the bloodstream to heart disease. In 1972 the NRC's Food and Nutrition Board joined the AMA's Council on Food and Nutrition in issuing a cautious warning that "the average level of plasma lipids in most American men and women is undesirably elevated." However, they shrank from recommending any across-the-board changes in the national diet. They recommended "maintaining desirable body weight" and said "those deemed at risk" should consume less saturated fat to lower their cholesterol intake.[59] The American Heart Association had no such compunctions. In 1974 it intensified its anticholesterol campaign, calling on the entire nation to cut back its meat consumption by one-third.[60]

By the mid-1970s, said a Norwegian survey, an astounding 98.9 percent of the world's nutrition researchers believed that there was a connection between blood cholesterol levels and heart disease—a consensus practically unheard of in nutritional science.[61] But the consensus disintegrated when it came to questions of exactly what the connection was and how or even whether it could be remedied. Even more uncertainty reigned on other life-style issues, such as the purported relationship between diets high in fats and cancer and the deleterious effects of the American sweet tooth. Much to the apparent disappointment of some members of the Senate nutrition committee, a host of antisugar witnesses, including John Yudkin, were unable to produce convincing evidence that sugar caused heart disease, diabetes, or even dental caries.[62] Warnings were also issued about too much salt, or sodium, in the national diet, but although popular "junk foods" and restaurant fast foods were loaded with it, it managed to escape the kind of opprobrium heaped on sugar. Evidently, it was not burdened with sugar's emotional and political baggage.[63]

The earlier scientists who tilled the fields of the vitamin-based Newer Nutrition had not had a particularly easy time demonstrating that not eating enough of certain tasteless, invisible things was unhealthy. But at least they could demonstrate the general principle by depriving rats of certain vitamins and taking pictures of the horrendous results. Those cultivating the emerging new paradigm—what Warren Belasco has aptly labeled "Negative Nutrition"—had a more difficult task: proving that eating too much of certain of things was harmful. Experiments with overfeeding animals are difficult and tend not to lead to the same dramatic results as deprivation studies. To most people, a fat rat looks disturbingly healthy. Although clinical examinations and biochemical analyses of humans were more sophisticated than in the past, they were still limited in what they could reveal. The human body's response to food is usually too complex to allow the health effects of one particular food to be isolated. Any adverse response might reflect past rather than present diet or might not appear for years. As for the new epidemiological studies, while useful, they still faced the same kinds of difficulties in measuring dietary intake as the poverty studies: All survey methods—from individual bookkeeping to dividing the national food supply by the country's population—had serious limitations. At best, experts could speak only of "associations" between diet and diseases, not causation.[64]

The Negative Nutrition also faced the problem of telling people to do things they really did not want to do. While the older paradigm told people that good health would come from eating more of some of their favorite foods—meat, milk, eggs, cheese, fruits, and vegetables—the Negative Nutrition did the opposite. They were now told not only to cut down on the total amount they ate but to cut down on and even cut out many of their historic favorites. A nation renowned since the nineteenth century for its love of the frying pan and, more recently, the deep fryer was now told to turn its back on fried foods. Health-conscious Americans now looked at a plate of ham and eggs—declared America's favorite restaurant dish in 1942—as a sodium-packed, cholesterol-laden, carcinogenic time bomb. Thick, well-marbled beef steaks and rib roasts, historically signs of the good life (the latter was a close second in the 1942 poll) were now a one-way ticket to the cardiac ward.[65] Apple pie à la mode, since the 1920s the all-American favorite dessert, now had to be spurned, along with most other sweets and the salty potato chips and other crispy snack foods that many Americans munched on incessantly. Indeed, it seemed that the experts not only frowned on all enjoyable foods, they disapproved of enjoying foods itself. "Feeding has changed from a necessity to a form of entertainment," complained an expert on heart disease.[66]

The ascetic Nader, well accustomed to self-denial, had no difficulty in taking up the cudgels of the Negative Nutrition. Jacobson and the others who had been attacking additives and processing also adjusted quite smoothly, for the new message allowed them to amplify their criticisms of the food industries. They could now blame obesity on the poor eating habits—particularly the sweet tooth—Americans developed as television-addicted children. Despite overwhelming evidence that infants are born with a sweet tooth, they assigned responsibility for it to the food industry. Children pestered their mothers for snacks, said one critic, because they were bombarded with TV ads by "the 'Cookie Man' or some other carbohydrate Pied Piper" extolling sugar's "instant energy."[67] Jacobson charged that the major food companies who spent millions marketing their excessively sweetened foods helped make refined sugar "perhaps the major villain" behind America's nutritional crisis.[68] A study coauthored by Jacobson called the contemporary American diet much worse than that of 1910 and said this was mainly because of increased consumption of sugar and fats, which it linked to the growth of the corporate-dominated processing and fast food industries.[69]

Negative Nutrition also fit in with growing disillusionment with the American medical establishment. The rise of more economical and equitable health care systems abroad undermined smugness about the superiority of American medical care and lent support to the New Leftist criticism that it ignored preventive medicine, such as nutrition counseling, in favor of more profitable forms of after-the-fact intervention. Critics such as Ivan Illich condemned most forms of modern medical intervention and called on people to take charge of their own health, mainly through improving their diets.[70] "Naturopaths" and "holistic" doctors came out of the woodwork with a bewildering array of diet therapies. In California thousands of cholesterol-concerned people paid handsomely to pass some weeks trying to shake off a passion for their favorite foods at the Pritikin Longevity Center. Its founder, engineer Nathan Pritikin, had refused to heed a medical specialist's advice to take drugs and avoid excessive exercise to lower his cholesterol level. Instead, he did so through exercise and a practically no-fat diet. This became the regimen at his center, where patients/clients went on a strict diet that allowed only infinitesimal amounts of animal fats and sugar and barred eggs, alcohol, tea, and coffee completely. Patronage was hardly affected when poor Pritikin committed suicide upon learning he had leukemia. Unlike Adelle Davis, he had never promised immunity from cancer. Indeed, patrons were buoyed by the news that an autopsy had revealed that "his arteries were as beautifully clear as those of a seven-year-old."[71]

The doubts about the medical establishment and the rise of the Negative Nutrition fit in well with the new, self-critical mood of the times. During the 1950s and early 1960s, the American cornucopia had symbolized God's blessing on the country. By the late 1960s, the generation raised in affluence was challenging the culture of "more is better." *Psychology Today* labeled a 1970 interview with Jean Mayer "Affluence, the Fifth Horseman of the Apocalypse."[72] The New Leftists and counterculturites raised new questions about the effects of food consumption on health, the environment, and the people who produced the foods, but they also reflected the traditional American propensity to judge people by their consumption habits. Now, in the sober aftermath of the race crisis, Vietnam, and Watergate, many other Americans—particularly middle-class ones—began to look at food through this moralistic prism and see their abundant food supply as anything but a blessing.

The steep rise in energy and food prices following the oil crisis of 1973 fueled a shocked realization of the extent of waste in the land of the gas-guzzling car. Americans seemed to be paying dearly for a society based on overconsumption—of fuel, minerals, forests, water, and, of course, food.

Rising food prices and new famines also brought renewed misery to the Third World, making the contrast with American abundance more striking than ever. The fact that obesity among the well-off was said to be America's greatest health problem seemed particularly shameful. "While a substantial proportion of the population is a victim of undernutrition," wrote Jane Brody in 1973, "a much greater proportion is suffering the ravages of overnutrition—obesity, heart disease, diabetes, and the like. Much of the nation's overweight problem is thought to be a function of an affluent society that encourages overeating."[73] A Harvard nutrition professor, lamenting the nation's "overabundant diet," called it "a happenstance related to our affluence, the productivity of our farmers, and the activities of our food industry."[74] Senator Charles Percy recalled the wisdom of Margaret Mead's 1968 statement that as a result of the previous decade's affluence "it was possible to say the major nutritional disease in the United States was overnutrition."[75] "Waste and gluttony are becoming passé" said a *New York Times* article on the day after Christmas in 1974. High food prices were spurring Americans to eat cheaper foods, use leftovers, and serve smaller portions. "I think we're all being schooled in a whole new way of thinking about food," said a San Francisco restaurateur.[76]

But what did all this mean for the core of the nation's postwar nutritional credo: the ideal of the "balanced meal" and faith in the Basic Four? Since World War II, when it started out as the cumbersome Basic Seven, food producers and processors had been aware of the benefits of this kind of nutritional advice: Any food, processed in any fashion, could fit into at least one category and could thus qualify as essential to a "balanced diet." In 1956 the USDA had reduced it to the more easily recalled Basic Four so that it would be a more effective tool in stimulating food consumption. Thanks to the industry and USDA, it became the centerpiece of the nutrition message in the schools and the media. When the industry's Food Council, feeling the heat from rising concern over denutrification, mounted a "nutrition-awareness" campaign in 1970, it naturally chose as its theme "Eat the Basic Four Foods Every Day."[77] It need hardly have bothered, though, for the public had already absorbed the message. By 1970 over 85 percent of Americans agreed that "anyone who eats balanced meals can get enough vitamins in his regular food"—something that perplexed those who fretted about the concurrent wave of vitamin-mania.[78]

Government and the media felt comfortable with advice supporting the "balanced diet" and the Basic Four because it allowed them to avoid playing favorites. The USDA's chief of nutrition research explained it quite succinctly in 1977: "Since there is a certain level of ignorance in the nutrition field, we say you should eat something of everything—something out of each category of food; and two, since there are a number of different power groups, you should eat some of everybody's products."[79] In 1972, when *Good Housekeeping* was faced with mounting inquiries from readers about "those so-called health foods," the Basic Four credo allowed it to respond without alienating any of its advertisers. The best way to maintain good health, it said, was "to eat a well-balanced diet of food from the Basic Four categories . . . simple, everyday foods." This meant dishes such as "Creamy Scrambled Eggs," corned beef hash, hamburger, fish sticks cooked in ketchup, and "Puffy Cheesewich."[80]

But while other government agencies held steady on this course, the Senate nutrition committee veered off toward the Negative Nutrition. Over almost five years of hearings on diet and disease, its staff lined up witness after witness to testify that the American diet was a major contributor to the nation's main "killer diseases," heart disease, cancer, stroke, and diabetes. Finally, the staff came up with a blueprint for implementing the Negative Nutrition. In effect, this special report, *Dietary Goals for the United States*, called for a complete about-face in government nutrition policy. Whether promoting wartime enrichment, the Basic Seven, the Basic Four, or cod liver oil, all previous government efforts had centered on getting people to eat more of what was thought to be good for them. *Dietary Goals* emphasized eating less of what was thought to be bad for them—the core of the Negative Nutrition.

In introducing the report, Senator McGovern echoed the critics' gloomy assessment of where the American diet had been heading. "Our diets have changed radically within the past fifty years," he said, "with great and often harmful effects on our health. . . . Too much fat, too much sugar and salt, can be and are directly linked to heart disease, cancer, obesity, and stroke, among other killer diseases." The steady decline in the proportion of the national diet made up of grains and their carbohydrates—long regarded as a welcome sign of affluence and progress—was now decried. Americans should reduce the proportion of fats, particularly saturated fats, in their diets by almost one-third and cut consumption of sugar and salt even more. To accomplish this—and here was the most controversial part—they should eat less meat, butterfat, eggs, and other foods high in cholesterol, replacing them with more poultry, fish, fruits, vegetables, and whole grains. They should also substitute nonfat milk for whole milk.[81]

If the report's thrust were correct, even vegetarians and natural foods buffs would have to make major dietary adjustments. Francis Lappé relied on large amounts of eggs and milk to provide the essential proteins in her diet. Adelle Davis, another whole milk buff, now seemed to have gotten many things wrong and rapidly went out of fashion. Jean Hewitt's best-selling *The New York Times Natural Foods Cookbook* (1971) now prompted visions of rapidly clogging arteries and heart failure: Alongside the recipes using whole grains, pulses, organic foods, and other acceptable foods were cholesterol-laden ones such as that for calves' brains with scrambled eggs or potato chowder made with lots of bacon and whole milk.[82]

But most natural foods advocates adjusted to the new mandate relatively easily. The real uproar came from beef, dairy, and egg producers. While their lobbyists worked the halls of Capitol Hill, experts sympathetic to them attacked *Dietary Goals* in the media and in testimony to Congress. Gilbert Leveille, chairman of the food science and nutrition department at Michigan State University, said, "The American diet is better than it has ever been." The higher rates of degenerative diseases such as heart disease and cancer resulted from the reduction or elimination of other causes of illness and death. "The concept that dietary modification will prevent or delay atherosclerotic heart disease remains a hypothesis and not a fact," he said. "It hardly seems a sufficient basis for the recommendation of major dietary changes for the entire population."[83] The recommendation that dietary cholesterol be reduced was "premature," said Dr. Robert Olson, the head of biochemistry at St. Louis University Medical School. The epidemiological evidence for it was entirely "circumstantial." Even if it were true, the dietary changes recommended for the entire nation would benefit only 10 percent of the population, males between twenty and fifty.[84] The AMA opposed *Dietary Goals* on the grounds that there was not enough scientific evidence to support "such universal dietary goals" and cited their "potential for harmful effects."[85] It issued a special report reaffirming its faith in the Basic Four, a varied diet, and "moderation." It warned specifically against "avoidance," the underpinning of the Negative Nutrition.[86]

The NRC's Food and Nutrition Board also dissented. It began a methodical two-year scrutiny of the scientific literature, which ultimately echoed the AMA's qualms and added some more of its own. It too disagreed with the claim that the nation's diet and health were declining, and it warned that changing the national diet might jeopardize the past century's steady improvement in these fields. It called the evidence that diets high in saturated fat caused heart disease inconclusive and contradictory and was very skeptical of the purported links between dietary fat and cancer. Although conceding that high levels of serum cholesterol and low-density lipids were risk factors for heart disease, it warned that "it has not been proven that lowering these levels by dietary intervention will consistently affect the rate of new coronary events." The benefits of lowering fat intake to the level recommended by *Dietary Goals* had "not been established," it said, yet the entire population was being asked to make dietary changes that might, at best, be beneficial to only a small minority. Frederick Stare supported this critique vigorously, warning against blanket condemnations of cholesterol and sugar in particular.[87]

But support for *Dietary Goals* and the Negative Nutrition mounted steadily. In 1979 the surgeon general issued a report, *Healthy People*, calling for a national health strategy to "emphasize the prevention of disease" that should include eating fewer calories and less saturated fat, cholesterol, salt, and sugar. It also recommended that Americans eat more fish, poultry, complex carbohydrates, and legumes—more or less the same regime as that proposed by *Dietary Goals*.[88] A joint report of the Departments of Agriculture and Health came out cautiously on the side of *Dietary Goals*. It tried to escape the wrath of powerful interests by advising Americans to "eat a variety of foods" and condemning only sugar and salt—easy marks—by name. Otherwise, they were to avoid, not specific foods such as beef or eggs, but "too much fat, saturated fat, and cholesterol."[89] Later that year, even the FNB seemed to soften its opposition to recommending general dietary change along Negative Nutrition lines. Its revised Recommended Daily Allowances suggested that the proportion of calories one derived from fat be cut to 35 percent (the national average was over 40 percent; *Dietary Goals* recommended cutting it to 30 percent) and proposed an upper limit for sodium intake.[90] By 1981 the National Cancer Institute, the Society for Clinical Nutrition, and even the American Medical Association had come around to recommending national dietary changes along Negative Nutrition lines. Sensing the sea-change in attitudes toward food and health, the National Institutes of Health now began to direct a larger portion of their research funds toward studying the links between diet and cancer and other diseases. Bob Bergland, Carter's secretary of agriculture, even resurrected the long-spurned idea of the 1930s that nutrition and health, not selling food, should be the goal of federal farm policy—though at some future time, he cautioned, and he was out of office before the future arrived.[91]

Those opposed to *Dietary Goals* turned out to have more political than scientific clout. It had taken political courage on the part of McGovern—almost 80 percent of whose state's farm income derived from meat animals and their feed grains—to go along with the recommendation that Americans "decrease consumption of meat and increase consumption of poultry and fish." But McGovern had not anticipated the brouhaha the report would cause. A flood of calls from his constituents and some private visits from livestock industry representatives chipped away at his determination, convincing him that a switch in wording from *less* meat to *lean* meat would be advisable.[92] The suggestion itself had come from Senator Percy of Illinois, the ranking Republican member, who had developed "serious reservations" about the report after hearing from his state's powerful beef and dairy interests.[93] In December the committee agreed on a revised *Dietary Goals*, which, instead of calling for cutting down on meat, told Americans to "decrease consumption of animal fat, and choose meats, poultry, and fish which will reduce saturated fats intake."[94]

Nick Mottern, the committee staff member who was the chief architect of the report, had refused to go along with the changes and was asked to resign. McGovern, he reported, had said that he did not want to "engage in a battle with [the meat] industry that he could not win."[95] But potent as the beef lobbyists were, it seemed that the greatest obstacle to implementing the new goals lay in the economics of the food industry itself—something that had contributed to a vast outpouring of advertising for its most unhealthful products. The food market had begun to stagnate around 1970; population growth had slowed, and individuals can eat just so much food. Supermarket shelves were now packed with so many items (more than twelve thousand different items in an typical large one) that introducing new food products was an ever more awesome—and expensive—task. Processors also complained that suspicion of processing made them reluctant to strike out in new directions. "The food industry is so over-regulated," grumbled the chairman of Kraftco, "that there can be no real innovation."[96] Whatever the reason, he was right in that the 1970s brought no great breakthroughs in convenience—nothing like the frozen foods of the 1950s and 1960s or the microwave ovens of the 1980s—to impel a burst of new, high-value-added products. (100% Natural Cereal was the *only* successful new product for Quaker Oats from 1970 to 1978.)[97]

There were also fewer possibilities for expansion through the traditional route, gobbling up smaller competitors. A burst of agglomeration in the late 1960s and early 1970s left a mere one hundred corporations in control of almost all of American food manufacturing. Some tried to lessen their dependence on food processing by diversifying, into shoes, luggage, clothes, furniture, housing, even jewelery. General Mills bought the companies that produced Play-Doh and Monopoly. Quaker Oats took over electric-train maker Louis Marx and Fisher-Price toys, while Nabisco countered with Aurora Toys.[98] Ultimately, however, they could not escape dependence on food. Yet there they almost invariably found themselves locked into situations in which each sector was dominated by three or four established giants, whose market shares could expand only at the expense of the other behemoths.[99] In the absence of new product lines, grabbing market share from competitors meant one main thing—advertising.

Even by 1969, Ralph Nader had pointed out, the largest food manufacturers spent about 18 percent of sales on advertising, compared to 3 percent by auto producers—no slouches in that department themselves.[100] Moreover, food (and beer and soap) advertising dollars translated directly into profits to a far greater extent than in all other major industries.[101] This was particularly true of the most profitable food products, which critics charged were usually the most highly processed and innutritious, the very items that seemed most responsible for the high proportion of sugar, salt, and cholesterol in the American diet. The staff report that accompanied *Dietary Goals* said 28 percent of total television advertising was for food and beverages and that the bulk of the time purchased—including 85 percent of that on weekends—went to sell foods high in saturated fat, cholesterol, sugar, and/or salt. This was particularly alarming, said *Dietary Goals*, because television was likely the main source of nutrition information for the 30 percent of the nation—mainly low-income—who were classed as functional illiterates.[102]

The staff report called on the government to counterbalance food advertising by mounting an extensive television campaign to spread the new message, but this was wishful thinking.[103] If the media were to carry the new message to the masses, the food companies themselves would have to play the leading role. In fact, some of them were already doing this. The edible oils producers already had a head start on alerting the public to one aspect of the Negative Nutrition: cholesterol-phobia. The three out of four Americans who, pollsters reported in 1977, were worried about the cholesterol in their diets almost certainly owed much of that concern to the efforts of margarine and vegetable oil producers, who had been marketing their products as low in saturated fats since 1962.[104] After 1981, when a new administration did make "caveat emptor" the name of the game by lifting most restrictions on health claims in food advertising, the rest of the industry would join in. Then, as they discovered the manifold commercial opportunities in the Negative Nutrition, they would indeed become major forces in its dissemination. But they would do this in the same unbalanced and distorted fashion—probably spreading more misinformation than useful nutritional knowledge—as they had done previously with vitamins and the Basic Seven and were currently doing with their "low cholesterol" and "natural" products.

By the time *Dietary Goals* was issued in 1977, there were indications that many Americans were already altering their diets in that direction. As a whole, they were eating 10 percent less sugar per person than in 1965, almost one-third more chicken, 20 percent more fish, fewer eggs, one-third less butter, more margarine and potatoes, and a bit more vegetables. Of the foods to be avoided, only beef consumption had risen markedly, fueled by the continuing boom in fast foods.[105] Yet it was already clear that the new self-denying ordinances were affecting mainly the middle and upper classes. The working class and the poor, in particular, remained relatively unaffected. While the upper and middle classes fretted about whether to eat "junk food," the working class and the poor continued to munch on crispy snacks and sweets with little apparent guilt. As the middle class began to regard the historic markers of material success—the thick beef steak or slab of rib of beef—with fear and trepidation, poorer folks continued to regard them as signs of the good life. Cholesterol tests, fitness tests, stress tests, blood sugar tests, subcutaneous fat measurements, and digital scales scattered throughout the home—these were the marks of the middle and upper classes.[106] Paradoxically, while the middle class worried that their excess poundage would fell them, obesity was much more prevalent among the less-concerned poor, particularly among women.[107] Indeed, while poor adult males tended to weigh about the same as better-off males of their age and height, poor women weighed much more than better-off women and seemed hardly as affected by the Negative Nutrition's admonitions about the health perils of obesity as were middle-class women.[108] Asked about the various reasons for their food choices in 1978, forty poor Hispanic women in New Mexico selected health only 4.6 percent of the time. Sensory, economic, and preparation reasons were given much more often, as were reasons of religion, race, and culture. Weight loss for appearance's sake was not even mentioned.[109]

Yet the Negative Nutrition was only one of a number of demands pulling at middle-class women, often in contradictory directions. The rise of "gourmet" cooking, increased restaurant-going, the fast foods boom, and the substitution of snacking for formal meals tended to undermine the best of intentions to abide by the rules of the Negative Nutrition. The growing quest for eternally youthful looks and the progressively leaner definition of what that meant encouraged weight-loss regimes far removed from considerations of healthy eating. When these dissonant sounds clashed with each other, as well as with the ever-changing variations on the Negative Nutrition theme, middle-class America did indeed seem to be resounding with what Claude Fischler has called a "dietary cacophony."[110] Few people could pick out a cadence they could march to with any consistency. "Food has replaced sex as a source of guilt," said a 1978 article in *Psychology Today*.[111]

CHAPTER 14

✗ ✗ ✗ ✗ ✗ ✗

Darling, Where Did You Put the Cardamom?

In October 1966 *Ladies' Home Journal* ran a lush photo spread on "How America Entertains." The people it featured, mainly urban socialites raising money for charities, were hardly a cross-section. But there, among the tanned, carefully coiffed, and well-dressed beneficiaries of the mid-1960s economic boom, one could see French food at the pinnacle of its status in postwar America. In Minneapolis, Atlanta, and even Boston, the refrain was the same: The hosts were particularly proud of their French dishes, especially if they had been discovered in one of New York's most fashionable French restaurants. The San Francisco dinner featured *boeuf en gelée;* the Minneapolis midnight supper included "Pâté Brasserie," a chicken liver, ham, and orange pâté, which the hostess said came from the Brasserie restaurant in New York. The Atlanta hostess said that her most "luscious" dessert was "inspired by" one at Manhattan's Four Seasons restaurant. Even the genteel Boston hostess who put on an English-style tea at her gracious, antique-filled Beacon Hill mansion served the tea sandwiches as "*Pain Surprise*"—in a long loaf of scooped-out French bread—and let it be known that the special desserts, petits fours and tartlets, were from a Four Seasons recipe.[1]

Three years earlier, an assassin's bullets had put the rangy Texan Lyndon Johnson into John Kennedy's place in the White House. Acutely conscious of the disdain in which he was held by old-line members of the so-called Eastern Establishment, when Kennedy's chef announced he was leaving LBJ made sure that suitable publicity accompanied his hiring Henri (later "Henry") Haller, a French-trained Swiss, to replace him.[2] (No such fanfare accompanied his bringing in a black American chef to cook his southwestern favorites in the White House "family" kitchen and installing a Fresca-spouting fountain just outside the Oval Office.)

Richard Nixon, who succeeded Johnson in January 1969, had spent the past seven years—during his involuntary sabbatical from elective office—living at Fifth Avenue and East Sixty-second Street on New York's plush East Side. There he thrilled to life on what he called "the fast track," with its visits to the neighborhood's elegant French restaurants.[3] Perhaps because of this exposure to the extremes of Soulé-style snobbery, Nixon was even more conscious of the importance of serving fine French food than Johnson, who eventually took to entertaining foreign dignitaries at his Texas ranch with beer, bourbon, and barbecue. In any event, he kept Haller on to prepare French food for both public and private occasions.[4] The Californian president even ordered that California wines not be served at White House dinners, particularly if foreigners were present.[5] (Sophisticates found his regular lunch dish, cottage cheese and ketchup, rather appalling, but so did he. He despised cottage cheese, which he ate for weight control, and put ketchup on it to make it palatable.)

In March 1970 Craig Claiborne, marveling at the new ascendancy of French food, labeled the 1960s as the decade in which haute cuisine had finally come of age in America.[6] A new breed of superstar French chefs now showed up regularly in America, suitcases full of truffles and other such exotica, and fanned out to give cooking demonstrations, prepare charity dinners, appear on talk shows, sign cookbooks, and promote their private-label wines and foods. Julia Child, who had worked so hard to promote French food in America, found the attendant food snobbery disturbing. In November 1973, writing to her French friend Simone Beck about a particularly orgiastic series of articles on the great chefs of France by *New York* magazine's Gael Greene, she said, "Food is getting too much publicity, and is becoming too much of a status symbol and 'in' business."[7]

Yet ten months earlier an event fraught with symbolism had taken place: Le Pavillon had closed. True, Soulé had been dead for six years, and it is by no means unusual that the public fails to accept successors to great restaurateurs. However, the last owner-manager, Stuart Levin, a man much experienced in the restaurant business, saw that there was more to it than that. "The era of the grande luxe restaurant is over," he said. "There are simply not enough patrons to keep a restaurant of this stature in the style it should be kept. . . . In the society in which we live there are no longer any requirements for this kind of cuisine."[8]

Levin was correct in seeing that the days of traditional grande luxe cuisine were virtually over, but this was not because, as he implied, Americans were no longer willing to pay top dollar for fine food. Rather, eating "this kind of cuisine" had ceased to be particularly distinctive. One reason was that, while Levin was referring to haute cuisine, in fact the "gourmet" food served at most of America's expensive French restaurants in the 1960s was not all that haute. Most relied on dishes such as escargots, onion soup, sole meunière, coq au vin, and calves' liver, which were part of the standard bourgeois repertoire. Even Soulé, who served a number of classic haute cuisine specialties, served many bourgeois dishes as well: his core menu was dishes such as onion soup, filet of sole with mustard sauce, frogs' legs *provençale*, and *caneton aux cérises*. His most famous dish was *quenelles de brochet*. In prewar years this difficult-to-prepare dish had been something of an haute cuisine mainstay, but after Claiborne raved about it—and particularly after the commercial kitchens' Robot Coupe/Cuisinart made it a snap to prepare—it too became commonplace.[9]

In the trans–Hudson River hinterland a genre of high-priced French or "Continental" restaurants had sprung up, wowing the natives with duck *à l'orange*, steak *au poivre*, *boeuf bourguignon*, and other standards of the bourgeois kitchen—preferably, if possible, flambéed—but by the mid-1970s, the foreign travel boom had exposed many of their customers to better versions of these dishes in France itself, or even in each others' homes. Sophisticated *New Yorker* readers now chuckled knowingly as Calvin Trillin speculated that the continent in "Continental" might be Australia. Plummeting airfares allowed their own offspring—even their own hairdressers' children—to enjoy French cuisine at its source. (A character in "Life Lessons," director Martin Scorsese's segment of the movie *New York Stories*, says: "I heard these two kids in a restaurant the other day. One said 'What's chocolate pudding?' The other said, 'It's good. It's a lot like chocolate mousse.' ") So by the time Pavillon closed, one had to travel—and eat—much further afield than Paris or Nice to evoke much interest back home.

One of the reasons the American elite of the 1950s and 1960s had originally thought French cuisine the world's finest, recalled Claiborne, was their "naiveté." They were "not aware of great Italian cooking, or of the great cooking of other nations."[10] That this was beginning to change had been reflected in the charity supper in Dallas covered by the *Ladies' Home Journal* in 1966. It featured "three world-famous cuisines"—French, Indian, and "Oriental"—and all three received equal billing.[11] A new breed of restaurant critics such as Claiborne and Greene, who waxed just as ecstatic over "Northern Italian" or Chinese cooking as French, helped matters along. This was particularly important toward the end of the decade, when many newspapers began to feature weekly restaurant reviews, often written by people who were more culinarily intrepid than their audience. On January 1, 1970, two months before rhapsodizing over the previous decade's improvement in New York's French restaurants, Claiborne noted that Chinese restaurants had been "multiplying like loaves, fishes, and bean sprouts" and that the city had recently gained fifty new Japanese restaurants. "The public," he said, "seemed to take to sashimi and sushi and the likes of shabu-shabu with what could be regarded as passion."[12] Who would have dreamed, a mere ten years earlier, that trend-setting Americans would be eating raw fish?

Although travel to Europe helped stimulate appreciation for French food, other foreign destinations played only a minor role in fueling the foreign food boom. Japan has never been a major American tourist destination. The cuisines of China became the rage well before American tourists were even allowed to travel there. Nor was immigration to America a major factor. The great influx of European immigrants had dried up over forty years earlier. Their cuisines, like their personages, had met a generally cold reception, and for the most part their food habits gave way to, rather than altered, the dominant ones. It is no accident that the major exception, Italo-American food, entered the American mainstream only after the great wave of immigration from Italy subsided. Even fifty years later, the status of what was labeled "Northern Italian" food was enhanced by sophisticates' knowledge that it was disdained by most Italo-Americans, still enmeshed in the bountiful pasta-and-spicy-tomato-sauce syndrome, who were regarded as quite déclassé.

In fact, both empire and immigration, each of which exposes nations to new foods and cuisines, seem to have a similar impact on their food. While the people of the dominant culture may adopt some of the foods of the subordinate culture—the Spanish in America, the British in India, and the French in North Africa come to mind—they do so by divorcing them from those aspects of the culture they regard as inferior, particularly the foods of its lower orders. They also tend to marry colonial foods, flavors, or methods to foods that are of high status in the imperial centers, giving us curried beef, shrimp, and even lobster, for example. To a certain degree, this is a natural human reaction. As omnivores, we are understandably cautious about the dangers of new foods and have learned to incorporate them into established cuisine by preparing them with familiar tastes—what psychologists call "markers"—and methods.[13] But choosing which foods, flavors, and cooking methods are worthy of incorporation into one's diet is a process very much involved with status and distinction. The chefs of France have drawn much more inspiration from the cooking of far-off Japan than what was being eaten in their own country's teeming North African *bidonvilles* or its squalid Vietnamese ghettos.[14]

Indeed, the adoption of new food tastes is probably facilitated by an absence of low-status people from whose homelands they originate. The dearth of East Asians in nineteenth-century New England likely encouraged, rather than hindered, the measured use of curry powder. A taste for curried dishes among the English working class was well established by the early 1960s, before the massive influx of immigrants from the subcontinent stoked ethnic tension on factory floors and urban terraces. Italo-American food became popular in parts of America far removed from its Little Italies, while many of those who lived next door to Italian ghettos derided it. Since the 1930s Greek immigrants have played an important role in the American restaurant business, but mainly in places serving anything but Greek food. (The owner of Khansson's restaurant in Corrales, New Mexico, which the *New York Times* called "perhaps the best Mongolian restaurant between the two coasts" in 1976, was Greek.)[15] The fact that most Chinese were concentrated on the coasts likely helped, rather than hindered, the steady march of chop suey and chow mein through Middle America. When Chinese food did sweep the sophisticates on the coasts it was not the cuisine of the overwhelming majority of Chinese-Americans, who were southerners of Cantonese origin, but cooking claiming to be from the north—"Mandarin" or "Peking" and "Shanghai"—and, later, inland provinces such as Szechuan and Hunan. Only after they became jaded with these did big-city Chinese food mavens rediscover the food of Canton, which in Peking, Shanghai, Szechuan, and Hunan is usually acknowledged as China's finest.[16]

The trajectory of Mexican cuisine in America is a case in point. Although Mexican food (and here one is not speaking of "Tex-Mex" or "Cal-Mex") is at least as varied and interesting as that of the Indian subcontinent, it was not among the three cuisines (French, Indian, and Oriental) chosen by the Dallas socialites whose 1966 dinner was covered in the *Ladies' Home Journal*. The reason was obvious: The Dallas fund-raisers were surrounded by Mexicans, including many of their servants, whose status was by no means high in their eyes. No distinction derived in the Southwest from having visited Mexico, whose food was the butt of jokes about "Montezuma's Revenge." When Mexican food's turn at titillating palates finally did arrive, much of the impulse came from New York City.[17] Yet the very same New Yorkers who trooped to raffish Fourteenth Street or the tumble-down market stalls in Spanish Harlem's La Merced in search of *chiles anchos* and *pasillas* generally disdained the food of the Puerto Ricans and other Caribbeans who were part of their own underclass. As we shall see in the next chapter, Mexican food's later success in the snack and fast food markets was based very much on its purveyors' ability to disassociate it from its Mexican connotations.

But while travel played only a minor role in the specifics of the non-French ethnic foods boom, it did stimulate a growing cosmopolitanism, which made people receptive to foreign foods in general.[18] The fact was that, although most Americans did not define it as such, the United States had become the center of an impressive, worldwide, informal empire. As in Rome at the peak of its influence, a certain amount of distinction rubbed off on those whose tables manifested the cosmopolitan nature of the empire. This was particularly true of those who had been sent abroad to help run the overseas dominions. Many thousands of Americans in government, the armed forces, private industry, and the media returned from abroad with their gustatory horizons expanded. "Everybody's becoming more sophisticated about eating, especially Air Force families that have spent time in other countries," said Bobbie Snyder, the proprietor of a gourmet food shop and restaurant in Great Falls, Montana, in 1976.[19] It was also true of many of the millions who each year went abroad as tourists. Their number tripled during the 1960s, passing the four million mark in 1970, and has expanded more or less steadily ever since.[20]

The ethnic food boom was also connected to the rebirth of cooking as a status symbol during the 1960s. Again demography—particularly the baby boom—played a major role. In the 1960s cute little baby boomers who had been the focus of the previous decade's child-centered culture turned into self-centered teenagers unwilling to sacrifice their individual wants to traditional family expectations. As parental illusions about the satisfactions of child-rearing evaporated in the *Sturm und Drang* of domestic tensions, middle-class parents turned to other forms of self-fulfillment. Often they looked in the same direction as their youngsters—inward. They too began to reject traditional family-centered values in favor of devotion to their own personal "growth" and "self-realization," helping to feed what Christopher Lasch aptly called "the culture of narcissism." Their expectations oiled by affluence (median family income tripled between 1950 and 1970), the middle class searched for new experiences to widen their horizons and fuel personal growth.[21] Could there be a more enjoyable way to broaden oneself than by developing an appreciation for—and talent at preparing—exotic foreign cuisines?

The search for new food tastes also fit in with a turn toward sensuality. A heightened appreciation for the pleasures of the table marched almost in lockstep with the increasing popularity of notions that all and sundry should enjoy a wider variety of intense sexual experiences. In the mid-1960s *Vogue* magazine, previously devoted mainly to fashion as a sensual/sexual experience, began a steady expansion of its food and travel coverage.[22] The venerable *Ladies' Home Journal*'s food coverage expanded almost in tandem with ever more frank discussions of the joys of sex (indeed, the latter was the title of a 1972 best-seller, whose subtitles were "A Gourmet's Guide" and "Cordon Bleu Guide").[23] By the end of the decade it was running more than 450 recipes per year—many of them ethnic—more than triple the 1960 number.[24]

Of course, by then ethnic food had become the subject of almost as much food snobbery as French food. Claiborne's first articles on foreign foods, in the late 1950s and early 1960s, had stimulated some interest, but little in the way of a following. Then, he recalled, in the mid-1960s his articles about "unknown foods" began to evoke "a certain amount of snobbery." Suddenly, when he wrote of pesto, "people were dying to become the first ones to discover where to buy fresh basil."[25] His 1963 *Herb and Spice Cook Book*, which was full of foreign recipes—*baba ghanouj*, *tapenade*, Peruvian ceviche, tandoori chicken, *cassata*—calling for herbs and spices rarely used in American kitchens, turned into a runaway success, going through many printings in the next five years.[26] "At this moment," wrote a disgruntled traditionalist in 1970, "scores of harried women are tramping through the East Side in search of a spice imported from Mozambique and available only in Nieman-Marcus, Dallas, and one Armenian food shop on Third Avenue."[27] "It's a real challenge and status symbol," Jean Thwaite, food editor of the *Atlanta Constitution* reported in 1977, "to come up with something your company hasn't tasted before, something they don't even know how to pronounce."[28]

The ethnic food boom was also abetted by the revival of ethnic consciousness. The rise of black nationalism in the late 1960s had brought a new appreciation for the southern black culinary tradition, now called "soul food." On New Year's Eve well-off urban blacks ostentatiously eschewed steak and roast beef for the ham hocks, peas, and collard greens of their rural forebears. One entrepreneur set up shop importing prepared soul food directly from North Carolina to affluent blacks in New York City.[29] In the early 1970s, as Italians, Poles, Chicanos, Amerindians, and others began to manifest pride in their heritages, they too tried, with greater or lesser success, to cultivate their culinary roots. Dying food stores in withering urban immigrant ghettos received life-saving transfusions from third-generation suburban ethnic shoppers. Shops in Boston's Italian North End, Philadelphia's South End, and Detroit's Greektown and even on Bergenline Avenue in Union City, New Jersey, sprang back to life.[30] When upscale nonethnics began exploring these far-off places, "gourmet" food stores—hitherto reliant on canned escargots, pâté de foie gras, preserved kumquats, and other such delicacies—shifted gears and began selling Italian pasta and olive oil, Greek feta cheese, Central European sausages, pirogi, and *hoi sin* sauce. "New gourmet stores are sprouting like Mung beans—which many also sell," said *Time* in 1977.[31]

Meanwhile, thanks in part to the lingering influence of the New Left and counterculture, natural and health food enthusiasts were discovering the lively spicing many of the world's peasants used in preparing healthy whole grains, pulses, and vegetables.[32] In 1971, when *Mother Earth News*, the best selling of the "old-timey" back-to-nature magazines, ran an article on how to cook Chinese food at home ("inexpensive, nutritious, delicious, and fun"), it was written as if this were a wondrous new discovery. ("It's a real trip just preparing the meals!") Within two years, however, recipes for ethnic dishes such as "Basque Food Delights" were regularly featured in its new glossy, urban-oriented sister publication, *Lifestyle*.[33]

The people initially affected by these trends were mainly middle- and upper-middle class. Many were former counterculturites and New Leftists who had rejoined the mainstream, bringing with them their reverence for the artisanal and a disdain for foods that were not fresh, natural, and "authentic." Indeed, the most distinction accrued to those who came closest to authenticity. The most-traveled, the best-read, the most discerning consumers knew the difference between true French bread and long loaves that were merely shaped like it, between South Asian dishes made with a special combination of spices to suit each dish and ones made from commercial curry powder. They scrupulously cooked meals from the same ethnic cuisine from start to finish, and they bought woks, bamboo steamers, *couscousiers* and pasta machines (the latter reluctantly, for they knew that the best kind was rolled by hand). The trend in cookbook publishing, said *Time* in 1977, was "towards more esoteric books on specialized foreign cuisines. . . . The best cooks are learning Indian, Indonesian, Indo-Chinese and Chinese (especially Szechuan and Hunan) and Japanese recipes."[34]

"Every thirty-eight seconds a cheese shop opens in Manhattan," the *New Yorker* reported two years later. "Croissants, pâtés, fresh pasta, imported preserves and biscuits are as ubiquitous as cat food; arugula, raspberries, chives, and mint never go out of season. The days when we innocently spoke of 'gourmet' food seem as distant as the days of 'casseroles.' There is scarcely a child left in town who doesn't know how to eat an artichoke or wield his little snail tongs, or an adult who can't handle phyllo pastry sheets and eviscerate squid."[35] While still expensive, the food processor, that year's smash-hit kitchen utensil, seemed to expand the possibilities and the culinary horizons exponentially. Now, ambitious home cooks were told, a mere flick of the wrist could put pasta dough, pâtés, quenelles, *taramasalata*, and authentic Cantonese *jao tze* on one's fingertips. It was, said *Time*, a "gourmet breakthrough."[36]

But the breakthrough was still a limited one, confined mainly to some of the upper and middle classes. For the most part, Americans assimilated foreign foods into their diets the way most other people do: They used them as "sauces" for their still-familiar "core" foods—chip dips, sandwich fillings, and so on—or domesticated them with familiar markers such as ketchup or mustard. They toned down spicing that was too piquant for their tastes and served dishes in the American context and order, rather than in the foreign one—Mexican frijoles as a side dish to the main course, rather than as a separate course following it; soup as a first, rather than a last, course in a Chinese meal, and rice served with, rather than toward the end of, a Chinese banquet.[37]

Meanwhile, high-status French cooking, reeling from the cholesterol-phobic perception that it meant bathing everything in butter and cream, had received a shot in the arm from the nouvelle cuisine. A new wave of chefs rejected the older haute cuisine tradition, which transformed foods through elaborate processes, calling it a kind of falsification. Instead, they emphasized simplicity of preparation—grilling, roasting, or poaching—to bring out the natural taste of fresh ingredients. These were served, not in Escoffier's old *espagnole* or *béchamel*-based sauces, but in reductions of their own juices. Fresh, raw, or barely cooked vegetables were esteemed as never before.[38]

Initially, the American echoes of the nouvelle cuisine resounded mainly in expensive restaurants in a few major metropolitan areas. In 1972, when Alice Waters launched her trail-blazing Berkeley restaurant, Chez Panisse, she was inspired by hearty French home cooking of the kind described so lovingly by Elizabeth David in *French Country Cooking*, particularly the idea of having her cooking reflect the fresh products of the changing seasons. As unimpressed by vegetarianism and cholesterol concerns as David, she searched out local suppliers of the finest butter, cheese, and suckling pigs, as well as excellent vegetables. By the end of the decade, however, her cooking techniques had been modified by nouvelle cookery, particularly in the treatment of vegetables. The nouvelle affection for the hot grill was amplified by the discovery of the unique aroma of southwestern/Mexican mesquite wood. Her food, Waters thought, had evolved into "a celebration of the very finest of our regional food products."[39] It had also spurred the development of "California cuisine," whose emphasis on grilled foods garnished with the freshest of raw or barely cooked vegetables, fruits, and edible flowers helped inspire a surge of interest in other American regional products and cooking. Now, said Jane and Michael Stern,

> modern menus provided obsessive accounts of each ingredient's pedigree. . . . No more "steak au poivre."' Instead diners were enticed with "entrecote cut from a two-year-old, corn fed, Montana steer of Freesian-Holstein lineage, dry-aged five weeks, broiled over mesquite and grape vine clippings,

> served with a natural Gilroy-garlic gravy of *Morchella semilibera* mushrooms harvested by the Syfert family in the woods outside Boyne City, Michigan."[40]

Home cooks tried to follow suit, at least in heightening their appreciation of good produce. Suddenly supermarkets across the country were expanding their fresh produce sections with a widening variety of products. "Crisp fresh vegetables," said *Time*, "which used to be as scarce as lap-wings' eggs, have become a mainstay of any well-planned menu . . . lightly steamed, stir-fried, Chinese-style, or tossed raw in oil and vinegar."[41]

The appreciation for regional cuisine and foods struck a resonant chord among Americans without strong ethnic ties.[42] "On their way to *gourmandise*," said the rather overexuberant *Time*, "American cooks have rediscovered the glorious raw materials and inimitable provincial dishes of their own country. Newly appreciated are such home-grown marvels as Long Island duckling, Maine lobster, Maryland lump crabmeat, Chesapeake oysters, Gulf shrimp and pompano, Louisiana crawfish, California abalone and Columbia River salmon. Back in style are New England boiled dinners, Kentucky burgoos, Florida conch ceviche, New Orleans gumbo—and soups, chowders, breads and pies of every stripe and spice."[43] Calvin Trillin stimulated sophisticates' interest in often-scorned regional specialities with his witty pieces in the *New Yorker* extolling down-to-earth American foods consumed in the most unchic surroundings. People who ten years earlier had smacked their lips over escargots and *boeuf bourgouignon* now argued over where one could find the best ribs, crab cakes, chili, or barbecue. In contrast to LBJ, Georgian Jimmy Carter saw no need to disguise his down-home food tastes when he moved into the White House in 1977. When after his election he let it be known that he expected to see grits and other southern dishes on the White House menus, Julia Child voiced her strong support for going even further and "trotting out American regional dishes for official visitors."[44]

Meanwhile, Paul Prudhomme, a massive young entrepreneur in the kitchen, was beginning to combine sophisticated New Orleans Creole cuisine with zesty back-country Cajun cooking. In 1979 he began to dish out "Louisiana cooking" in K-Paul's Louisiana Kitchen in New Orleans. A masterful promoter, the genial Prudhomme and his restaurant were soon featured on PBS, the three major networks, and in innumerable newspapers and magazines. He rapidly became a regular on the talk show and charity dinner circuits, marketed a line of spices and sauces bearing his name, and began writing a cookbook (de riguer for star restaurateurs of the 1980s). By 1983, thanks in part to his promotional genius, regional American cooking had risen so high in status that President Ronald Reagan (no doubt prompted by his wife Nancy's sensitive antennae regarding style and status) invited him to cook for the Group of Seven heads of state at the 1983 economic summit meeting in Williamsburg, Virginia.[45] Soon the nation's fire departments, which had briefly stood down while the Continental orgy of flambéing abated, were again on high alert as acrid smoke billowed out of restaurants across the nation where cooks tried to imitate his signature dish, blackened redfish, by searing pompano, bluefish, shrimp, pork, and, it seemed, practically anything else they could fit into a dry, superheated frying pan.

The greater variety ethnicity and regionalism added to the American menu obviously helped counter the social, economic, and technological forces that had been tending to standardize and homogenize the national diet. So did the revival of cooking and eating as status symbols. Brillat-Savarin's aphorism "Tell me what you eat: I will tell you what you are" seemed more apt than it had been for fifty years, as the social classes again diverged in their food tastes and expectations. In particular, the national community of shared values that had persisted for twenty-odd years after World War II—years in which even those in the highest economic and social brackets appeared to share the straightforward national food tastes—seemed to break down, as food again became an important sign of distinction. Unlike previous eras of food snobbery, high status could now be derived from preparing the right foods—not just consuming them. A Palm Beach reporter noted in 1977 that "there's nothing more chic right now than a small gourmet party prepared by the hostess, instead of her staff."[46] Hundreds of cooking schools sprang up (Boston alone had twenty-nine of them in 1977) to teach the well-heeled hostess how to pull off this kind of thing.[47]

The widening of the class gap seemed to accelerate in the aftermath of the 1973–1975 oil shock, the spread of the Rust Belt, and other industrial woes. As well-paying blue-collar jobs disappeared and were replaced by lower-paying jobs in service industries, many on the lower-middle rungs lost ground, scotching their hopes of moving up to a middle-class style of life. But an increasing number of well-trained and/or well-off young people in the top 20 percent income bracket were beginning to reap the benefits of what some called the new information-based economy.[48] For them food became more than simply the mark of travel and sophistication it had become in the 1960s; it was central to one's "lifestyle." A reverence for anything "fresh" was a badge of this new attitude toward life. Like the counterculturites, the new "foodies" sought to eat things that were as close to their living, natural state as possible. "One of the nostrums of fashionable eating," the Sterns have written, "was that everything ought to be undercooked, therefore more purely itself. Vegetables and pasta were always served painfully *al dente;* bloody rare duck breast and pink-fleshed chicken were considered proper; salmon was sautéed on one side only, uncooked tuna fish was pounded into oceanic carpaccio, and even beans were cooked so briefly they broke when bitten."[49] An appreciation for French nouvelle, "Northern Italian," ethnic, American regional, and, of course, healthy foods also distinguished the upscale from those in the lower orders, who were thought to be wedded to unimaginative, overprocessed, unhealthy, and fattening "junk" foods, all of which were dizzyingly high on the food chain. Even their own mothers' cooking—now defined as the bizarre manipulation of 1950s-style processed foods—became the butt of humor rather than, as in previous generations, something to be revered.

Perhaps nothing exemplified the revival of class differences better than what happened to pasta. By the 1960s, as we have seen, Southern-Italian-inspired dishes based on pasta and tomato sauce were essentially classless parts of the national diet. But in the late 1970s, old standbys like spaghetti and meatballs went out of fashion among higher-class diners, who were expected to favor more esoteric kinds of fresh pasta, sauced with ever more exotic foods. Southern Italian inspiration was shunned in favor of dishes purporting to come from the more sophisticated and stylish north of Italy. Soon Italy itself was left behind. Alice Waters said her 1984 pasta cookbook was "more Provençal than Italian in inspiration, highly personal in execution." (Indeed, dishes such as buckwheat noodles with caviar and *crème fraîche*, or pasta with truffles, chanterelles, and endives roamed as far from Aix-en-Provence as from the Mezzogiorno.)[50] Meanwhile, a popular commercial for Alka-Selzter showed an overweight, distinctly unfashionable-looking Italian-American moaning "Mamma-mia, dattsa some spicy meatball" and reaching for relief.

By then graduates of Waters's kitchen and a host of imitators were fanning out on both coasts and into mid-America, mesquite supplies in hand, visions of Waters's nasturtium butter in their heads. There they peeled grapes and kiwis and nurtured their edible flora, using them to decorate tiny portions of grilled or poached main courses. Garnished by "baby" vegetables, these floated on, not under, painfully thin sauces on enormous plates of dazzling colors and shapes. The well-heeled patrons paying so much for so little appeared to have found a way of eating much more bewildering to the vast majority of Americans than that of their legendary forebears of the turn of the century, who sat down to tables that groaned under the weight of a plethora of delicacies. The food market, industry analysts noted approvingly, was becoming increasingly "segmented," with expanding "upscale" niches representing welcome hope for higher profit margins.[51]

One of the reasons for the accelerating pace of segmentation was the difficulty of staying ahead of the masses—in this case the mainstream middle class—whose tastes were being affected by media coverage of upscale food trends and the foreign travel boom. The collection of fund-raising community cookbooks recently assembled at Radcliffe College reflects this well. Historically, most of these compendiums of recipes women served in their homes had been solidly "American." From the 1930s to the 1960s, they reflected the pride housewives felt in mixing together strange new concoctions of convenience foods. Condensed soups, processed cheese, cracker crumbs, potato chips, and Jell-O often held pride of place. Then, in the mid-1970s, dishes that had swept upper-middle-class New York City in the 1960s began to appear on their pages. Admittedly, the "French Onion Pie" in the Cohasset, Massachussetts, Garden Club's *Cohasset Entertains* (1978) is made with eggs, milk, and canned onion rings; the "Seafood Casserole" is still layered with white bread and Velveeta cheese; and the "Creamy French

Dressing" is made with canned condensed tomato soup, cider vinegar, and lots of sugar. But sprinkled throughout are recipes for dishes such as *blanquette de veau* ("Super Elegant"), *saltimbocca alla Romano*, quiche (with canned mushrooms and shredded Bonbel cheese), filet of sole and asparagus in a white wine sauce thickened with egg yolk and cream, and minestrone; while not rigorously "authentic," these represent a real departure from the tastes and techniques of traditional American home cooking.[52]

The cookbook put out that same year by the volunteer committee at the Franklin County Hospital in Greenfield, in the more rugged western part of the state, is also schizophrenic. On the one hand, it is full of American bowdlerizations of foreign recipes with American-invented names such as "Chelupez"—frozen "tacos" (meaning tortillas) baked with frozen onion and American (i.e., processed) cheese—and "Lamb Khayam"—lamb cooked in wine, yogurt, Worcestershire sauce, and curry powder. But a dish made from the recipe for "the best moussaka" in Athens would not be rejected in one of that city's better tavernas, and the pistachios, almonds, and dates in "Shurin Chicken" would not contradict its claim to Iranian origin. Most striking, though, considering that the cookbook was for people living some distance from a major metropolitan area, are the number of recipes calling for items such as shallots, bulgar wheat, phyllo pastry, feta cheese, canned jalapeño chiles, and imported black olives.[53] "Supermarkets coast to coast now stock such one-time exotica as game pâtés, Beluga caviar, imported mustards, goat and sheep cheese, leeks, shallots, bean curd, pea pods, bok choy, capers, curries, coriander and cornichons," said *Time* magazine in 1977.[54]

By then large food processors had sensed that the ethnic bandwagon was heading somewhere profitable and were taking up its reins. They had begun to acquire medium-sized ethnic food manufacturers—particularly of Italian and Oriental foods—during the diversification drive of the late 1960s. Prince spaghetti, Prego tomato sauce, Chun King Chinese vegetables, El Paso Mexican products, and a host of others were gobbled up by decidedly nonethnic corporations such as Standard Brands, Beatrice Foods, and Borden's. The tough sledding in the aftermath of the 1973 oil shock—the soaring food costs, price controls, stagnant markets, and consumer resistance to processed foods—led them to seek expanded shares of the torpid market by mounting national campaigns for these newly acquired ethnic products. They also developed a new variation on the old strategy of product line extensions: ethnic variations on well-worn themes. Why risk developing completely new products when one could add Italian, French, or Oriental variations to already well established lines? All that was needed were some inexpensive flavorings—tomato sauce for Italian, soy for Oriental, imitation wine for French, chili powder for Tex-Mex—and a change in packaging. (Not all consumers were impressed. "How many different kinds of beans can you eat?" a supermarket shopper asked. "I don't need sixteen different brands of baked beans or twenty-five kinds of pinto beans.")[55]

Ethnicizing processed foods also helped address a problem inherent in processing itself, the loss of flavor. This had normally been dealt with by adding salt, sugar, or monosodium glutamate, but mounting pressure to cut down on these encouraged a turn toward stronger flavors from the ethnic repertoire. It was for this reason, for example, that manufacturers of crispy snack foods who had reduced the salt content in their products developed Mexican (nacho cheese), Italian (pepperoni), and British/Canadian (vinegar) flavors for them.[56] Food processors scoured the earth's cuisines searching for new flavors. In 1973 the industry journal *Prepared Foods* began to suggest a different one each month—Portuguese, Swedish, Jamaican, West African—for flavor-mining.[57] As usual, the industry found fool's gold more often than it struck the real thing. Naarden International Flavors, Inc., for one, came up empty with a peanut butter–based coating for corn snack foods, which it claimed "combined the popularity of ethnic food and peanut butter." It came in two flavors: one a sweet and spicy Indonesian satay flavor, the other a rather far-out Hungarian flavor, with paprika, onion and, beef broth mixed into the peanut butter base.[58]

Foreign baked goods were more successful, but often after having undergone major transformations. French visitors were nonplussed to see breakfast croissants become "croissanwiches."[59] Thanks in large part to this innovation, by 1985 Americans were spending seven hundred million dollars a year on crescent-shaped rolls.[60] Bagels emerged from big-city Jewish ghettos to find a sizable niche in the mass market, but only after a thorough reconstruction. They were normally twisted by hand to make a good-sized hole and boiled before baking—both labor-intensive processes. Now, a way was found to extrude them from machines and steam, rather than boil, them. Their holes were eliminated to facilitate making "pizza bagels" or sandwiches, and they were considerably lightened to accommodate mainstream tastes. The end result neither looked, felt, or tasted like a traditional bagel—crisp on the outside, doughy on the inside—but remained, for reasons philosophers may ponder, a bagel.[61]

While class differentiation increased, gender differences became more muted, largely as a result of the still-increasing proportion of married women in the labor force. By the mid-1970s over 40 percent of them were employed outside the home; by 1982, a majority of them were.[62] The wholesale entry of middle-class married women into the work force was particularly striking. Nearly 60 percent of the women entering the labor force between 1960 and 1977 were married to men with *above*-average incomes.[63] Indeed, from about 1973 until 1990, any real gains in middle-class living standards—more expensive housing, transportation, vacations, and so on—came mainly as a result of the rise of two-income families in that class. Except for the top fifth, who benefited from Reagan-era tax cuts and other kinds of government largesse, most American families' real incomes remained about the same. Indeed, for a majority of American families, working wives were now essential to prevent a sickening falloff in living standards.[64] Thus, Barbara Ehrenreich has noted (with a touch of exaggeration), by 1978 "in the popular media the full-time housewife had sunk to approximately the level once occupied by single women. She was more and more likely to be portrayed as the object of pity."[65] Men were now expected to contribute to housework, and the old idea that home cooking was an exclusively female pursuit became passé. Husbands were no longer confined to the backyard barbecue ghetto, and the kitchen was no longer the mysterious female sanctum sanctorum, where displays of masculine competence were signs of effeminacy.[66]

Yet the changing attitudes hardly bespoke a clear feminist victory, for cooking's rising status in itself made it an acceptable male pursuit. In a way, this represented a reversal of the process of "feminization," in which traditionally male occupations lose status when they become predominantly female. As more males mastered home cooking, its status rose; as its status rose, it became more acceptable for males to do it. This was reflected in the media: As the food pages of major newspapers stopped being mere filler around supermarket ads and drew closer to the journalistic mainstream, accomplished journalists such as Raymond Sokolov, William Rice, and John Hess joined Craig Claiborne in what had previously been a decidedly low-status journalistic ghetto. The media now featured prominent celebrities such as actors Danny Kaye and Walter Matthau and Hollywood director John Frankenheimer as outstanding home cooks: standing at their woks and restaurant-gauge stoves, being creative in the kitchen. Home cooking, which had been regarded as quintessentially feminine—an expression of women's nurturing, emotional, and inutuitive nature—could now assume what was regarded as a decidedly masculine cast. "Cooking and law are quite similar," said a Manhattan attorney who had been a star pupil at two cooking schools. "With both, there's the challenge of problem-solving, logic and reasoning."[67] The new machismo in the kitchen also brought the decline of the "man-pleasing" dish.[68] While gender differences over quantity remained, foods themselves, like clothing and hair styles, became more "unisex."[69] But this one tendency toward a more homogeneous diet could hardly counterbalance the powerful trends toward segmentation, diversity, and choice. The contemporaneous boom in eating out, for one, helped to present Americans with an array of food choices that was more dizzying than ever.

CHAPTER 15

Fast Foods and Quick Bucks

Mentioning that I am writing about modern American eating almost invariably evokes the comment: "Oh, the rise of fast foods—McDonald's and all of that." This is particularly true abroad; from the Champs Elysées and Piccadilly to the Ginza and Tienanmen Square, McDonald's and Kentucky Fried Chicken are regarded as the quintessential symbols of how Americans eat—successors to those hallmarks of the postwar years, the can opener and home freezer. It is obvious why this is so. For one thing, until recently most served only distinctively American foods: hamburgers, batter-coated fried chicken, milk shakes, and cola drinks. The teenaged help maneuvering among the gadgetry behind the counters with the precision of well-drilled football teams also seemed fitting reflections of the American obsession with time- and labor-saving. Only in America, it seemed, could anyone aim—as Burger King initially did—to serve customers a complete meal in fifteen seconds. The franchise system upon which they grew seemed to represent a happy marriage between the two forms of capital thought to have made America great: the hard-working, risk-taking individual franchisee and the organizational skills, know-how, and promotional capabilities of the giant corporation.

Of course, none of these things was new in itself. Americans' reputation for eating quickly stretched back to the early years of the republic. There were self-service restaurants doling out quick lunches to white-collar workers who ordered them at a counter and took them to a table to eat in New York and Chicago in the early twentieth century, where their boisterous bustle earned them the nickname "smash-and-grab places." Other kinds of self-service restaurants, such as cafeterias, have roots almost equally deep. Nor, as we have seen, were franchising and drive-ins new in the 1960s. As for technology, while Burger King's famed "Insta-broiler," which turned out cooked hamburgers in gravy on rolls at the rate of four hundred an hour, was indeed impressive, such contraptions were also common in other kinds of 1950s restaurants.

But by the late 1960s, the drive-ins with call-boxes and Rube Goldberg-esque serving machines had faded away, and "automatic restaurants" such as Pat Boone Dine-O-Mat, Buy-O-Mat, and White Tower's Tower-O-Matic, with their frozen, microwavable foods, hardly left a mark; they had been supplanted by the new, limited-menu, self-service fast food chains.[1] These burgeoning outlets had triumphed because to consumers they seemed to combine all of the advantages of lightning-fast service, drive-in convenience, and the economies of mass production techniques. To their owners, they also held the promise of a well-disciplined labor force immune to the vagaries of traditional restaurant workers. (One reason the McDonald brothers fired their attractive female carhops and converted to a drive-in format was that they suspected that the girls were more interested in selling their bodies than the hamburgers. They developed their militarized production system, using teenaged boys, to—in the words of one McDonald—free themselves from "drunken fry cooks and dishwashers.")[2]

But there was more to the new chains' success than this ability to put all of these virtues into a new package. At the outset, at least, they also tapped into something that had fueled the rise of earlier chains: American concern for restaurant hygiene. Ever since the germ theory of disease swept America in the late nineteenth century, smart restaurateurs had profited from Americans' fretfulness over the cleanliness of what emerged from the world hidden behind the swinging kitchen doors. The nation's first successful chain, the Fred Harvey System, which arose with the germ theory, had been based on confidence in the cleanliness of the food served in often woebegotten railroad stations by the immaculate "Harvey Girls." Soon thereafter John R. Thompson of Chicago had built one of the great chains of urban restaurants on the basis of white tile, bright lights, and gleaming dining rooms. After the First World War, companies like White Castle, White Tower, and Horn and Hardart did very well by following in this path. When Walter Anderson and Edgar Ingram founded the White Castle hamburger chain in the 1920s, they sought, in Ingram's words, to "break down a deep-seated prejudice against chopped beef" and modeled their stands on Chicago's water tower, making them "white for purity."[3] This was also, as we have seen, the key to White Tower's success, as well as a reason for the nation's confidence in Duncan Hines. Ray Kroc, the mastermind behind the rise of McDonald's as a national chain, hoed the same row, although with unusual vigor. Personally fastidious, on weekends he himself joined in cleaning his own Des Plaines, Illinois, outlet, hosing down the garbage cans and scraping chewing gum off the green cement in front of the store with a putty knife.[4] Like their predecessors, the structures under the golden arches featured sparkling white tile, but Kroc also added great swaths of glass, putting the food preparation into a virtual fishbowl. "You could see the cleanliness," said Edward Schmitt, McDonald's president. "How many restaurants are there in the United States where you can look into the food facilities and preparation area?"[5]

In fact, as we have seen, Schmitt was crediting his boss with innovations that had long preceded him. One could watch the counterman prepare one's hamburger in scrupulously clean surroundings at White Tower too, pay as little for it as at McDonald's, and even have it made to order. Yet White Tower and the others withered while McDonald's flourished. Why? Here is where Kroc's real genius lay: He had an unerring sense of how to exploit the new demographics. By 1960 most White Towers, White Castles, Horn and Hardardts, and other fast food chains were still in the inner cities, often in fraying surroundings that bespoke a migrated white working class. Kroc, aiming straight at the families that had created the baby boom, would not go near the center city. He concentrated instead on the new commercial strips of the suburbs. "Where White Tower had tied hamburgers to public transportation and the workingman . . . McDonald's tied hamburgers to the car, children, and the family."[6] As a McDonald's president subsequently explained, targeting the suburbs "was a conscious effort to go for the family business. That meant going after the kids."[7]

Aiming at baby boom families meant that McDonald's turned its back on those enthusiastic drive-in patrons of the mid-1950s, teenagers. Deliberately setting out to create an "antijoint" climate, Kroc went to extraordinary lengths to prevent McDonald's restaurants from becoming teenage hangouts. He banned jukeboxes, vending machines, and even telephones from his outlets and, like the McDonald brothers, refused to hire females. (Teenaged female staff, he said, "attract the wrong kind of boys.")[8] The outlets' cleanliness was also intended to mark them off from the conventional drive-ins, which often sat amidst swirling clouds of discarded napkins and teenage litter. Franchisees were also enjoined to burnish the family-friendly image by supporting local family-oriented community functions.[9] Discovering that children determined where three out of four families ate, McDonald's set out to make visits to its outlets "fun experiences," with special giveaways and packages for children's meals. By 1970 the many millions the corporate giant spent on child-oriented television advertising and local promotions had succeeded in making its clown mascot, Ronald McDonald, identifiable by fully 96 percent of American children, making him second only to Santa Claus on that score.[10]

Kentucky Fried Chicken also enjoyed unprecedented success (in the late 1960s its sales volume exceeded that of McDonald's) by cultivating a family image to attract suburban baby boomers. It was founded in 1954, the same year Kroc made his famous deal with the McDonald brothers to franchise the system they had pioneered in California, by Harlan Sanders, another dropout (Kroc dropped out of high school; Sanders left after grade seven). Prominent in its marketing was the avuncular, white-haired, white-goateed, white-suited "Colonel" Sanders amiably presiding over happily munching children.[11] As was so often the case, Burger King played catch-up in the image game. Its little cartoon "Burger King," attired in royal robe and crown, tried to assure children that it too was a fun place to eat, but Ronald—all image and no personality—and the genial Colonel (who was actually quite a hothead) could not be matched.[12]

While not yet a bona fide fast food, pizza was soon giving the fast foods a run for the consumer's money. By the mid-1950s, thanks to the popularity of spaghetti and tomato sauce, a taste for a white farinaceous base slathered in thick and salty tomato sauce had become an integral part of the American palate. The country was therefore well primed for the invasion of pizza. Yeast bread dough flattened and baked with a tomato sauce topping had appeared in Naples as early as the seventeenth century, not too long after the tomato was introduced into Europe from America. By the late nineteenth century, it had become common to add mozzarella cheese to the topping too, but the dish remained mainly a local Neapolitan speciality: a poor people's food that evoked little interest outside the town and decreasing interest within it. It was given a new life, however, in bakeries in some of the East Coast's Little Italies, which began turning out versions of it soon after the turn of the century.[13] By the 1920s it was a common sight, if not a major seller, at Italian festivals in New York City.[14]

In the 1950s, however, pizza suddenly burst onto center stage. In part this was because it fit so well in the culture of the times. It was regarded as an ideal family food, equally acceptable to all ages and both sexes. Its taste hardly departed from the tried and true, yet its form could be readily accommodated to the era's newer, more casual ways of eating: children's parties and snacking in front of the television set. The informal, communal way it was eaten in restaurants made it particularly popular with teenagers, and by the mid-1950s boisterous "pizza parlors" dotted the main streets of Italian neighborhoods, their oversized booths for six or eight crammed with voracious young eaters, while others lounged by the entrance waiting for take-home orders. "A trip to Wooster Street for Vinnie's or Sal's thin crusted, aromatic cheese and sausage pizzas was a gustatory event," recalled an ex–New Haven aficionada of Jewish origin. "No short cuts, no tricks, just the mouth scorching mozzarella and the olive oil dripping down our chins, the delicate tomato sauce and the pile of filthy napkins on worn formica tables."[15] Pizza also became the hottest restaurant item of the 1950s because, unlike most pastas, it was not particularly affected by delays between cooking and eating. This made it ideal for the two main growth sectors in the television-battered restaurant industry, drive-ins and take-home places. By 1956 it had shunted aside hot dogs as the most popular item in both.[16] By the late 1960s, Americans were consuming two billion pizzas annually, using six hundred million pounds of mozzarella cheese and eight hundred million pounds of tomatoes in the process.[17]

By then the major problem that stood in the way of pizza becoming a real fast food—the crust—was on its way to solution. To make a traditional crust, after the breadlike dough had risen it had to be weighed, cut, stretched, and flattened, often by tossing it in the air. The sauce was then slathered on, toppings were added, and it was baked for about twenty minutes, bringing total preparation time after the dough had risen to almost thirty minutes. Greek-Americans, who had been opening storefront pizzerias on the East Coast since the early 1950s, now discovered that the day's supply of pizza dough could be rolled out in the morning and refrigerated in ten-inch pans, ready to be quickly topped and baked in the same pan when ordered. While it made for a flatter, thinner crust, this drastically reduced labor costs and made the mealtime rush more manageable.[18] But the new method was also a boon for franchised chains such as Wichita-based Pizza Hut, which turned pizza into a true fast food. They developed a method for freezing the dough in a central commissary, from where they sent them to their outlets to be slathered with toppings and quickly baked on conveyors in special infrared ovens.[19]

By the mid-1960's McDonald's had already set many an American to salivating, not over its food, but by becoming the first franchised food operation to "go public" with a stock issue. The immense amount of cash this put in Kroc's pocket aroused many others to dream of a double killing; huge profits might be made on both up-front franchise fees and a stock flotation. So frenzied did the atmosphere become that sharp promoters, such those behind Lum's, a chain whose main innovation was a hot dog steamed in beery liquid, made enormous profits by floating stock even before they had sold a significant number of franchises. Some had little more to offer than a celebrity's name—Johnny Carson, Joe Namath, Mickey Mantle—and a pie-in-the-sky prospectus. In 1968 even John Y. Brown, Jr., the mastermind behind the rise of Kentucky Fried Chicken, was gulled by a debonair, smooth-talking Englishman into having KFC pay an outrageous sum for the shares of his H. Salt Fish and Chips chain, a company with remarkably poor prospects in America. (Among other things, KFC discovered—too late—that many Americans did not know that chips are french fries.)[20]

Inevitably, in the early 1970s most of the high flyers came down to earth, many with a thud. Minnie Pearl's Fried Chicken's stock sank from $23.00 to 12½ cents before disappearing under a different name.[21] Even KFC, still the industry leader, dropped from $55 a share in April 1969 to $23 a year later.[22] However, about fifty franchising chains staggered along, hoping to be acquired, merged, or gobbled up. Many were: Burger King took over Carrol's and Burger Chef; Gino's merged into Roy Rogers. At the same time, food processors looking to diversify were tempted to expand into this related field. Pillsbury's had acquired Burger King in 1967. Now, Hueblein took over KFC, Ralston Purina bought Jack-in-the-Box, General Foods began a disastrous fling with Burger Chef, and Pepsico took over Pizza Hut. The same forces of concentration that had transformed food production, processing, and distribution had caught up with the new fast food industry.

The small entrepreneur was soon being squeezed out at the lowest level, that of the individual franchisee, as the cost of franchises in the more successful chains climbed to well out of reach of those without substantial resources. Blocs of franchises in the lucrative chains were controlled by very large investors. After 1976 Pizza Hut would not even sell franchises to individual investors.[23] The hope of getting a franchise in a successful chain became a pipe dream for the proverbial "ambitious little guy." In 1980 a prospective franchisee in a not-too-promising Mexican chain such as

Taco Time or Taco John's had to come up with $35,000 (including a $10,000 to $15,000 franchise fee), plus $100,000 to $150,000 for land and a building.[24] Increasingly, parent companies themselves took up new franchises and repurchased older ones.[25] At the same time, to compete with the giants, with their huge promotional budgets, by opening an independent outlet became a generally foolhardy proposition.[26]

So, despite the paeans of industry leaders such as Kroc to the spirit of entrepreneurship, the fast food chains helped snuff out a traditional entry point into the ranks of the self-employed for aspiring but capital-poor working people. For at least two generations the dream of renting a storefront, purchasing some used restaurant equipment, establishing a meager line of credit with some food suppliers, and opening a small restaurant had inspired many ambitious and hard-working Americans. Now, these dreams were headed in the same direction as previous rural aspirations to go into food-processing or urban hopes of opening neighborhood grocery stores.[27]

Kroc, the Colonel, and the other leaders in the fast food race had made their bundles by following the baby boomers as they progressed through childhood and adolescence. In the later 1960s, when they were teenagers, McDonald's—spurred by a labor shortage and new federal antidiscrimination laws—even began hiring females. In the 1970s, as the teenagers turned into young adults, the chains followed with a wholesale shift to "family-style" restaurants. Inviting brick and shingles replaced glitzy exteriors; cozy mansard roofs replaced soaring angular ones; glaring white tile and fluorescent lights were replaced by warm wood, posters, greenery, and softer lights. The old concept of discouraging patrons from lingering and forcing them to eat off the premises was the first to go. Burger King began installing tables in 1967; McDonald's followed in 1968.[28] In the 1970s the seats were made softer and the tables enlarged. The policy of confining the labor force as much as possible to behind the counter, where contact with the patrons was polite but curt, was altered. Now, friendly hosts greeted patrons and oversaw the queues, buspeople roamed about, and in some cases there was even table service. New lighter entrees such as fish were introduced, and salads and salad bars began to appear. Industry veterans shook their heads as McDonald's began serving fish, Burger King croissants, Bonanza chicken, and Long John Silver's wine. Pizza chains began serving hot sandwiches, submarines, pasta, wine, and even seafood.[29]

The chains also sought to adjust to what appeared to be the breakup of the traditional family structure. Many of the young adult boomers postponed or even rejected marriage in favor of more informal ways of living together. Those who did formalize their unions did not produce children at anything near the 1950s rate. However, they did take to divorcing at an all-time high rate, adding to the number of people living alone, with other singles, or in single-parent families. When their numbers were combined with a byproduct of the pathology of urban slum life—soaring rates of teenage pregnancy and fatherless families—the resulting statistics seemed to portray those living in traditional two-parent, two-and-a-half-child families, with mother home doing the cooking, as museum pieces. In March 1977 the Labor Department reported that the concept of a family in which "the husband is the only breadwinner and the wife is a homemaker out of the labor force and there are children" now applied to only seven out of every hundred households. Half of all households were now composed of single men or women or married couples without children. Almost all of the people in this category, male and female, worked outside the home and had little time or inclination for home food preparation. "When I was married I ate out maybe twice a week," a divorced male Kansas City executive told a reporter in 1977. "Now I eat out five nights a week. I like to be with people. I seldom go to a supermarket."[30]

The fast food chains responded by abandoning their exclusive concentration on the suburbs and opening outlets in and near the glass office towers through which hundreds of thousands of these white-collar workers hurried each day. Special breakfast and dinner items were introduced in the hopes of enticing them to stop by on their way to and from work. They mounted campaigns to entice females in particular to drop in at any time of day; McDonald's asserted, "You Deserve a Break Today." Statistics of all kinds attested to their success. By 1977 the proportion of the American food dollar spent on food outside of the home had risen to over 35 percent, up ten percentage points from 1954. Most of this increase was accounted for by fast food establishments.[31] By 1983 there were already three times as many fast food outlets as in 1963—more than 122,500. Their thirty-four-billion-dollar intake amounted to about 40 percent of all public eating place sales.[32]

Market research indicated that many fast food patrons were peripatetic diners who ate on impulse and had little brand loyalty. Many of them also ate out so often that monotony became a threat.[33] The slightest differences in menu, service, price, or promotion could thus be all-important. Some fast service restaurants therefore sought diversity by adding ethnic touches to their menus. As was the case with processing, this need not involve much real change. "Foreign and ethnic foods are all the rage these days," said *Fast Service* magazine, but all that was needed to "ethnicize" a menu was to change some of the spicing. "To make something German, don't hire a German cook, just give your roast beef a topping of German-style sauerkraut, that is, canned sauerkraut with some caraway seed added. Mix oregano, basil and garlic with canned tomatoes, add chicken, and you have an unusual Italian hero sandwich. For Chinese, add one or more of: ginger, anise seed, garlic, onions, red pepper, fennel seed, cloves, or cinnamon."[34]

Some entrepreneurs tried to develop fast food chains based entirely on one regional or foreign-sounding cuisine, but none could rival the giants serving American-style food. Even Oriental food, so popular in full-service restaurants, fell flat as fast food, largely because almost everything in its repertoire was unsuitable for flash freezing. Pizza chains were no exception to this generalization, for by then pizza had for the most part lost its Italian connotations. Indeed, part of Pizza Hut's success derived from a mid-1970s campaign to "Americanize" its image, which included replacing the pizza-tossing Italian chef on its logo with an outline of a roof.[35]

Only Mexican chains achieved significant success, but this was mainly because they were able to piggyback on pizza. What they called tacos, burritos, and so on were essentially a variation on pizza: a flattened dough topped with tomato-flavored meat sauces, some vegetables, and cheese. The pancakes were smaller, folded, and often crisper, but the step from pizza to tacos was hardly more daring than the one that had led Americans of the 1950s from pasta to pizza. Mexican fast food also received a boost from the national mania for snack food. The popularity of corn chips spurred supermarket sales of taco shells—most of which are really large, folded corn chips—and sauces and fillings to go with them, helping to create a core of consumers who were then willing to risk Mexican fast food.[36]

Still, the Mexican connotation was a handicap. When Pepsico took over Taco Bell in 1978 and began expanding it from its southwestern base, it mounted a campaign to assure customers that its tacos were no more spicy or un-American-tasting than hamburgers.[37] Others followed suit. "There is no problem with consumer acceptance of Mexican food," a Taco Time spokesman assured a trade journal, for it was "seasoned to American taste."[38] In 1982, as Taco Bell planned a further expansion campaign, market researchers warned against promoting the food as Mexican. In particular, they said, Americans distrusted Mexican restaurants as dirty. As a result, Mexican allusions were banished from Taco Bell's decor and menus, and its original symbol—a sleeping Mexican in a sombrero—was replaced by a bell.[39] The places of origin of Taco Bell's competitors—Eugene, Oregon (Taco Time), Wyoming (Taco John's), Ogden, Utah (Taco Maker), Topeka, Kansas (Taco Casa), Chicago (Pepe's), and Mississippi (Pedro's Fine Mexican Foods)—hardly augured any more commitment to a Mexican identity, and most followed suit.[40] Thanks to this bowdlerization process, by the late 1980s Taco Bell had led the Mexican sector into a 5 percent share of an otherwise overwhelmingly American-style business, but by then it was questionable whether anyone but Mexicans should have considered it foreign food.[41]

Full-service restaurant chains were more successful in ethnicizing their menus. The more formal ones had initially taken quite a beating from the fast food chains, whose informality fit in better with the mood of the late 1960s.[42] The perennial labor problem seemed to lock all full-service restaurants into increasing dependence on preportioned, mainly frozen, generally nondescript dishes. By 1975 giant food service corporations turning out bland dishes in huge central facilities had a virtual hammerlock on the $2.8 billion restaurant-supply market. The larger full-service chains merely emulated them. In 1975 one of them, Quality Inns International, tried to raise quality and cut costs by "cooking from scratch," but the bold experiment soon ran aground on the shoals of labor costs, the very factor that had led to frozen foods' triumph in the first place. When the more skilled kitchen workers needed for the new/old system demanded wages as high as six dollars an hour, a Quality spokesman announced, "We're just not paying that kind of money," and the firm returned to the frosted fold.[43] A New York woman who applied for a job as a cook at the Stouffer's chain was told that they did not need cooks, only "thawer-outers."[44]

In the mid-1970s, however, the food service corporations began expanding their repertoire by offering frozen French-sounding dishes to "Continental" restaurants. (Calvin Trillin now quipped that the continent in question must be Antarctica.) Then, in the later 1970s, as restaurateurs realized that patrons were alarmingly peripatetic in searching for ever more varied eating experiences, they began turning out ethnic and American regional-sounding dishes. These dishes not only struck the requisite note of informality, but they also promised greater profit margins, because most used more starch and less meat. Now restaurateurs facing what a restaurant industry magazine called "the seeming inconsistency—if not downright madness" of divergent patron tastes could, at the flick of a teenager's wrist, serve any or all of frozen quiches, crepes, lasagne, canneloni, fettucini, or burritos alongside the usual array of steaks, burgers, and chicken fingers.[45] Bennigan's, Pillsbury's chain of "casual restaurants," featured croissant sandwiches and tacos as well as Cajun-style burgers and barbecued back ribs.[46] "Theme" restaurant chains, such as Victoria Station, whose Casey Jones–clad cooks worked in the open amidst old cabooses and gas lamps, also fostered the illusion of diversity.[47] Some industry consultants, such as Lippincott and Margulises in New York and S&O Consultants in San Francisco, specialized in developing not new recipes or foods but new themes. Working for Pillsbury's Steak & Ale subsidiary, S&O devised a chain called Orville Bean's Flying Machine and Fixit Shop—decorated with the fictitious inventor's contraptions such as steam-driven roller skates and an electric fork—as well as Juan and Only's, a Mexican restaurant with eclectic displays of old records, photos, and junk. (The themes themselves soon became as throwaway as the decor; to last two years was considered a success.)[48]

Much of the success of fast food and theme restaurant chains was based on their ability to attract working women. Working wives in particular ate out—morning snack, lunch, and dinner—much more than homebound ones. "I eat out at least three times a week," a thirty-year-old medical technologist told the *New York Times* while waiting for her moderately priced meal amidst the wood, greenery, and glass of a Los Angeles Great American Soup Company restaurant in April 1977. "Nobody feels guilty any more if they don't cook every meal at home. There is money to eat out because so many people are working."[49] By 1985 the average nineteen- to fifty-year-old woman ate 38 percent more of her food (measured in calories) in cafeterias than she had in 1978, 60 percent more in full-service restaurants, and an impressive 120 percent more in fast food outlets.[50]

Meanwhile, supermarket industry leaders had watched in stunned disbelief as the nuclear family, the core of their market, seemed to disintegrate before their eyes and Americans shunned their own kitchens in favor of other defrosters.[51] In 1960 Americans had spent twenty-six cents of every food dollar away from home; by 1981 thirty-eight cents of every dollar escaped the grocery trade.[52] While restaurants were the major culprits, take-home outlets also posed an increasing threat. But supermarkets counterattacked by expanding their own offerings of fully prepared foods to take home. Delicatessen counters now commonly featured barbecued chickens, spare ribs, lasagne, and a variety of pasta and other salads, while salad bars sold large arrays of washed, chopped, and sliced vegetables and dressings. The food-processing industries rolled in behind them, developing new products that could be just as easily microwaved at home as by teenagers in a restaurant. As a result, the variety of foods displayed in the supermarkets increased exponentially, as did the size of the stores themselves. The range of choices now available to the shopper made that which had seemed so impressive in the 1950s pale in comparison. At the same time, the choice of where to shop expanded, as chains of smaller markets such as 7-Eleven arose, specializing in prepared and convenience foods aimed at singles or families with working mothers.

The end result of all of this was an incredible extension of the food choices facing Americans. Although they still clung to certain "core" foods and traditional taste "markers," Americans were accompanying them with an ever-increasing variety of "sauces" and eating them in ever less formal fashions. Family meals seemed on the wane, even where traditional family structures survived. Snacking, on the other hand, was becoming a continuous process, indulged in practically at all times in all places. Indeed, it was calculated that three-quarters of all Americans derived at least 20 percent of their energy needs from snacks.[53] (This might explain the success of the song "Junk Food Junkie," which climbed to the top of the pop music charts in 1976.)[54] There were even predictions that among certain middle-class groups, particularly singles, "grazing"—snacking in different locales—would soon almost entirely replace formal meals. When this explosion of options regarding what, where, and when to eat was set against the recurring scares over food and health, the result verged on dietary chaos, at least for the middle and upper classes. Possibly the only thing that could have made matters even more confusing was a competing concern over weight loss, and this, of course, is exactly what was happening.

CHAPTER 16

Paradoxes of Plenty

The Reagan family had hardly unpacked its bags in the White House when Nancy Reagan began choosing new china for her new home. The enormous cost—$209,508 for 220 gold-embossed place settings—seemed to set the pattern for the rest of the decade, as the Reagans helped make ostentatious wealth more fashionable than it had been since the 1920s. So did their inauguration, a sixteen-million-dollar, multi-venue extravaganza tarnished only somewhat when the building's derelicts and bag ladies invaded the Union Station party and helped themselves to the lobster bisque, shrimp merlin, escargots, and veal laid out for the invited guests.[1] An economic downturn in the early 1980s hardly affected the upper-income beneficiaries of the tax cuts Reagan pushed through soon after assuming office, and the long shadows of foreign challenges to Americans' self-perception as the wealthiest people on earth could still be ignored. Newly rich financial wizards and media culture heroes took new pride in living "Life in the Fast Lane." It was to be the last of the century's ages of abundance: a time—like the 1920s—when American business leaders and their cheerleaders in the White House seemed to have found, in the "magic of the marketplace," the keys to everlasting prosperity.

Nancy Reagan, surrounded by new Sun Belt wealth, presiding over glittering events in her fashionable size 4 outfits, seemed to epitomize the decade. Unlike her immediate predecessors as First Lady, who left the details to their social secretaries, she took a particular interest in the food chef Henry Haller prepared for these functions, asking him to make the platters fancier, the color combinations more striking, and the portions smaller.[2] Her increasingly cadaverous appearance began to make "Babe" Paley's famous aphorism "you can't be too rich or too thin" look like a sick joke, but the remark seemed to neatly encapsulate the new elite's aspirations. A lust for wealth displaced older ideas of public service in Washington, drove considerations of responsibility to clients, stockholders, and the public from Wall Street boardrooms, and turned Ivory Tower college campuses into centers for glorifying greed.[3] Indeed, there was no such thing as "too rich" in a culture that viewed wealth as the ultimate sign of achievement and elevated a way of getting it—"the deal"—into an art. Success was defined in terms of consumption, the more conspicuous the better, including consumption of food. "May you have caviar wishes and champagne dreams," was host Robin Leach's signoff for the popular television program "Lifestyles of the Rich and Famous." Yet, perversely, there also seemed to be no such thing as "too thin," as corporeal ideals reached the thinnest extremes ever. The critics of the 1930s and 1960s who had bemoaned starvation amidst plenty now saw it occur in a form they never anticipated. For the first time in American history, a substantial number of deaths were directly attributable to starvation—but it was of a voluntary kind, the result of a rash of eating disorders, mainly among middle- and upper-class females. That people who could afford more than enough food would starve themselves to death reflected, in part at least, the rather bizarre turn ideas about food and body image had taken.

Yet while it may have reached unprecedented extremes, the ascendance of a slender ideal was not in itself an aberration. Nor was the fact that it coexisted with the booms in luxury dining, casual restaurants, fast foods, grazing, and snacking. As Claude Fischler has noted, "societies of abundance are tormented by the necessity to regulate feeding." This leads to a paradox: "They are at one and the same time impassioned over cuisine and obsessed with dieting."[4] Still, it is difficult to think of a society in which the paradox was more pronounced than in the United States of the 1980s.

On the surface, the most obvious culprit for at least the feminine side of the compulsion to be thin seemed to be "fashion." Yet for all that has been written about changing fashions in dress and body type, no one has come up with a compelling explanation of its vagaries, particularly in the twentieth century. Contradictory conclusions almost inevitably emerge from speculation about such things as the relationship between short skirts and women's liberation. For example, the short skirt styles of the 1920s and 1970s, which are often regarded as signs of liberation, can just as easily be interpreted as regressive, emphasizing women as sex objects. Few would argue that the revival of short skirts in the late 1980s had anything to do with liberation. On the other hand, the longer hemlines of the 1930s and 1940s are often associated with strong female figures such as Joan Crawford and Katharine Hepburn, who in many ways embodied a more independent image of women than the flappers of the 1920s or their miniskirted successors of the 1970s and 1980s.

Changing body ideals are equally difficult to link to wider social changes. As was indicated earlier, it is sometimes said that fulsome body types tend to be admired in societies where food is scarce while thinner ones are favored where it is abundant. But, although this generalization may apply to some remote cultures, it seems hardly relevant to modern industrial societies. As we have seen, even the Great Depression did not bring the amplification in ideal body type that might be expected; the urge to slenderize continued to plague many middle-class females. Conversely, the shapely ideal rose again during the long period of sustained prosperity after the war.

Indeed, in the mid-1950s women cinched in their waists with tight belts to emphasize bosom and hips to a degree second only to those who aspired to "hourglass" figures during the *belle époque*. Zaftig actresses such as Jane Russell, Anita Ekberg, Marilyn Monroe, and Jayne Mansfield were able to achieve stardom practically on the basis of the size of their bosoms alone. It is easy to see why only 5 percent of the women polled by Gallup in June 1955 said they were on diets.[5]

Yet for reasons that are still unclear, in the early 1960s the beauty and fashion pendulum began to swing back toward the thin ideal. A statistical analysis of the measurements of *Playboy* centerfolds and Miss America pageant contestants in the 1960s and 1970s has charted this, showing how both groups of women became considerably thinner over that period. The *Playboy* models also changed shape significantly, becoming much more "tubular," with smaller bosoms, larger waists, and smaller hips. The thinner trend was particularly remarkable, said the researchers, because it went in the opposite direction of the actual changes in young women's bodies reported by life insurers over that period: Their average weight actually increased. "It is worth noting," the study concluded, "that just over 5% of female life insurance holders between the ages of 20 and 29 are as thin as the average Miss America Pageant winner between 1970 and 1978." What was even more disturbing to the authors, who treated eating disorders in Toronto, was that "*Playboy* centerfolds and Miss America pageant contestants hardly represent the bony-thin body frame that is typically promoted by the fashion and advertising industries. It is obvious that the prevailing shape standards do not even remotely resemble the actual body shape of the average woman consumer."[6] By the early 1980s, study after study and poll after poll showed that a large majority of American women had come to think they were too fat.[7] Fully 76 percent of respondents in a 1983 *Glamour* magazine survey, for example, categorized themselves that way. Yet 45 percent of this group actually weighed less than the 1959 recommendations of the Metropolitan Life Insurance Company, which advocated a degree of slenderness that was soon abandoned as unhealthy.[8]

Yet where "the prevailing shape standards" came from is still not clear. Feminists often blamed the (male-dominated) media and fashion industries for creating another impossible hoop through which women were forced to jump. That the media abetted the weight-loss mania is indisputable. Sitcom heroines, television commercials, ads in the print media—all provide visual evidence for the growing obsession with slenderness, particularly after about 1968. A survey of six women's magazines counted 70 percent more articles on dieting in the years 1968 to 1979 than in the previous ten years.[9] Studies showed that women had come to believe that slenderness was the most important aspect of physical attractiveness. An analysis of primetime television showed that women with thin bodies were also associated with favorable personality traits.[10]

Why did this message pour forth at that particular time? On one level it seems linked with the emergence of the baby boomers as a major engine of the economy and culture. The media often apotheosized them as the repositories of Truth and Beauty—particularly the latter—and their youthful standards of beauty came to dominate all others. Now, both young girls and their mothers aspired to the same youthful body type.[11] As Fischler points out, this was unique in history. In most societies there are at least three stages of womanhood—a young, nubile one; a fecund one; and a mature, aging one—which women's appearance is expected to reflect. "In no civilization before our own," he says, "has the same ideal body and clothing type been imposed at these different ages and social roles."[12]

But there was something else that was also unique to our times. Although young nubile women had generally been considered sexually attractive in the past, they were usually rounded in shape. Now, the desirable young look was an angular one, taut and wrinkle-free, the image of an immature, prepubescent girl. Noting that adult women who managed to achieve this ideal would stop menstruating, the Toronto eating disorder therapists remarked that "it is ironic that the current image of sexual attractiveness in women is a shape associated with both loss of reproductive functioning and sexual appetite."[13] But perhaps there is no irony there, for the asexuality of this look posed no problem to many of the trend-setting men in the media and fashion industries. (Andy Warhol, the homosexual guru of "pop art," and his emaciated "star" Edie Sedgwick come to mind.) Of course this runs counter to some feminists' notion that demeaning fashions are imposed on women by lustful heterosexual males. Indeed, if anything, fashion seemed less concerned than ever before about catering to these instincts. As we have seen, the *Playboy* and Miss America body types—which approximated the conventional male ideal—were far fuller than what most women idealized. Most men seemed puzzled by, rather than attracted to, Twiggy, the fashion phenomenon of the late 1960s, who was 5′7½″ tall and weighed but ninety-one pounds.[14] These sexual differences persisted into the mid-1980s, when studies continued to show that males' ideal female body image was considerably heftier than that of females.[15]

This is not to ascribe the fashionable female image to some kind of gay conspiracy. Not only was there no conspiracy, but there were other important factors, aside from fashion, at play. As part of the era's turn toward the "natural," women rejected the layers of clothing and the stiff undergarments that had helped hide or support the fleshier parts of their bodies, exposing their real figures to unprecedented public examination. Miniskirts and minimalist clothing might even have reflected, as Roberta Seid suggests, the generation's extremism in other areas—particularly politics. But, as Seid also points out, perhaps the most important force underlying the compulsion to lose weight was the deep current of "fat-phobia" that had been steadily gaining strength in much of the industrialized world since at least the 1920s.[16]

Historically and anthropologically, fat people have been regarded in contradictory ways: as either predatory gluttons, voraciously grabbing more than their fair share of food, or as benign gourmands enjoying of one of life's great pleasures. As fears of food shortage and famine receded, so did the first—malign—view of fat people, although it still emerges from the collective unconscious occasionally, as in the negative feelings psychologists report people express when shown photos of fat people. But the second—benign—view also changed, particularly with the rise of dieting in the twentieth century. Increasingly, as we saw regarding the War on Hunger, being fat signified an inability to control one's impulses.[17] Middle-class fat people, it was agreed, must be deeply unhappy. How else could they feel about having lost their self-control and become slaves to their cravings? One wonders if the students of the 1980s readily understood Shakespeare's Julius Caesar when he said, "Let me have men about me who are fat; sleek-headed men who sleep o' nights." How could self-indulgent fat people enjoy the sound sleep of the guiltless? Conversely, why Caesar thought Cassius's "lean and hungry look" made him "dangerous" must also have been puzzling. They would think Cassius would be quite content to look like a rock star.

That males as well as females were affected would support the idea that much more than fashion or the media was involved. As we have seen, beginning at least as early as the 1910s, and particularly in the 1960s, an increasing number of nutritional experts blamed many of America's most serious health problems on obesity. Although they disagreed on many other scores, establishment nutritionists such as Frederick Stare and Jean Mayer were in solid agreement on this. In the 1970s more radical critics joined the antifat campaign, denouncing the corporate-dominated food industries for foisting fatty, oversweetened, nutritionally vapid food on unresisting Americans. It was "an establishment that wanted us passive, blissed out in front of televised sports, too impacted by beer and junk food to prevent the robbery of our health and country," said socially conscious *New York Times* sportswriter Robert Lipsyte.[18] These ideas fit in well with the campaign against commercials for sweetened cereals, sweet snacks, and "junk foods" on children's TV, much of which was based on the (fallacious) idea that overeating in childhood boosted the number and size of fat cells and caused obesity in adulthood. "The total number of fat cells we carry into adult life is totally dependent upon eating habits established during childhood and early adolescence," said a typical critic.[19] The fight against overweight also went along well with the larger turn toward the Negative Nutrition, for it emphasized getting overindulgent Americans to cut down and cut out. However, unlike other calls of the Negative Nutrition, fat-phobia itself threatened few powerful food interests. If anything, it opened new windows of commercial opportunity. Consequently, few experts were ready to challenge some of the questionable "facts" upon which it was based.[20]

With no one lobbying for avoirdupois on Capitol Hill, there were no objections to the 1977 Senate Nutrition Committee's *Dietary Goals* calling it a national evil to be extirpated. Other government agencies, such as the Departments of Agriculture and Health, soon joined the campaign. In 1985 a Consensus Development Panel of the National Institutes of Health declared thirty-two million Americans, about 28 percent of the adult population, to be overweight and warned that the available evidence indicated this adversely affected health and longevity. Not only was it psychologically burdensome, they said, but it was clearly associated with hypertension, high-risk serum cholesterol, non-insulin-dependent diabetes, certain cancers, and other medical problems.[21] The 1988 *Surgeon General's Report* declared "overeating" to be a problem for *most* Americans. They were admonished to "achieve a desirable weight" by cutting consumption of foods high in calories, fats, and sugars, limiting alcohol intake, and exercising more.[22] "Every authority, every institution in our society," wrote Seid in 1989, "urges us to fight our fat."[23]

While the crusade against overweight was, in itself, nothing new, the extent to which it affected males was.[24] As we have seen, the dieting crazes of the 1930s were overwhelmingly female phenomena. For almost thirty years thereafter, most males continued to be blithely indifferent to losing weight. In 1955 74 percent of them told Gallup they were satisfied with their weight or wanted to put on more poundage. A mere 3 percent were dieting at that moment, and only 17 percent said they had ever seriously tried to lose weight.[25] The Metrecal and other diet fads of the early 1960s mainly involved females. However, in the late 1960s and early 1970s, an increasing number of middle-class men began to work seriously at losing weight. By the mid-1970s dieting for weight loss had lost its mainly feminine connotation, and by the 1980s, particularly as it became intimately linked with exercise, to watch one's weight was becoming a sign of prudence among middle- and upper-middle-class men. To lunch on poached fish and/or a salad and reach for a Diet Coke after a workout connoted a healthy concern for one's well-being. Clever commercials featuring burly sports heroes who glowered menacingly in proper macho fashion before reaching for a "lite" beer even helped make calorie-reduced foods acceptable among working-class men. Not surprisingly, a 1986 study indicated that more than half of the seventy-eight million Americans using low-calorie foods had only begun doing so during the past five years and that the bulk of these recent converts were male.[26]

Yet while conceptions of the ideal male body shape slimmed down, particularly among the middle and upper classes, it hardly approached the slimness of that prevailing among women. One reason seems to be that, while there was certainly no shortage of modern-day Ponce de Leóns, men did not face the same pressure to maintain their youthful looks as women. Men's looks could still reflect their stage in life. Indeed, they were thought to get more "distinguished" with age. A few wrinkles on the face were regarded as denoting pleasing cragginess rather than, as with women, decrepitude. Also, while tautness of body was highly regarded, men are burdened with a smaller percentage of body fat in the first place. Nevertheless, there was more than enough cause for concern. Most middle-class men had sedentary occupations, and "love handles" or even "spare tires" had a tendency to show up in their midriffs by age thirty, a landmark most baby boomers had passed by the early 1980s. Studies now showed that, like their wives, lovers, and daughters, most middle- and upper-class men over thirty now regarded themselves as overweight. While they were ultimately not prepared to sacrifice as much as females to battle it (the proportion of female to male dieters remained about two to one), they were still affected by the new weight-control ethic.[27] In this, they were part of an international trend. By the 1980s, says Fischler, "a large proportion of the population of the most developed countries dreamed of being thin, saw themselves as fat, and suffered from the contradiction."[28]

At first, the food industries reacted to rising fatphobia in the usual way, by changing the labels and advertising of old products to emphasize their noncalorific properties. But then commercial lightening struck, starting a spectacular outpouring of newly formulated "lite" foods.[29] Calorie-reduced foods had originally been aimed mainly at diabetics. The approval of the artificial sweetener cyclamate in the early 1950s had widened their scope and encouraged a migration from drugstores to supermarket shelves. Changing the signs over these sections from the off-putting "diabetic foods" to "dietetic foods" also encouraged sales, but subsequent gains were led mainly by soft drinks. By the mid-1970s, however, it was becoming apparent that—for reasons Adam Smith would have been hard-put to explain—consumers were willing to pay more for less; that is, they would shell out premium prices for "lo-cal" products with reduced food value.[30] At first, large processors of well-established brand names turned to the time-honored method of product-line extensions. These were particularly profitable in the beverage industry, whose diet segment boomed after aspartame was approved for use in soft drinks in 1983. But introducing "lite" prepared foods remained a problem, for frozen prepared foods, their most obvious form, had developed an inferior TV-dinner image. Indeed, sales of frozen dinners and pot pies had been in steady decline since 1972.[31]

All this changed in 1981 with the introduction of Stouffer's Lean Cuisine. Stouffer's had spent five years developing a line that would overcome the general perception that frozen dinners meant meager portions, poor quality, and little variety. Convinced, correctly, that it had finally done so, it then hired "Lean Teams," consisting of a nutritionist and well-known athletes, to tour the country promoting its fourteen-day diet plan, "On the Way to Being Lean." Within a year of its introduction it had almost single-handedly led sales of lo-cal items to jump from 7 to 17 percent of the frozen entree market.[32] Heinz then weighed in with its Weight-Watcher's line, and grocery freezer shelves were soon packed with premium-priced, tempting-looking entrees and dinners, emblazoned with the vital calories-per-serving information necessary to regulate dosage. As *Prepared Foods* noted the next year, Lean Cuisine's success "spread far beyond the frozen food case. It demonstrated dramatically that today's 'upscale' consumers will spend a little more for food products in tune with their lifestyles."[33] A new age of "frozen food for the Jacuzzi generation," not the "old TV types," had dawned, said the *Washington Post*.[34]

Ironically, among the main corporate beneficiaries of the diet craze were snack food manufacturers, whose business had also been in the doldrums for much of the 1970s. In 1983 *Prepared Foods* noted that dessert consumption had been hard hit by the fact that one-third of the nation's females and one-sixth of its males were dieting, but "snack foods are taking up much of the food treat slack." Indeed, what seemed to be happening was that weight-conscious dieters were filling in the voids created by meal-skipping with more snacks.[35] Housewives trying to lose weight were passing up breakfast and lunch, as well as cutting down on dinners, noted a food industry consultant, but they were making up for it with snacking. Like most other Americans, she said, they snacked "from morning to bedtime" and even after—an average of about twenty "food contacts" per day.[36] A similar psychology allowed the diet fetish to coexist with the continuing boom in "gourmet" dining. The "gourmet" food industry did very well on consumers who made up for weekday deprivation by overindulging in "gourmet" and specialty foods on weekends; "shape up, pig out," they called it.[37]

The publishing industry was a minor beneficiary of fat-phobia. By 1984 there were three hundred diet books in print in the United States. They were fixtures on every best-seller list; five million copies of Dr. Irwin Stillman's *Quick Weight Loss* alone had been sold.[38] No daytime TV talk show was complete without an interview with a peripatetic diet-book author whose work had been excerpted and summarized in countless newspapers and magazines. Most of their regimens were up-to-date-sounding variations on age-old nostrums—low protein, high protein, low fat, high fat, low carbohydrate, high carbohydrate, fruitarian, vegetarian, grapefruit, and so on. Stillman's popular diet, which allowed only protein-rich foods like meat, eggs, and cheese, could be traced back to the all-meat diets, such as the Salisbury Diet, which followed on the late-nineteenth-century discovery of proteins, as could the ill-fated Dr. Herman Tarnower's low-carbohydrate, high-protein Scarsdale Diet.[39] There were some new twists—Weight Watchers borrowed quasi-religious techniques from Alcoholics Anonymous, for example—but the main new departure was the emergence, in the mid-1980s, of exercise as a major weapon in the attack on body fat. The Jane Fonda Diet gave way to the Jane Fonda Workout. (Only then did Ms. Fonda confess that her weight-reduction efforts had also been accompanied by bouts of bulemia.)

When all was said and done, though, the millions of middle-class men and women who periodically dieted, jogged, and exercised seem to have had little impact on what was perceived to be the national problem with overweight. In 1989 the government's most comprehensive health and nutrition monitoring survey concluded that in the past two decades there had been "no decline in the prevalence of overweight."[40] Indeed, one analysis of its statistics concluded that in that period the average young woman (age twenty-five to thirty-four) had even gotten fatter.[41] What also stayed the same was who fought the hardest in the battle against weight. The higher the class, the greater the effort to reduce, so it seemed, particularly among women. College-educated young women weighed less than those of the same age whose education had stopped at high school. One study even showed that married women's weight was inversely proportional to their husbands' income.[42] Curiously, though, black women weighed more than their white counterparts at all economic levels, opening another area for "nature versus nurture" speculation.[43]

The studies also evidenced the continuing disparity between the nutrition and health of rich and poor. Indeed, the gaps between the extremes of wealth widened during the 1980s. While the richest fifth experienced significant improvements in living standards and the middle class more or less held its own, mainly by increasing reliance on two incomes, the bottom 15 percent sank downward.[44] The deterioration of the living standards of the poor could be seen, as it had been during the Depression, in the proliferation of breadlines, soup kitchens, and food banks. It could also be measured, as it could not during the Depression, in statistics. These seem to indicate that by 1979 two government programs, food stamps and expanded medical and Social Security benefits for the aged, had substantially improved the lives of the poorest 20 percent of Americans.[45] Indeed, even the Field Foundation, whose 1967 study of hunger had spurred the War on Hunger, acknowledged in 1979 that federal food programs had eliminated most of the gross malnutrition in America's backward rural areas and urban slums.[46] However, during the 1980s the ranks of the poor were swelled by an influx of single parents. The evaporation of decently-paid blue-collar jobs also put downward pressure on those on struggling to stay on the lower rungs of the employment ladder, forcing many into poorly paid part-time service jobs or onto the welfare rolls. In Pennsylvania skilled steelworkers formed the bulk of the clientele for the two hundred soup kitchens that sprang up during the decade.[47] A very visible wave of homeless people took to the streets, evoking images of the darkest days of the Depression.[48] The century-long decline in infant mortality rates—still linked, among the poor, to nutrition levels—slowed, and the United States sank to twenty-second in the world on that score. Black infant mortality rates returned to where they had been earlier in the century, double those of whites.[49]

It now seemed that the antipoverty advocates who in 1969 had ignored Daniel Moynihan's warnings and abandoned support for the guaranteed income plan to concentrate on food stamps and other targeted programs had probably made a grave error. When the political winds changed direction, targeted programs such as food stamps could be cut back or gutted much more easily than income supplement programs such as those that helped many of the aged climb from poverty. This is more or less what happened to the food programs in the 1980s. In the guise of targeting only those most desperately in need, the Reagan administration worked persistently to cut back eligibility for food stamps. Simultaneously, it expanded the previously discredited commodity distribution programs, for the old surplus-disposing reasons.[50] Little did the antipoverty activists who struggled for further expansion of food benefits in the late 1970s realize that those years would soon be regarded as a kind of golden era in the struggle against hunger and malnutrition.

But the health of the poor, whether crowded in inner-city ghettos or scattered in remote country hamlets, was hardly a concern of most Americans in the 1980s, particularly the upper-income people benefiting most from the apparent miracle of "Reaganomics." They and the middle class continued to be buffeted by the contradictions of affluence. On the one hand, a continuing swirl of high-status foods and eating places beckoned. On the other, the dieting and exercise manias sent stern signals to cease and desist. In France the nouvelle cuisine helped mute the clash. When food processors were allowed to introduce "light" products there later in the decade, the cuisine took on a decidedly nouvelle hue. The two terms—*nouvelle* and *light*—became interchangeable on grocery shelves. But real nouvelle cuisine, classical in its simplicity, was just too restrained for the upscale trend-setters of Reagan's America. Classicism gave way to baroque, and even rococo, as a new breed of "maverick chefs" used nouvelle cuisine as a springboard for ever more dizzying dives into novel combinations of exotic ingredients.[51] Austrian-born Wolfgang Puck took California by storm with dishes such as ravioli stuffed with lobster mousse, pizza with smoked salmon and caviar, and oysters dusted in curry-flavored flour, sautéed in butter, and placed on a purée of cucumbers and cream. Anne Greer wowed flush Dallasites with veal chops sauced with a papaya purée and surrounded by a "halo of dark cordon sauce," red-chili pasta tossed with spinach and goat cheese, and warm chicken, avocado, and papaya salad with a soy sauce and ginger dressing.[52]

But while the infrastructure of ex–graduate students and lawyers producing goat cheese in Vermont and organic baby vegetables and edible flowers in Sonoma County expanded healthily, the enormous system producing foods for the mass market was beginning to stumble here and there. Only slowly did the realization begin to dawn that, as in other fields, American food processors were no longer in the technological lead. In the 1950s and 1960s, processing industry journals had routinely hailed any American manufacturers' advances as "firsts" in the world. By the 1980s innovation in U.S. equipment-manufacturing (the object of 90 percent of food industry research and development is raising productivity) was stagnating, while European processors, spurred by the expanding Common Market and overseas sales, had leapfrogged the Americans.[53] Yes, American chemists still had their triumphs: In 1982 *Prepared Foods* hailed the National Starch and Chemical Company for coming up with clamless stuffed clams and meatless meat raviolis and lauded Nabisco for cleverly using red-dyed dehydrated apples in its strawberry parfait. But increasingly, articles on companies using new processes routinely began "first time in the United States."[54] Even in packaging, the field where America had always reigned supreme, the widespread use of new methods such as *sous-vide* and aseptic cartons allowed the Europeans to push ahead of the Americans in maintaining the quality of preserved foods.[55]

A series of ever more spectacular debt-financed takeovers—R. J. Reynolds of Del Monte and then Heublein, Dart of Kraft, Beatrice of Esmark (Swift, Hunt-Wesson), Pillsbury of Green Giant, climaxed by the fall of venerable Standard Brands and Nabisco to R. J. Reynolds in a thirty-two-billion-dollar deal organized by junk bond specialists Kohlberg, Kravis, and Roberts—did not help.[56] Inevitably, as in other industries plundered by the corporate raiders, executives became fixated on next quarter's bottom line at the expense of long-term planning and research and development, putting even more distance between the American food megaconglomerates and the surging Europeans. Significantly, it was not their research, development, or even production facilities that made the food giants attractive takeover targets. It was their brand names—Del Monte, Jell-O, Ritz—emblazoned at enormous cost over many years in consumers' minds as symbols of quality and confidence. The number of brands that are leaders in their markets today and were also the top brands in 1925 is striking testimony to the enduring value of these names: Campbell's soups, Swift's bacon, Nabisco crackers and cookies, Kellogg's cereals, Del Monte canned fruit, Crisco shortening, Lipton tea, Coca-Cola soft drinks, Wrigley chewing gum, Hershey chocolate, and Life Savers mints.[57] Thus, when Swiss-based Nestlé's acquired Carnation for eight billion dollars in 1984—until then the largest non-oil-and-gas deal in history—Carnation's unimpressive R and D program was hardly a consideration. As in the other takeovers, its brand names were the lure. Asked about the benefits they were experiencing from the takeover, Carnation executives sounded like managers in Third World outposts as they hailed Nestlé's longer-term outlook and impressive research and development effort, most of which was done at its laboratories in Vevey, Switzerland.[58]

Concentration also continued to be the name of the game in the chain restaurant business. By the mid-1980s it had become, in the words of an industry magazine, "top heavy." In 1987 the ten largest chains accounted for 57 percent of the top one hundred's sales volume, and the top twenty-five pulled in over 75 percent. Those below were by no means mom-and-pop operations. Orange Julius, with 580 outlets and eighty million dollars in sales, was number ninety-eight.[59] As the industry matured, established brand names also proved invaluable. It cost half as much in advertising money to sell a McDonald's hamburger as one from Burger King.[60]

Yet foreigners muscled into even this quintessentially American sphere, albeit with mixed results. Pillsbury and its fractious Burger King subsidiary were likely set back by being swallowed by Britain's Grand Metropolitan, a debt-laden brewery-and-pub-based conglomerate that had little to offer in terms of capital, technology, or organization. The British conglomerate Imperial Tobacco could not turn Howard Johnson's around, although its Canadian subsidiary, Imasco, had more luck with Hardee's. Perhaps the most telling sign of the passing of American supremacy came after the great age of mergers was over. Among the debt-ridden companies unable to raise capital in the recession of the early 1990s was Restaurant Associates, an innovative leader that had helped raise the level of American dining a notch or two in the 1960s with restaurants such as the Four Seasons in New York City. In 1991 it too slipped into foreign ownership, leaving people to ponder the significance of Mamma Leone's, that boisterous New York Italian restaurant, now being Japanese-owned.

Even before the recession that pushed Restaurant Associates into foreign hands began to batter the nation, the pendulum was already swinging back from its 1980s' extremes. As the Reagans glided through the final series of glittering events before leaving the White House in January 1989, the economic house of cards they had helped construct was falling apart. Huge budget deficits handcuffed the federal government, starting a domino effect on state and local governments facing cutbacks in Washington aid. Within two years of the Reagans' return to California, despite the cutbacks in eligibility, one in ten Americans was poor enough to qualify for and receive food stamps, whose cost now consumed half of the Department of Agriculture's entire budget. In November 1991 other Hollywood stars, in an eerie echo of the five-cent meals served at the Waldorf-Astoria and other posh hotels in December 1930, prepared to attend the Hollywood Hunger Banquet. There, the likes of Dustin Hoffman and Cybill Shepherd were expected to draw lots to see whether they would be seated among the small group served an elegant three-course meal or at wooden tables with a larger group eating rice and beans with tortillas, or with the largest group of celebrities, on the floor, to eat rice and water with their hands.

The collapse of the junk bond market and the multibillion-dollar savings and loan scandals brought many of the entrepreneurial culture heroes of the Reagan years into disrepute. As in the 1930s, *sua culpas* filled the air as the nation bemoaned the previous decade's profligacy, cursed its misplaced faith in Wall Street, and sought to return to the traditional verities—the comforts of home, family, and "the simple life." Executives at Kraftco noticed a steady rise in sales of Kraft Macaroni Dinner: a sure sign of recession and the search for economical home-style food, they said.[61] Americans had "tired of trendiness and materialism," said *Time* in April 1991, and were "rediscovering the joys of home life, basic values and things that last."[62]

Restaurants were among the first to reflect this, for the reaction against the Reaganesque excesses was also reinforced by the pinch on discretionary spending. Suddenly, as in the 1930s, home cooking—now called "comfort food"—became the name of the game. Expensive French restaurants, facing shoals of empty white-napped tables, scrambled to downsize and downprice their menus by serving "bistro food"; sophisticated "Northern Italian" restaurants replaced their tiny portions of angel's hair pasta with hearty platters of rigatoni and baked ziti, installed brick ovens, and became rustic-looking trattorias. Robust "Mediterranean" food rose in fashion, while "California cuisine" was proclaimed a dead duck, or at least a dying swan. "California cuisine has stopped," said Joachim Splichal, chef and co-owner of Patina in Los Angeles. "I'm moving away from the salad approach to cooking." "I don't even know what California cuisine is," said Michel Richard, chef and owner of the Citrus restaurant in the same city. "Maybe it's a lot of uncooked vegetables. I don't like that. . . . I think people are tired of grilled food too." Instead of exotic salads and miniscule portions of artfully decorated grilled foods, they served filling dishes such as roast pork, mashed potatoes, and puréed lentils and beans.[63]

As in the 1930s, there was also a revival of middle-class interest in home cooking—or, at least, eating at home. Advancing age, family obligations, and slower income growth were transforming the high-flying "yuppies" of the 1980s into stay-at-home, self-proclaimed "couch potatoes." "After years of takeout," said the *New York Times* in September 1989, "many are returning home." There they confronted their ignorance of even the rudiments of home cooking. Publishers therefore rushed out a wave of basic "cookbooks for the 90s" such as Julia Childs's *The Way to Cook.* These were profusely illustrated to meet the needs of a generation who responded better to visual than literary cues, with recipes that were spicier than the earlier standards and paid due respect to considerations of health.[64]

At the other end of the gastronomic pole, the microwave revolution was eating into patronage at fast food restaurants. One reason was speed. "Fast food can no longer live up to its name," said one food industry consultant. "There are never any lines at home. With the microwave you just reach into your freezer and pop it into the oven and zap! it's done."[65] The increasing variety of take-home food also hurt, although some of it was from fast food outlets. By 1987 fully seventy-one million American households—eight out of ten—regularly purchased take-out foods to be consumed at home, spending 15 percent of the national food dollar on these pizzas, chickens, and submarine sandwiches.[66] When Gallup asked a sampling of people eating at home in September 1989 what they were eating, almost half were sitting down to frozen, packaged, or take-out meals. Partly as a consequence, a gloomy pall settled over the fast food restaurant industry; domestic expansion slowed to a crawl, and chains resorted to vicious price-cutting.[67]

There were even signs that the weight pendulum was beginning to swing back. The apparent epidemic of eating disorders among middle-class girls prompted criticism of impossibly thin ideals.[68] Psychotherapists and others began to turn out books—such as *Don't Diet*—questioning the efficacy and desirability of dieting.[69] For the first time in almost two decades diet books were *not* regulars on best-seller lists. Fat people took a leaf from the pages of other denigrated groups and organized pressure groups to fight invidious stereotypes of them.[70] They added their bit to the mania for "political correctness" by attacking "lookism" and warned against using language that demeaned them. The importance of heredity in weight gain and loss finally came to be recognized—that fatter people seem to inherit the ability to burn calories more efficiently than thinner ones—offering hope that individuals would not have to take moral responsibility for their excess poundage.[71] Serious questions were also raised about how detrimental excess weight really was to health. As with cholesterol, it now began to appear that only a small proportion of the population was at risk from overweight. Straying from the average or recommended weight for one's height and bone structure seemed to have deleterious effects mainly for those at the extremes of the charts, the most overweight *and* the most underweight. Indeed, by the end of the decade some experts were acknowledging that to be 10 to 15 percent over the recommended weights was not only not unhealthful, it might even be health-promoting—an echo of age-old notions of the healthfulness of what the French call *embonpoint*, a pleasing stoutness.[72]

Women's fashion seemed to change in tandem, as curves began to creep back into the women's magazines and bosoms became noticeable on couturiers' runways. The exercise and fitness boom had helped modify the completely bony look of the early 1980s with a new admiration for muscularity. Although still taut and wrinkle-free, this in turn began to give way to a grudging respect for curves. Cosmetic surgeons, whose liposuction and breast reduction techniques had presented a popular alternative to painful hours in the gym and on the jogging path, began to do more breast amplifications. From 1985 to 1991, the weight of women featured in *Playboy* centerfolds rose by one pound per year. The cover of the June 1991 *Vogue* magazine pictured a model who would almost certainly have been considered too buxom for it five years earlier—in a pose that clearly revealed this.[73] Men's fashion seemed to be moving in the same direction, as reports that a "fuller shape" would soon be fashionable emerged from fashion centers in early 1991.[74]

Yet the pendulum hardly threatened to swing back even to where it was in the 1950s. If considerations of fashion were the only ones, then the possibilities would be practically infinite: There could be a revival of the 1890s' body ideals or even those of the Rubensian era of the early seventeenth century. But the other factors that fed this century's fat-phobia seemed to brake the swing. Though modified, the ideas that obesity is unhealthy and that exercise is good are too ingrained to simply go away, and they are still underpinned by the compulsion of a society of food abundance to regulate eating by frowning on signs of overindulgence in it. Santa Claus may remain corpulent, but unless Doomsday scenarios of worldwide famine come to pass, future generations of children may well wonder why he is so jolly. For women, the decline of the matronly look would also seem to be related to the steady expansion of their role in workplaces outside of the home—the change from a society in which woman is regarded primarily as reproducer to one in which she is also a producer.[75]

Similarly, it was likely that the American diet would continue to reflect the growing impact of foreign cuisines and tastes, defanged and domesticated though they might be. The nature of the food industries and their market seemed to ensure this. During the 1980s the seventy-six million Americans in the baby boom generation continued to age and—in marketing jargon—segment. That is, they subdivided into singles, single-parent families, two-income families, and traditional one-income families, with differing responses to new products and sales pitches. The elderly were emerging as a distinctive—and important—market, with their own dietary agenda.[76] The young formed other lucrative segments. Although labeled the "baby bust" generation because they were relatively small in number, they still had an enormous amount of disposable income, particularly as part-time employment (much of it in food retailing) became more common. As a result, said a food processors' journal, the days when food companies "treated consumers as a homogeneous mass with many mutual needs and wants" were over. The old "crossover marketing techniques" could no longer carry new products past most consumer thresholds. "We are in an age of multidimensional marketing: a division of food shoppers into various segments and sub-segments." The mass market had become micro, said the head of Campbell's.

This opened the door wider for ethnic foods, for, while most ethnic foods and tastes would fall flat in the mass market, a goodly number could do well among certain segments of the market, particularly the more upscale ones. Ethnicity thus became one of a battery of pitches, including health, lo-cal, convenience, freshness, status, and value, to be mixed and matched in attempts to tap into specific segments. Campbell's combined ethnic, convenient, and lo-cal with a bit of status in its upscale L'Orient line of frozen dinners. Others put items from different cuisines in the same dinner—cannelloni with mornay sauce, Salisbury steak with Italian tomato sauce. Still others tried pushing the ethnic frontier further than ever. The Silverbird Company of New York City even came out with a line of native American foods—Navajo Fry Bread, Buffalo Burgers, Blue Corn Pudding, and Rabbit Stew.[77] Fifteen or twenty years earlier the idea of carving out a chunk of the market for this kind of food would have seemed ridiculous, but then so would the thought that *Good Housekeeping* would include an Ethiopian restaurant on its list of the nation's one hundred best, which it did in 1988, or that by 1990 Birmingham, Alabama, would count over sixty Chinese restaurants.[78]

Nor did the pendulum seem to be swinging back against food and health scares. Every available statistic chronicled the effects of cholesterol-phobia.[79] Americans had been replacing whole milk products with low-fat dairy ones since the early 1960s and had been cutting back on butter and eggs and increasing their intake of margarine and vegetable oils since the late 1960s. Per capita consumption of beef had continued to rise until 1975—mainly because of the rise of fast food hamburger chains and the popularity of backyard barbecuing—but it then began to plummet. Men between the ages of nineteen and fifty, those deemed most at risk from high cholesterol, were responsible for much of this decline, reducing their intake by about 35 percent. They also cut back on whole milk and eggs by about one-quarter while boosting their intake of low-fat milk and fish by 50 percent. Women of the same age reacted even more negatively to beef, cutting consumption by almost half, and also halved the amount of eggs they ate. Hamburger-based fast food chains scrambled to expand the available alternatives, but they still felt the pinch. In 1991, when McDonald's shocked Wall Street by turning in particularly mediocre results, analysts ascribed the grim results in part to the qualms of aging baby boomers about eating beef.[80]

A rather bizarre offshoot of the turn against beef was a radically altered image for bovines. Previously, cows had been regarded as docile, pleasant animals, peacefully grazing, lowing, and sleeping until called to genially do their duty to their human masters. For years Borden's had the happy Guernsey "Elsie" on its labels, and Carnation boasted that its products came from "contented cows." Now, when the health-conscious looked at bovines they saw hundreds of pounds of life-threatening cholesterol production. Francis Lappé and the radicals of the early 1970s had already shown how ecologically wasteful they were, consuming much more protein than they produced. Now, the destruction of the Brazilian rain forest to make room for cattle ranches was linked directly to the American demand for fast food hamburgers, as was the displacement of farmers by ranchers in other parts of the Third World. As if that were not enough, environmentalists also charged that their relentless burping, flatulence, and droppings produced enormous amounts of methane gas, which depleted the ozone layer and was a major contributor to the "greenhouse effect." The very people struggling to save dolphins, horned owls, and other endangered species now seemed intent on driving cattle into extinction.

Of course, food companies went through the usual orgy of reformulation, repackaging, and product-line extension to try to catch the cholestrol-phobic winds. Enterprising book publishers supplemented earlier Negative Nutrition cookbooks such as *The Don't Eat Your Heart Out Cookbook*, *The Long Life Cookbook*, and *The American Cancer Society Cookbook* with titles such as *The Count Out Cholesterol Cook Book*, *Cholesterol-Control Cookbook*, and *Dr. Dean Ornish's Program for Reversing Heart Disease*.[81] But one of the more unlikely groups that sought to profit from it was the American Heart Association—the organization that, it will be remembered, was one of the first to begin warning of its dangers in the early 1960s. Someone—perhaps one of the old-timers who remembered the 1930s AMA seal of approval or the World War II WFA symbol—persuaded it to organize a system whereby, in return for a fee, it would allow foods deemed healthy for the heart to display a special seal.[82] But the food industry was not nearly as enthusiastic over this scheme as it had been over its predecessors. The food conglomerates were now so humongous that, while they all had divisions that might benefit from this, they also had ones that would look bad. Moreover, the FDA did not take kindly to the plan, seeing in it the possibility of opening a Pandora's box it had originally been created to close, that of misleading health claims for food.[83] The petulant AHA was soon forced to withdraw the proposal.

But the cholesterol problem turned out to be much more complex than originally thought. The public were now told that three kinds of fatty acids affected it—one good, one bad, and one of uncertain effect. People who had followed expert advice and dutifully substituted margarine for butter now learned that much of the margarine was made of hydrogenated oils containing a kind of fat—transmonounsaturated fatty acids—that actually increased the risk of coronary heart disease.[84] Corn and most other vegetable oils, the darlings of the 1960s' cholesterol scare, also sank in experts' estimates, replaced on the altar by olive oil, which was now said to reduce cholesterol levels, but only if cold-pressed. Meanwhile, oat bran came and went as a miracle cholesterol reducer, but rice bran and rice oil, which most people did not even know existed, seemed poised to take over the limelight. Contradictions abounded. Wine labels were made to carry health warnings, yet studies showed that drinking it likely raises the level of "good," high-density lipids in the bloodstream.[85]

And on it seemed to go. Some studies indicated that increasing the intake of cholesterol itself had no effect on blood cholesterol levels. On the other hand, broiling and grilling meats, highly recommended by every expert concerned with weight loss and cholesterol, were now said to create carcinogenic nitrosamines. Only poaching seemed safe—hardly good news for steak lovers. Government agencies abandoned the old Basic Four. The surgeon general and even the USDA now emphasized getting the right percentage of energy from fats, warning that only a certain percentage of these fats could be bad-for-you fats. Nevertheless, experts attacked their recommendation that only 30 percent of calories be derived from fats as at least 50 percent too high; it should be 20 percent or lower, they said.[86] A group called the Physicians' Committee for Responsible Medicine called on them to endorse a new meatless Basic Four, a demand that aroused suspicion that perhaps the founding director's animal rights agenda shaped its views on human health.[87] Meanwhile, in April 1991 the USDA had timidly introduced a replacement for the famous pie chart illustrating the Basic Four. It was a pyramid of different categories of food, emphasizing grains, legumes, fruits, and vegetables, intended to make its 1980 "Dietary Guidelines" easier to learn. When it quickly withdrew them, apparently in response to farm and food industry pressure, critics again charged—as they had since the 1930s—that it "placed agribusiness first, public health second."[88]

When combined with the continuing choruses imploring Americans to lose weight and the changing drumbeats of status-makers—all amplified by misleading food advertising campaigns—the result seemed to be Fischler's "dietary cacophony" played with the volume turned up as high as it could go. Perhaps it will end in the paralysis of "gastro-anomie"—a hypothetical condition in which, bombarded by nutritional and culinary messages from all sides, people lose all sense of dietary norms and rules.[89] But that is an extreme scenario. The fact is that, as Fischler says, "homnivores" cannot live without food rules, norms, and restrictions. The cacophony may well produce a degree of deafness in the general public, manifested in a certain wariness about the daily pronouncements of experts on food and nutrition. (One hesitates to use the term "grain of salt.") After all, for over one hundred years nutrition experts have been telling Americans to subordinate taste to concerns about the economy and healthfulness of food. Yet while it may not be as important a consideration as it used to be, taste still plays an major role in their food choices.[90] Moreover, despite the toll in flavor and variety taken by the industrialization of food, in their restaurants and markets Americans still have access to an unprecedented array of high-quality, tasty food. Indeed, the nation now stands—as it never did before—on an equal plane with Europe's finest in terms of the opportunities for good eating, no matter how defined.

There remained another ray of hope. A common thread linked most of the scares: the idea that it was necessary and possible to have a national nutrition policy. This aspiration stretched back to the first fond hopes for the New Nutrition in the 1890s, when nutritionists thought food reform would head off social upheaval. It was nurtured by the World War I conservation effort, stimulated by the malnutrition concerns of the Great Depression, and underlay the National Nutrition Conference of June 1941, which aspired to codify the Newer Nutrition. From there sprang the half-baked Recommended Daily Allowances—which, as we have seen, inflated the recommended nutrient intakes for the majority in order to accommodate the presumed needs of a minority—and the official Basic Four–"balanced diet" line, which was based on the assumption that everyone should eat plenty of all kinds of foods. The Negative Nutrition, which looked askance at many of the most highly recommended of these foods, represented a *volte face* in terms of approach but was nevertheless grounded in the same futile search for the Grail of national nutrition rules. Salt and sugar, as we have seen, while perhaps harmful to a small minority, were condemned out of hand for all. In 1988 the surgeon general persisted in recommending a drastic reduction in fat intake for the large majority of Americans in the face of evidence—since reinforced—that it would benefit only a minority. Indeed, in June 1991 a study concluded that if every American heeded the advice to cut fat intake to 30 percent of total calories, the benefits would be so limited that the average life span would increase by at most several months.[91] A later study estimated the gain at only several minutes.

The same seems to apply, as we have seen, to the calls for nationwide weight loss. Indeed, an analysis of the Framingham study statistics indicated that dieting might even be harmful to health. Men who dieted and then regained the weight they lost had much higher (from 25 to 100 percent) rates of death from heart disease than those who did not. Given the usual 95 percent recidivism rate in dieting, this would mean that, if followed in the usual zealous but fitful manner, the admonitions that the nation as a whole lose weight might lead to much more ill health than they might prevent.[92]

There may be grounds for optimism, however, in the increasing use of the phrase "for those at risk." Slowly, it seems, the experts are beginning to acknowledge that the relationship between nutrition and health may be modified by so many other factors—individual biological makeup in particular—that few rules are universally applicable. Hopefully, this will lead both to a rethinking of the role of government in telling people what to eat and—more important, for they are still the main source of nutritional information—more restrictions on food vendors' ability to alarm and deceive the public by distorting the health benefits of their products. The "truth-in-labeling" law passed in 1991, which seeks, as David Kessler, the new head of the FDA put it, to shift the industry's creative efforts from the marketers' "word processor to the laboratory," would seem to be a healthy step in this direction, although past experience must temper optimism in this regard. Nevertheless, modified ideas about what constitutes a healthy diet may take hold, allowing Americans to—in a variation of the old saw—have their good food and eat it too. This would certainly represent a giant step toward resolving the paradox of a people surrounded by abundance who are unable to enjoy it. But then again, these are difficult times for optimists.

Abbreviations for Frequently Cited Periodicals

AC	*American Cookery*
AJPH	*American Journal of Public Health*
BHG	*Better Homes and Gardens*
BW	*Business Week*
FE	*Food Engineering*
FF	*Fast Food*
FFF	*Fountain and Fast Food*
FFR	*Food Field Reporter*
FP	*Food Processing*
FT	*Food Technology*
GH	*Good Housekeeping*
JADA	*Journal of the American Dietetic Association*
JAMA	*Journal of the American Medical Association*
JHE	*Journal of Home Economics*
LHJ	*Ladies' Home Journal*
NN	*Nutrition Notes*
NR	*Nutrition Reviews*
NYT	*New York Times*
NYTM	*New York Times Magazine*
OGF	*Organic Gardening and Farming*
PF	*Prepared Foods*
RB	*Restaurant Business*
RH	*Restaurant Hospitality*
RM	*Restaurant Management*

SNL	*Science News Letter*
TGM	*Toronto Globe and Mail*
USNWR	*US News & World Report*
WP	*Washington Post*
WSJ	*Wall Street Journal*

Notes

Prologue

1. *NYT*, Aug. 6, 1930.

2. *NYT*, April 1, 1931; Bonnie Fox Schwartz, "Unemployment Relief in Philadelphia, 1930–1932: A Study of the Depression's Impact on Voluntarism," *Pennsylvania Magazine of History and Biography* 42 (Jan. 1969), rpt. in Bernard Sternsher, ed., *Hitting Home: The Great Depression in Town and Country* (Chicago: Quadrangle, 1970), 65.

3. Mary Kay Smith, "Dark Days of Depression: Lonoke County, Arkansas, 1930–1933," *Red River Valley Historical Review* 7, no. 4 (1982): 18; Janet Poppendieck, *Breadlines Knee-Deep in Wheat: Food Assistance in the Great Depression* (New Brunswick: Rutgers Univ. Press, 1986), 31.

4. Charles S. Johnson, *Shadow of the Plantation* (Chicago: Univ. of Chicago Press, 1934), 100–101.

5. *NYT*, May 4, 1930.

6. Ibid.

7. *NYT*, Sept. 30, 1930.

8. Steven Mintz and Susan Kellogg, *Domestic Revolutions: A Social History of American Family Life* (New York: Free Press, 1988), 134.

9. Russell Nixon and Paul Samuelson, "Estimates of Unemployment in the United States," *Review of Economic Statistics* 22 (Aug. 1940): 106–7.

10. Poppendieck, *Breadlines*, 27; Federal Writers' Project (Roi Ottley and William Weathery), "The Depression in Harlem," rpt. in Sternsher, *Hitting Home*, 106; Richard Hooker, *Food and Drink in America: A History* (Indianapolis: Bobbs-Merrill, 1981), 308.

11. Irving Bernstein, *The Lean Years: A History of the American Worker, 1920–1933* (Boston: Houghton Mifflin, 1960), 363.

12. B. T. Quinten, "Oklahoma Tribes, the Great Depression, and the Indian Bureau," *Mid-America* 49 (Jan. 1967), rpt. in Sternsher, *Hitting Home*, 200.

13. C. Roger Lambert, "Hoover, the Red Cross, and Food for the Hungry," *Annals of Iowa* 44 (1979): 534.

14. Robert S. Lynd and Helen Lynd, *Middletown in Transition: A Study in Cultural Conflicts* (New York: Harcourt Brace, 1937), 105.

15. Poppendieck, *Breadlines*, 25.

16. Ibid.; *NYT*, Aug. 5, 1932.

17. Lynd and Lynd, *Middletown in Transition*, 105.

18. Lambert, "Hoover, the Red Cross, and Food," 538–39.

19. In 1932 the national office abandoned even this tack, leaving local chapters to come up with the funds if they wished to continue employing dietitians. Adelia Beeuwkes et al., eds., *Essays in the History of Nutrition and Dietetics* (Chicago: American Dietetic Association, 1967), 235.

20. Lambert, "Hoover, the Red Cross, and Food," 533–34.

21. *NYT*, Jan. 25, 1932.

22. *NYT*, Dec. 2, 1932.

23. Smith, "Dark Days of Depression," 20–21.

24. Another problem was that the Red Cross barely existed in most of the parched farmlands, dusty rural villages, and remote mountain towns where the suffering was most acute. Elizabeth Etheridge, *The Butterfly Caste* (Westport, Conn.: Greenwood, 1971), 196–97; Poppendieck, *Breadlines*, 28–30.

25. Lambert, "Hoover, the Red Cross, and Food," 537–38.

26. Poppendieck, *Breadlines*, 21–22; James Patterson, *America's Struggle Against Poverty, 1900–1980* (Cambridge: Harvard Univ. Press, 1981), 26.

27. Harvey Levenstein, *Revolution at the Table: The Transformation of the American Diet* (New York: Oxford Univ. Press, 1988), 120.

28. She recommended devoting one-fifth of the food budget to dairy products, one-fifth to meat, one-fifth to fruit and vegetables, and two-fifths to dry groceries and staples. Yet her recommendations for milk consumption amounted to twenty-six quarts per week. At sixteen cents a quart, noted the *New York Times*, this would "take some pretty sharp shopping." *NYT*, Feb. 2, 9, 1930.

29. Lucy Gillett, "Growth and Activities of Nutrition Service in AICP-CSS," July 1948, typescript in Community Service Society, Papers, Division of Archives and Manuscripts, Columbia University Libraries, New York, N.Y. (hereafter CSS Papers), Box 68. Mulberry Health Center, Semi-Annual Report, 1933, and Annual Report, 1934–1935, CSS Papers, Box 60.

30. *NYT*, May 15, 1932.

31. *NYT*, Jan. 9, March 8, 1932. The technique the USDA advised was the same one the Iowa home economist brought to New York, dividing the food dollar into five parts: 25 cents for milk or cheese; 20–25 cents for fruit and vegetables; 15–20 cents for butter, lard, other fats, and sweeteners; and 15–20 cents for meat, fish, and eggs. *NYT*, Dec. 21, 1931.

32. *NYT*, Dec. 22, 1932.

33. *NYT*, Dec. 13, 1932.

34. James Gray, *Business Without Boundaries: The Story of General Mills* (Minneapolis: Univ. of Minnesota Press, 1954), 180.

35. *NYT*, May 15, 1932.

36. Poppendieck, *Breadlines*, 25; Schwartz, "Unemployment Relief," 78; Patterson, *America's Struggle*, 57.

Chapter 1. Depression Dieting and the Vitamin Gold Rush

1. Eunice Fuller Barnard, "New Styles in Diet as Well as Dress," *NYTM*, Nov. 1, 1931.

2. Eunice Fuller Barnard, "So, a New Fashion Is Here," *NYTM*, May 4, 1930.

3. In fact, hardly any anthropological studies support this.

4. Veronique Nahoum, "La belle femme, ou la stade du miroir en histoire," *Communications* 31 (1979): 22–32.
5. Barnard, "So, a New Fashion."
6. Thelma Rose, "Scientific Eating Campaign at Stephens College," *JHE* 27 (Nov. 1934): 564.
7. Wingate M. Johnson, "The Rise and Fall of Food Fads," *American Mercury*, April 1933, p. 477.
8. Carl Malmberg, *Diet and Die* (New York: Hillman-Curl, 1935), 91–92.
9. Ibid.
10. Hillel Schwartz, *Never Satisfied: A Cultural History of Diets, Fantasies, and Fat* (New York: Free Press, 1986), 192–93.
11. Esther L. Smith, *The Official Cook Book of the Hays System* (Mount Pocono, Pa.: Pocono Hay-Ven, 1934), xix; "The Wonders of Diet," *Fortune*, May 1936, p. 90.
12. "Wonders of Diet," 90–91; *NYT*, April 14, 1938.
13. *NYT*, Dec. 31, 1934.
14. *FFR*, Oct. 9, 1933.
15. James Gray, *Business Without Boundaries: The Story of General Mills* (Minneapolis: Univ. of Minnesota Press, 1954), 181.
16. Helen S. Mitchell and Gladys M. Cook, "Facts and Frauds in Nutrition," *Massachusetts Agricultural Experiment Station Bulletin* 342 (April 1937): 9–13.
17. U.S. Federal Trade Commission, Stipulation no. 01620, Agreement to Cease and Desist, Welch Grape Juice Company, May 28, 1937.
18. Schwartz, *Never Satisfied*, 192.
19. Mitchell and Cook, "Facts and Frauds," 24.
20. Ronald M. Deutsch, *The Nuts Among the Berries: An Exposé of America's Food Fads* (New York: Ballantine, 1967), 185, 200–206.
21. *NYT*, June 1, 1934.
22. Barnard, "So, a New Fashion."
23. "Wonders of Diet," 86.
24. In fact, acidosis is an extremely rare ailment—having nothing to do with acid in the stomach—which sometimes appears in diabetics when they are in a coma.
25. Deutsch, *Nuts Among the Berries*, 170–74.
26. Barnard, "So, a New Fashion."
27. Kenneth Roberts, "An Inquiry into Diets," *Saturday Evening Post*, Oct. 15, 1932.
28. FTC, Stipulation no. 01620; Harvey Levenstein, *Revolution at the Table: The Transformation of the American Diet* (New York: Oxford Univ. Press, 1988), 153–54.
29. *SNL*, Nov. 10, 1934.
30. *NN*, May 1933.
31. Clarence Lieb, "The 'Compatible Eating' Fad," *Hygeia* 14 (Aug. 1936): 685.
32. Henry C. Sherman, "Foods for Health Protection," *JHE* 26 (Oct. 1934): 493–95.
33. U.S. Federal Trade Commission, Stipulation no. 02180, Cease and Desist Agreement of Standard Brands, Inc., July 28, 1938.
34. *GH*, Jan. 1935, p. 147.
35. *NYT*, April 20, 1937.
36. *NYT*, April 22, 1937.
37. *NYT*, Sept. 11, 1933.

38. *NN*, Sept. 1937.

39. *LHJ*, Aug. 1934, p. 59; *GH*, Aug. 1936, p. 144.

40. *NYT*, Dec. 30, 1934. The belief was still current seven years later, when a University of Pittsburgh biologist said that the reason three times as many draftees were being rejected for faulty teeth than had been the case in 1917 was that as infants they had been fed evaporated or pasteurized milk, which was deficient in vitamin C. *SNL*, May 3, 1941.

41. *NYT*, May 12, July 11, 1937.

42. *SNL*, Dec. 20, 1930; *NYT*, Sept. 8, 1937.

43. M. Daniel Tatkon, *The Great Vitamin Hoax* (New York: Collier-Macmillan, 1968), 28.

44. Elmer McCollum and Nina Simmons, *Food, Nutrition, and Health* (New York: Macmillan, 1928); General Mills, "Outline of the Career in Advertising of Marjorie Child Husted," Feb. 1950, typescript in Marjorie C. Husted Papers, Schlesinger Library of Women's History, Radcliffe College, Cambridge, Mass.; *FFR*, March 26, 1934; Frederick J. Schlink, *Eat, Drink, and Be Wary* (New York: Covici, Friede, 1935), 14–19.

45. Schlink, *Eat, Drink, and Be Wary*, 14–15.

46. An appropriation of $250,000 to set up a special institute for research on food and nutrition was a further evidence of their realization that the doubts about processing must be dispelled. *NYT*, Nov. 30, 1938.

47. *JHE* 24 (June 1932): 528.

48. Levenstein, *Revolution*, 156.

49. Mitchell and Cook, "Facts and Frauds," 9–24; Schlink, *Eat, Drink, and Be Wary*, 212–15. In many cases the defense of canning was technically correct, but only if the liquid in the can was consumed. Home economists thus had to recommend that it be laboriously boiled down and used as a sauce, which undermined the convenience that was often the reason for using using canned foods in the first place.

50. *NYT*, April 3, 1933; Schlink, *Eat, Drink, and Be Wary*, 27–31.

51. *FFR*, July 10, 1939.

52. Schlink, *Eat, Drink, and Be Wary*, 30.

53. *LHJ*, May 1934, pp. 105, 137.

54. Gove Hambidge, "This New Era in Foods," *LHJ*, May 1929, pp. 26–27.

55. *GH*, Aug. 1934, p. 18.

56. James Harvey Young, *The Medical Messiahs* (Princeton: Princeton Univ. Press), 161–63.

57. Frederick J. Schlink and Arthur Kallett, "Poison for Profit," *Nation*, Dec. 21, 1932, pp. 608–10. They did not ignore denutrification completely. Schlink's book was none too flattering about either the reversal of McCollum's position on white flour or the subservience of home economists to business interests. It discussed both in a chapter entitled "The Misinformers."

58. Arthur Kallett, *100,000,000 Guinea Pigs: Dangers in Everyday Foods, Drugs, and Cosmetics* (New York: Grosset and Dunlap, 1933), 30–31.

59. "Health Food Fad," *BW*, Feb. 29, 1936.

60. Edward Keuchel, Jr., "The Development of the Canning Industry in New York State to 1960" (Ph.D diss., Cornell Univ., 1970), 347–49; Ruth Atwater, "Information About Canned Food Products," *JHE* 24 (June 1932): 527.

61. Ruth DeForest Lamb, *American Chamber of Horrors* (New York: Ferrar and Rinehart, 1936), 183–84. Actually, it was never quite clear why the large canners

were so vehemently opposed to grade labeling. Arthur Kallett thought it was because "the scale of their output [did] not permit them to select from the best." Arthur Kallett, "Food and Drugs for the Consumer," *Annals of the American Academy of Political and Social Science* 73 (May 1934): 31–32.

62. James Whorton, *Before Silent Spring: Pesticides and Public Health in Pre-DDT America* (Princeton: Princeton Univ. Press, 1974), 236–37; Lamb, *American Chamber*, 333.

63. They argued that the act would make Tugwell, through the FDA, their prosecutor, judge, and jury and that the Federal Trade Commission already protected consumers from false advertising. Consumer advocates, however, pointed out that in December 1933 a court overturned an FTC order that Standard Brands stop advertising that Fleischmann's yeast had done wonders for singer Rudy Vallee's health. This was because the FTC, which was powerless to challenge the health claim, could only complain that it was misleading not to indicate that Vallee had been paid for his testimonial. "The Tugwell Bill and You," *BHG*, Feb. 1934, p. 38; *FFR*, Dec. 19, 1933.

64. The women's groups included the American Association of University Women, the League of Women Voters, the YWCA, and the Council of Jewish Women. Lamb, *American Chamber*, 320.

65. *JAMA* 99 (Oct. 1, 1932): 1195; *JAMA* 113 (Aug. 19, 1939): 681–82; *FFR*, Feb. 27, 1933.

66. In 1937 and 1938 they and toiletry advertisers, who held second place, each spent well over twice as much in national magazines than their closest competitors, the car makers. *FFR*, July 10, 1939.

67. Lamb, *American Chamber*, 320.

68. Ibid., 180.

69. David Montgomery, the head of the USDA's Consumer Affairs Department, who cooperated with the consumer movement in this, was ultimately forced out of that position by charges that he had leftist sympathies. See Montgomery Collection, CIO-Washington Files, Congress of Industrial Organizations, Papers, Walter Reuther Library, Wayne State Univ., Detroit, Mich.

70. The food industries had been partly mollified in mid-1935, when the bill made grade labeling voluntary, although they continued to support the drug manufacturers in opposing regulation of advertising. "Pure Foods Can Wait," *New Republic* 82 (May 31, 1935): 328–29.

71. Otis Pease, *The Responsibilities of American Advertising* (New Haven: Yale Univ. Press, 1959), 122–25; Ruth DeForest Lamb, "The New Food, Drug, and Cosmetic Bill and the Home Economist," *JHE* 30 (Oct. 1938): 128; Charles O. Jackson, *Food and Drug Legislation in the New Deal* (Princeton: Princeton Univ. Press, 1970), 175–200.

72. Conrad Elvehjem, "Seven Decades of Nutrition Research," in American Association for the Advancement of Science, *Centennial* (Washington: AAAS, 1950), 100; Martin Bell, *A Portrait of Progress: A Business History of the Pet Milk Company, 1885–1960* (St. Louis: Pet Milk, 1962), 145–47. By 1944, however, while 85 percent of the nation's evaporated milk was fortified with vitamin D, only 10 percent of the fluid milk was, mainly because small operators could not afford the new expense involved and were reluctant to charge the premium price fortified milk commanded. National Research Council, Food and Nutrition Board, Committee on Milk, "Report on Milk Fortification, Dec. 1, 1944," ms. in Margaret Mead Papers, Manuscripts Division, Library of Congress, Box F24.

73. John T. Edsall and David Bearman, "Historical Records of Scientific Activity: A Survey of Sources for the History of Biochemistry and Molecular Biology," *Proceedings of the American Philosophical Society* 123 (Oct. 15, 1979): 288–89.

74. Donald K. Tressler, Oral History Interview, Archives and Manuscripts Division, Cornell Univ. Libraries, Ithaca, N.Y., p. 15.

75. Daniel Maynard, Oral History Interview, Archives and Manuscripts Division, Cornell Univ. Libraries, Ithaca, N.Y., pp. 69–73.

76. *NYT*, Nov. 30, 1938.

77. Its products were sold, among other places, in special "Health Food Departments" in the Liggett's, Walgreen's, and Owl drugstore chains. *FFR*, Dec. 19, 1932.

78. Tom Dolan, "Food à la Concentrate," in U.S. Works Progress Administration, Federal Writers' Project, Records, "America Eats" Papers, Manuscripts Division, Library of Congress (hereafter "America Eats"), Box A830.

79. Arthur J. Brooks, "America Eats (Southwest Section)," Jan. 1942, p. 2, "America Eats," Box A833.

80. *SNL*, June 22, 1940; Richard Hooker, *Food and Drink in America: A History* (Indianapolis: Bobbs-Merrill, 1981), 310; W. H. Sebrell, "Nutritional Diseases in the United States," *JAMA* 115 (Sept. 7, 1940): 851–52.

81. *SNL*, Feb. 18, 1939.

82. *SNL*, June 22, 1940; *FFR*, April 17, 1939.

83. "Shotgun Vitamins Rampant," *JAMA* 117 (Oct. 25, 1941): 1147.

84. *FFR*, March 20, 1939.

85. *FFR*, Feb. 20, March 20, 1939.

86. *FFR*, Oct. 2, 1939.

87. *FFR*, March 20, 1939.

88. R. R. Williams, "Cereals as a Source of Vitamin B_1 in Human Diets," *Cereal Chemistry* 16 (May 1939): 301, cited in "Nutritionally Improved or Enriched Flour and Bread," *JAMA* 116 (June 28, 1941): 2853.

89. *JAMA* 113 (Aug. 12, 1939): 589, 681.

90. They had approved of adding vitamin D to milk only because that vitamin was not easily obtained in the normal American diet and milk was universally consumed.

91. Neurasthenia was the turn-of-the-century catchall diagnosis for ailments that were thought to result from a rundown nervous system, something that often struck well-off women unable to cope with the pressures of modern life. It is thus interesting that the Mayo doctors compared the symptoms of thiamin deficiency with two ailments with distinct cultural components, both of which were almost totally restricted to white middle-class females.

92. R. D. Williams et al., "Induced Vitamin B_1 Deficiency in Human Subjects," *Proceedings of Staff Meetings, Mayo Clinic* 14 (Dec. 1939), cited in R. D. Williams et al., "Observations on Induced Thiamin (Vitamin B_1) Deficiency in Man," *Archives of Internal Medicine* 60 (Oct. 1940): 785–89.

93. "Vitamins for War," *JAMA* 115 (Oct. 5, 1940): 1198.

94. *SNL*, April 12, 1941.

95. *SNL*, Jan. 11, 1938. Later, confronted with other doctors' rather devastating evidence that adding thiamin to patients' diets did not alleviate neurasthenia, he responded with a kind of pseudo-scientific soft-shoe routine, inferring that this must mean that the deficiencies had been long-standing and therefore would take longer

to cure. Russell M. Wilder, "Nutritional Problems as Related to National Defense," *American Journal of Digestive Diseases* 8 (July 1941): 242.

96. *SNL*, Feb. 8, 1941.

97. George R. Cargill, "The Need for the Addition of Vitamin B_1 to Staple American Foods," *JAMA* 113 (Dec. 9, 1939): 2146–51.

98. Ibid., 2147; *SNL*, Jan. 11, 1941.

99. They had protested that it would be impossible to ensure consistent levels of nutrients in different batches of flour, but Great Britain had already rejected this notion and promulgated compulsory enrichment regulations. So, when the National Defense Advisory Commission asked that a committee of specialists draw up blueprints for a new kind of flour that would be a better source of mental and physical stamina, the large millers could hardly refuse to cooperate. *SNL*, Jan. 11, 1941. Russell M. Wilder and Robert R. Williams, *Enrichment of Flour and Bread: A History of the Movement*, National Research Council Bulletin no. 110 (Washington: NRC, 1944), 1–7. Ironically, subsequently the British program was not implemented due to charges that it would unduly benefit vitamin manufacturers.

100. *SNL*, Feb. 8, 1941. It took about two years for (some) nutritionists to begin questioning what had happened. "It is a curious fact that enrichment of white flour and white bread was promulgated with little direct experimental evidence to demonstrate the value of such a proposal for the human being," said *Nutrition Reviews* in 1943. *NR* 1 (1943): 295–97, cited in Ross H. Hall, *Food for Nought: The Decline in Nutrition* (New York: Harper and Row, 1974), 29.

101. Henry A. Wallace, "Nutrition and National Defense," in U.S. Federal Security Agency, *Proceedings of the National Nutrition Conference for Defense, May 26–28, 1941* (Washington: USGPO, 1942), 37.

102. George Gallup, *The Gallup Poll, 1935–1971* (New York: Random House, 1972), vol. 1, p. 310.

103. "National Nutrition," *JAMA* 116 (June 28, 1941): 2854.

104. This was the number of people estimated to depend on wheat flour and grains for more than 25 percent of their calories. *SNL*, Feb. 8, 1941.

105. *NYT*, Dec. 30, 1941.

Chapter 2. The Great Regression

1. Robert S. Lynd and Helen Lynd, *Middletown in Transition: A Study in Cultural Conflicts* (New York: Harcourt Brace, 1937), 144.

2. For some, the burden of housework had even been lightened. Thanks to falling wages for domestic help, the proportion of them employing full-time help increased, from 9 percent to 16 percent. Winona L. Morgan, *The Family Meets the Depression* (Minneapolis: Univ. of Minnesota Press, 1939), 26–46, 97–99. The Depression also seems to have been psychologically beneficial for middle-class women. While its hardship scarred lower-class ones with feelings of inadequacy and insecurity, it fostered feelings of mastery and competence among middle-class ones, making them more assertive. Sheila Kischler Bennett and Glen Elder, Jr., "Women's Work in the Family Economy: A Study of Depression Hardship in Women's Lives," *Journal of Family History* (Summer 1979): 153–75; Glen Elder, Jr., and Jeffrey K. Liker, "Hard Times in Women's Lives: Historical Influences Across Forty Years," *American Journal of Sociology* 88, no. 3 (1982): 241–69.

3. *FFR*, Dec. 19, 1932.

4. *FFR*, March 20, March 27, 1933.

5. *FFR*, March 27, 1933.

6. By the end of World War II the one-plant fruit and vegetable canner had become extinct in California. Vicki L. Ruiz, *Cannery Women, Cannery Lives: Unionization in the California Food Processing Industry, 1930–1950* (Albuquerque: Univ. of New Mexico Press, 1987), 23.

7. Martin Bell, *A Portrait of Progress: A Business History of the Pet Milk Company, 1885–1960* (St. Louis: Pet Milk, 1962), 112.

8. *FFR*, Jan. 2, 1933.

9. U.S. Bureau of the Census, *Historical Statistics of the United States, Colonial Times to 1970* (Washington: USGPO, 1975), 331–32.

10. *Consumption of Food in the United States, 1909–52*, U.S. Dept. of Agriculture Handbook no. 62 (Washington: USGPO, 1953), 147; Valerie K. Oppenheimer, *The Female Labor Force in the United States* (Berkeley: Univ. of California Press, 1970), 35.

11. Abraham Hoffman, *Large-Scale Organization in the Food Industries*, Temporary National Economics Committee Monograph no. 35 (Washington: USGPO, 1940), 1–90.

12. Willard F. Mueller, "The Food Conglomerates," *Proceedings of the American Academy of Political Science* 34 (1982): 55.

13. Bread-baking provides an extreme example of this process, for the local nature of bread production meant that the giants that were created through mergers were simply networks of large bakeries, with no huge centralized production facility to give them economies of scale. The four large corporations that dominated the industry by 1930 rose on the basis of their ability to out-advertise and out-market—through discounting and the like—those competitors they could not buy out. William B. Panschar, *Baking in America: Economic Development* (Evanston: Northwestern Univ. Press, 1956), vol. 1, pp. 145–73.

14. Hoffman, *Large-Scale Organization*, 56–57.

15. *FFR*, May 8, 1933.

16. "Mr. Countway Takes the Job," *Fortune*, Nov. 1940, pp. 100–114.

17. Even when Wesson Oil, no midget, came up with a similar shortening, MFB, it concentrated on the wholesale restaurant market rather than take on the others on the grovery shelves. *American Restaurant*, 1935, passim.

18. Hoffman pointed out that they emerged in industries in which food was processed or otherwise manufactured and not in those in which agricultural products were sold more or less unprocessed to consumers. Hoffman, *Large-Scale Organization*, 77–88.

19. *NYT*, May 20, 1941.

20. "Flash" freezing was already established, but a dearth of freezers in groceries and homes inhibited its growth. See chap. 7.

21. *AC*, April 1931, Jan. 1935, Oct. 1941.

22. John W. Bennett, "Food and Social Status," *American Sociological Review* 8 (Oct. 1943): 561–69.

23. Margaret Cussler and Mary L. DeGive, *'Twixt the Cup and the Lip* (Washington: Consortium Press, 1952), 112.

24. Maud Wilson, "Time Spent in Meal Preparation," *JHE* 24 (Jan. 1932): 14.

25. W. Lloyd Warner, *The Social Life of a Modern Community* (New Haven: Yale Univ. Press, 1941), 61–62.

26. Robert C. Christopher, *Crashing the Gates: The De-WASPing of America's Power Elite* (New York: Simon and Schuster, 1989), 256–57.

27. Judith Goode, Janet Theophano, and Karen Curtis, "A Framework for the Analysis of Continuity and Change in Shared Sociocultural Rules for Food Use: The Italian-American Pattern," in Linda Brown and Linda Mussell, eds., *Ethnic and Regional Foodways in the United States: The Performance of Ethnic Identity* (Knoxville: Univ. of Tennessee Press, 1984), 66–88.

28. John W. Bennett, "Social Process and Dietary Change," in *The Problem of Changing Food Habits: Report of the Committee on Changing Food Habits, 1941–1943*, National Research Council Bulletin no. 108 (Washington: NRC, 1943), 122.

29. Jitsuichi Masuoka, "Changing Food Habits of the Japanese in Hawaii," *American Sociological Review* 10 (1945): 765. Although the study was not published until 1945, the research was done in 1933 and 1934.

30. Paul Radin, *The Italians of San Francisco: Their Adjustment and Acculturation* (San Francisco: n.p., 1935), vol. 1, p. 131.

31. Leonard Covello, ed., *The Social Background of the Italo-American School Child* (Totawa, N.J.: Rowman and Littlefield, 1972), 325–26.

32. Ruiz, *Cannery Women, Cannery Lives*, 13.

33. Emory S. Bogardus, *The Mexican in the United States* (Los Angeles: Univ. of Southern California Press, 1934), 36.

34. Noel Bunch, "Joe Dimaggio," *Life*, May 1, 1939, quoted in Richard Alba, *Italian-Americans: In the Twilight of Ethnicity* (Englewood Cliffs, N.J.: Prentice-Hall, 1985), 100.

35. Masuoka, "Changing Food Habits," 765.

36. Mrs. J. Riordan, "Americans Getting the Macaroni Appetite," *Macaroni Journal*, June 15, 1928. For more on this complex process, see my article "The American Response to Italian Food, 1880–1939," *Food and Foodways* 1, no. 1 (1985): 1–24, and the chapter I co-authored with Joseph Conlin, "The Food Habits of Italian Immigrants to America: An Examination of the Persistence of a Food Culture and the Rise of 'Fast Food' in America," in Ray Browne et al., eds., *Dominant Symbols in Popular Culture* (Bowling Green, Ky.: Popular Culture Press, 1990), 231–46.

37. The cooperative campaign fell apart amidst bickering between bulk and packaged producers and a wave of price-cutting that set the manufacturers at each others' throats. *Macaroni Journal*, March 15, 1932.

38. Mary Martensen, "Spaghetti Universally Popular," rept. in *Macaroni Journal*, March 15, 1933.

39. *Macaroni Journal*, Jan. 15, 1933, Jan. 15, 1935. Her sponsor, General Mills, was one of the largest producers of durum wheat for the wholesale market and also sold it retail as Gold Medal semolina.

40. Ibid., Sept. 15, 1932, April 15, 1933.

41. *Good Housekeeping Cook Book* (New York: Good Housekeeping, 1933), 33; Katherine Fisher, "Dining in Italy at Your Own Table," *GH*, Oct. 1931, p. 85; *AC*, Jan. 1930, p. 37.

42. *Good Housekeeping Cook Book* (1933), 76.

43. U.S. Dept. of War, Office of the Quartermaster-General, *Manual for Army Cooks, 1942* (Washington: USGPO, 1942), 180.

44. Erik Amfithreatrof, *Children of Columbus* (Boston: Little, Brown, 1973), 240.

45. Ida B. Allen, *The Service Cook Book* (New York: Woolworth's, 1933).

46. Lynd and Lynd, *Middletown in Transition*, 202.

47. Warren Susman, *Culture as History* (New York: Pantheon, 1984), 150–84.

48. Susan Ware, *Holding Their Own: American Women in the 1930s* (Boston: Twayne, 1982), 14–17.

49. Lawrence A. Keating, *Men in Aprons* (New York: Mill, 1944), ix.

50. Jane Humphries, "Women: Scapegoats and Safety Valves for the Great Depression," *Review of Radical Political Economy* 8 (Spring 1976): 106.

51. Mrs. Ralph Borsodi, "The New Woman Goes Home," *Scribner's*, Feb. 1937, pp. 52–56.

52. Ruth S. Cowan, "Two Washes in the Morning—Bridge Party at Night," *Women's Studies* 3, no. 2 (1976): 169.

53. Helen Mitchell, "The National Nutrition Outlook," *JHE* 33 (Oct. 1941): 537.

54. On the 1920s see chap. 13 ("A Revolution of Declining Expectations") in my *Revolution at the Table: The Transformation of the American Diet* (New York: Oxford Univ. Press, 1988).

55. Steven Mintz and Susan Kellogg, *Domestic Revolutions: A Social History of American Family Life* (New York: Free Press, 1988), 285.

56. Carol Reuss, "*Better Homes and Gardens:* Consistent Concern Key to Long Life," *Journalism Quarterly* 51 (Summer 1974): 292–93.

57. This is based on an informal reading of the magazines rather than a formal survey, but it is similar to the impression of Theodore Selden regarding French women's magazines after 1930. Theodore Selden, *France, 1848–1945*, vol. 2 (Oxford: Oxford Univ. Press, 1977), 557.

58. *FFR*, Sept. 18, 1939.

59. By 1940 the federal government was subsidizing food and nutrition courses in close to 7,500 of the nation's high schools, which taught it to over 430,000 students per year. In addition, an estimated 236,000 adults were taking home economics in 3,559 centers. M. L. Wilson, "Nutrition and Defense," *JADA* 17 (Jan. 1941): 15.

60. *FFR*, March 6, 1939.

61. *Delineator*, Oct. 1935, p. 13.

62. Cussler and DeGive, *'Twixt the Cup and the Lip*, 113–15.

63. None as yet had full-fledged "Food" sections, mainly for lack of enough food advertising. There were few supermarkets advertising weekly "specials," and food processors spent most of their advertising dollars on the women's magazines and radio.

64. They "include only a limited number of leading food products on an exclusive basis. This allows sufficient time for each manufacturer's message." *FFR*, Jan. 16, Sept. 25, Oct. 23, 1933.

65. "Mr. Countway," p. 114.

66. *FFR*, Feb. 13, 1933.

67. Helen Woodward, *The Lady Persuaders* (New York: Obolensky, 1960), 161.

68. General Mills, "Outline of the Career in Advertising of Marjorie Child Husted," Feb. 1950, typescript in Marjorie C. Husted Papers, Schlesinger Library of Women's History, Radcliffe College, Cambridge, Mass.; *NYT*, Dec. 28, 1986.

69. *GH*, Jan. 1935, p. 100.

70. *FFR*, March 6, 1939.

71. *FFR*, Sept. 18, 1939.

72. *GH*, March 1941, pp. 106–7.

73. *Hygeia* 18 (Aug. 1940): 728.

74. *NYT*, Oct. 21, 1934.

75. S. Warner, "Why Men Hate Spinach," *American Home*, March 1940, p. 54.
76. *Thoughts for Food: A Menu Aid* (Chicago: Institute Publishing, 1938), 70, 71–102, 248.
77. Anne Wetherill Reed, *The Philadelphia Cook Book of Town and Country* (New York: Barrows, 1940), 13, 14, 255.
78. Margaretta Brucker, "Best Sellers," *AC*, Jan. 1935, p. 302.
79. Bernard De Voto, "Notes from a Wayside Inn," *Harper's*, Sept. 1940, p. 447.
80. "La Societé des Gourmets de la Nouvelle Orléans," *AC*, May 1939.
81. *LHJ*, May 1934, p. 71.
82. Mary Grosvenor Ellsworth, *Much Depends on Dinner* (New York: Knopf, 1939), vii.
83. "Food for Husbands: Reader Recipes for a Man's Appetite," *Delineator*, Oct. 1935, p. 17.
84. Marjorie Husted, "The Women in Your Lives," typescript of speech, Oct. 13, 1952, Husted Papers, Folder 2.
85. *My Better Homes and Gardens Lifetime Cook Book* (Des Moines: Meredith, 1930).
86. *LHJ*, May 1939, p. 71.
87. *Good Housekeeping Cook Book* (1933), 100, 106, 90.
88. Ibid., 105.
89. The only recipe that included garlic was for chili con carne, which, aside from being the classic "man's dish," was also a traditional American (not Mexican) recipe; almost every version of it had historically called for garlic. Ibid., 59–67, 74, 136, 86.
90. Virginia Porter and Esther Latzke, *The Canned Foods Cook Book* (New York: Doubleday, 1939), vii.
91. Ellsworth, *Much Depends on Dinner*, 43–44.
92. John Tarbell, *The American Magazine: A Compact History* (New York: Hawthorn, 1969), 210–11.
93. Reuss, "*Better Homes and Gardens*," 293.
94. John C. Hudson, "The Middle West as Cultural Hybrid," *Pioneer America Society Papers* 7 (1984): 35–44.
95. Grace Smith, Beverly Smith, and Charles M. Wilson, *Through the Kitchen Door* (New York: Stackpole, 1938), 98–101.

Chapter 3. From Burgoo to Howard Johnson's

1. See my *Revolution at the Table: The Transformation of the American Diet* (New York: Oxford Univ. Press, 1988), 169–72.
2. Hazel Stiebeling et al., *Family Food Consumption and Dietary Levels: Five Regions*, U.S. Dept. of Agriculture Misc. Pub. no. 405 (Washington: USGPO, 1941), 1–6; M. A. Schaars, "How Butter Consumption Varies," *National Butter and Cheese Journal* 30 (March 1939): 8–9, 36.
3. Russell M. Wilder and Robert R. Williams, *The Enrichment of Flour and Bread: A History of the Movement*, National Research Council Bulletin no. 110 (Washington: NRC, 1944), 35.
4. This can be confirmed by glancing through the "Cookery" sections of the *Reader's Guide to Periodical Literature* for the 1930s.
5. Vella C. Reeve, "From Boston to Los Angeles via Motor Car," *AC*, Jan. 1935, pp. 344–49.

6. Catherine Mackenzie, "Food by the Wayside," *NYTM*, July 31, 1938.

7. "America Eats (Southwest Section)," in U.S. Works Progress Administration, Federal Writers' Project, Records, "America Eats" Papers, Manuscripts Division, Library of Congress (hereafter "America Eats"), Box A833.

8. Arthur Brooks, "America Eats (Southwest Region)," ms. in "America Eats," Box A833.

9. Walter Kipling, memorandum, Jan. 30, 1942, "America Eats," Box A830.

10. Ibid.

11. "The Birth of Burgoo," ms. in "America Eats," Box A833.

12. "Bad Advice Ruins Brunswick Stew," ms. in "America Eats," Box A833.

13. "We Refreshes Our Hog Meat with Corn Pone," ms. in "America Eats," Box A833.

14. "Oysters Are Good for Everything You Got," ms. in "America Eats," Box A833.

15. Named after the Big and Little Pee Dee rivers.

16. "Wind from the North, Fish Bite like a Horse," ms. in "America Eats," Box A833.

17. "Oysters Are Good."

18. Brooks, "America Eats (Southwest Region)"; Grace Smith, Beverly Smith, and Charles Wilson, *Through the Kitchen Door* (New York: Stackpole, 1938), 74.

19. R. L. Ryan, "America Eats (Washington): Annual Events," ms. in "America Eats," Box A833.

20. Montana Writers' Project, "America Eats (Montana)," ms. in "America Eats," Box A831.

21. "America Eats (The South)", ms. in "America Eats," Box A833, p. 47.

22. "America Eats (Illinois)," ms. in "America Eats," Box A831, p. 36.

23. Ibid., p. 61.

24. "America Eats (Montana)," pp. 50–51.

25. "America Eats (Montana)," p. 63; "America Eats (Southwest)."

26. Charles A. Finger to Henry Alsberg, Director, WPA Writers' Project, Oct. 23, 1937, "America Eats," Box A830.

27. Ryan, "America Eats (Washington)."

28. "America Eats (Montana)," p. 60.

29. "America Eats (Montana)," pp. 60–62; "America Eats (Colorado)," ms. in "America Eats," Box A830.

30. *FFR*, Oct. 9, 1933.

31. Smith, Smith, and Wilson, *Through the Kitchen Door*, 117.

32. Carroll Kennedy, "A Community Smelt Fry," in "America Eats (Washington)."

33. Arthur Van Vlissingen, Jr., "Fred Harvey," *Factory and Industrial Management* 78 (Nov. 1929): 1085.

34. I deal with these factors in more detail in *Revolution at the Table*, chap. 15, and "Two Hundred Years of French Food in America," *Journal of Gastronomy* 5 (Spring 1989): 67–90.

35. E. M. Fleischman, "Trends in Food Merchandising," *RM*, Nov. 1931, pp. 295–96.

36. *NYT*, Jan. 21, 1961.

37. Rae N. Sauder, "Announcing Mr. and Mrs. Milton Cross," *AC*, Jan. 1942, p. 267.

38. "The Wonders of Diet," *Fortune*, May 1936, p. 88.

39. Scudder Middleton, *Dining, Wining, and Dancing in New York* (New York: Dodge, 1938) 13–15, 20–36; Jerome Beatty, "Sherman Packs 'Em In," *American Magazine*, June 1941, pp. 44–45.

40. Caroline Bates, "Great Old Los Angeles Eating Places," *Gourmet*, Jan. 1991, p. 58.

41. Duncan Hines, *Duncan Hines' Food Odyssey* (New York: Crowell, 1955), 23–28.

42. Ibid.; Frank J. Taylor, "America's Gastronomic Guide," *Scribner's Commentator*, June 1941 pp. 14–18; Duncan Hines and Frank J. Taylor, "How to Find a Decent Meal," *Saturday Evening Post*, April 27, 1947; Duncan Hines, *Adventures in Good Eating* (Chicago and Bowling Green: Adventures in Good Eating, Inc., various dates).

43. Hines, *Odyssey*, 237; "Hines Abroad," *New Yorker*, June 24, 1954; *NYT*, March 16, 1959.

44. Taylor, "America's Gastronomic Guide," 16.

45. See, for example, the Linton's menu in *RM*, July 1931, p. 19.

46. Americans were estimated to have spent $3.569 billion on food and beverages consumed outside the home. By 1933 that figure had dropped to $2.158 billion. While this meant a sharp decline in revenues for restaurants, it did not really indicate a great shift in eating habits, for the amount spent on food consumed inside the house dropped just as precipitously, from $14.520 billion to $8.033 billion, leaving the percentage of food expenditures spent outside the home at about the same level, 20 percent. It continued at about this level until 1939, when it began to rise. *RM*, May 1953, p. 46.

47. *BW*, April 5, 1933; *RM*, Oct. 1931, p. 211.

48. *BW*, June 22, 1932.

49. *BW*, Oct. 19, 1940.

50. Philip Langdon, *Orange Roofs and Golden Arches: The Architecture of American Chain Restaurants* (New York: Knopf, 1986), 41–42; Clarence H. Lieb, *Main Street to Miracle Mile: American Roadside Architecture* (New York: Little, Brown, 1985), 202.

51. *BW*, Feb. 17, 1940.

52. Bernard De Voto, "Notes from a Wayside Inn," *Harper's*, Sept. 1940, p. 448.

53. Ibid.

54. Lieb, *Main Street*, 202.

55. Ibid.; Warren Belasco, "Toward a Culinary Common Denominator: The Rise of Howard Johnson's, 1925–1940," *Journal of Culture* 2 (1979): 506, 512–15.

56. See *RM*, Nov. 1938, pp. 38–39, for photos of an extraordinary variety of them.

57. "World's Finest Drive-In Restaurant," *RM*, July 1938, pp. 30–31.

58. Only fried frogs' legs with tartar sauce added a somewhat unusual touch, but then they were much more common in parts of the South and southern Midwest than they are today. Bayard Evans, "Profits Roll In on Rubber," *RM*, Aug. 1938, pp. 26–27.

59. *RM*, Oct. 1938, p. 78.

60. *RM*, Oct. 1931, p. 219.

61. The sauce was a tomato sauce with cream. *RM*, July 1938.

62. "America Eats (Southwest)," p. 36. The Brass Rail restaurant in New York seemed to owe much of its phenomenal success to covering practically all bases. It had a replica of the huge roasting jack in London's famed Simpson's-on-the-Strand

in its window and a "sea grill" in its basement Rathskeller, whose walls were covered in photomurals of sports events. *RM*, April 1936, p. 229.

63. De Voto, "Notes from a Wayside Inn," 448.

64. *FFR*, Oct. 1952, pp. 51, 98.

65. Smith, Smith, and Wilson, *Through the Kitchen Door*, 6.

66. "America Eats (Northeast)," ms. in "America Eats," Box A832. It is interesting that Automats, Swedish inventions that entered the United States through Philadelphia, did not catch on nearly as well anywhere else. Chicago was a particular disappointment for the company. Molly Miller, "Automatic Destruction," *New City*, Oct. 11, 1990.

67. It might be argued that because it was in Los Angeles it was sui generis, but most Americans would have recognized its name as a joke, whereas French, Chinese, or other people from cultures where restaurant-going is taken seriously would have been taken aback.

68. Langdon, *Orange Roofs*, 32–35; Paul Hirshorn and Steven Izenour, *White Towers* (Cambridge: MIT Press, 1981), 1–15.

69. Langdon, *Orange Roofs*, 32. The brochure might well have served as material for the movie *Metropolis*, Fritz Lang's contemporaneous view of a future mechanized Hell.

70. Diana Ashley, *Where to Dine in Thirty-nine* (New York: Crown, 1939), 82.

71. *Macaroni Journal*, Aug. 15, 1924.

72. Ibid., Apr. 15, 1933.

73. Rian James, *Dining in New York* (New York: John Day, 1930), 206.

74. Byron McFadden, "Let's Have Spaghetti," *GH*, March 1935, p. 89.

75. In 1937, for example, they owned 67 percent of the restaurants in New Haven, Connecticut, yet none of them seemed to specialize in Greek food. Lawrence A. Lovell-Troy, "Kinship Structure and Economic Organization among Ethnic Groups: Greek Immigrants and the Pizza Business," (Ph.D diss., Univ. of Connecticut, 1979), 2–38.

76. Mackenzie, "Food by the Wayside"; *AC*, May 1942, p. 162.

77. *RM*, April 1936, p. 245.

78. *RM*, Dec. 1931, p. 363.

Chapter 4. One-third of a Nation Ill Nourished?

1. *NYT*, Sept. 13, 1933. The Children's Bureau data are in Martha Eliot, "Some Effects of the Depression on the Nutrition of Children," *Hospital Social Services* 28 (1933): 585–98.

2. Martha Eliot, "Child Health, 1933–34," *Journal of Pediatrics* 4 (June 1934): 827–37.

3. Norman Thomas, *The Choice Before Us* (New York: Macmillan, 1936), 6.

4. Loren Baritz, *The Good Life: The Meaning of Success for the American Middle Class* (New York: Knopf, 1989), 119.

5. Robert S. Lynd and Helen Lynd, *Middletown in Transition: A Study in Cultural Conflicts* (New York: Harcourt Brace, 1937), 109.

6. Ouida J. Fulgham, "Roosevelt Feeds the Hungry," *Red River Valley Historical Review* 7, no. 4 (1982): 24–26; Arthur Schlesinger, Jr., *The Age of Roosevelt: The Coming of the New Deal* (Boston: Houghton Mifflin, 1959), 277–78.

7. *NYT*, Aug. 28, 1938.

8. Don F. Hadwiger, "Nutrition, Food Safety, and Farm Policy," *Proceedings of the American Academy of Political Science* 34, no. 3 (1982): 81; March 30, 1935. Had-

wiger places the incident in 1939, citing an interview done years later with Stiebeling, but she seems to have confused it with the 1935 uproar.

9. Bonnie Fox Schwartz, "Unemployment Relief in Philadelphia, 1930–1932," *Pennsylvania Magazine of History and Biography* 42 (Jan. 1969), rpt. in Bernard Sternsher, ed., *Hitting Home: The Great Depression in Town and Country* (Chicago: Quadrangle, 1970), 73–77.

10. "Hot School Lunches (Snohomish County, Washington)," in U.S. Works Progress Administration, Federal Writers' Project, Records, "America Eats" Papers, Manuscripts Division, Library of Congress, Box A833.

11. H. M. Southworth and M. I. Klayman, *The School Lunch Program and Agricultural Surplus Disposal*, U.S. Dept. of Agriculture Misc. Pub. no. 467 (Washington: USGPO, 1941), 17.

12. Janet Poppendieck, *Breadlines Knee-Deep in Wheat: Food Assistance in the Great Depression* (New Brunswick: Rutgers Univ. Press, 1986), 240.

13. R. S. McElvaine, *The Great Depression* (New York: Times Books, 1984), 80; Irving Bernstein, *The Lean Years: A History of the American Worker, 1920–1933* (Baltimore: Penguin, 1970), 362, 331.

14. Hadwiger, "Nutrition, Food Safety," 82.

15. He pointed out that death from industrial accidents had also declined, attributing the latter to unemployment, for "employment is a form of warfare." *NYT*, Jan. 31, 1933.

16. *NYT*, Sept. 19, 1933.

17. *NYT*, Sept. 13, 1933.

18. Ibid.

19. East Harlem Nursing Health Service, "A Progress Report, 1934," Community Service Society, Papers, Division of Archives and Manuscripts, Columbia University Libraries, New York, N.Y. (hereafter CSS Papers), Box 31, File 122-2.

20. They dropped from 98.6 per 1,000 live births in 1934 to 75.2 in 1937. Lindsley Cocheu and Sophie Rabinoff, "Effects of Changing Population on Health Conditions in East Harlem," CSS Papers, Box 31, File 122-2.

21. Bailey Burritt, foreword to Gwendolyn Berry, *Idleness and the Health of a Neighborhood* (New York: NYAICP, 1933), x.

22. Berry, *Idleness*, 66–67. In January 1933 the New York State commissioner of health declared that "even in the midst of plenty there is a starvation in the essential elements necessary for normal development" and called on every health and relief agency to ensure that children were taking cod liver oil over the winter and early spring, advice that was widely followed. *NYT*, Jan. 9, Apr. 2, 1933.

23. Eliot, "Child Health, 1933–34," 824–27.

24. For the origins and earlier application of these methods see chap. 9, "The Great Malnutrition Scare, 1907–21," in my *Revolution at the Table: The Transformation of the American Diet* (New York: Oxford Univ. Press, 1988).

25. In 1932, when only 2 percent of a large group of Detroit schoolchildren were declared malnourished in regular medical examinations, a reexamination of a smaller group of them using "careful clinical criteria" managed to declare 18 percent of them malnourished. Eliot, "Some Effects of the Depression," 588.

26. William Schmidt, "Newer Medical Methods of Appraisal of Nutrition Status," *AJPH* 30 (Feb. 1940): 165–69. Stature statistics are useful indicators of nutritional status, but only for similar populations over long runs.

27. Ada Moser, "Standard Food Budgets Used by Social Agencies," *JHE* 26 (March 1934): 132.

28. Hazel Stiebeling et al., *Family Food Consumption and Dietary Levels: Five Re-*

gions, U.S. Dept. of Agriculture Misc. Pub. no. 405 (Washington: USGPO, 1941); U.S. Bureau of Labor Statistics, "Food Expenditures of Wage Earners and Clerical Workers," *Monthly Labor Review* 51, no. 2 (Aug. 1940): 250–66.

29. Hazel Stiebeling and Esther Phipard, "Diets of Families of Employed Wage Earners and Clerical Workers in Cities," U.S. Dept. of Agriculture Circular no. 307 (Washington: USGPO, 1939). Canadian studies using these methods also came to alarming conclusions. One said that the diets of 90 percent of the *employed* poor in Halifax failed to meet minimum standards for maintaining health and that *none* could afford enough thiamin. A study of Torontonians with twice the Haligonians' income concluded that their diets were still terribly deficient. E. G. Young, "A Dietary Survey in Halifax," *Canadian Journal of Public Health* 32, no. 5 (May 1941): 236–40; E. E. McHenry, "Nutrition in Toronto," ibid. 30, no. 1 (Jan. 1939): 4–13.

30. Henry Boorsook, "Industrial Nutrition and the National Emergency," *AJPH* 32 (May 1942): 524.

31. *SNL*, May 31, 1941.

32. *NYT*, May 21, 1941; *SNL*, Feb. 8, 1941.

33. U.S. Senate, Select Committee on Nutrition and Human Needs, *Dietary Goals for the United States: Committee Print*, 95th Cong., 1st sess., 1977, p. 14.

34. New York Association for Improving the Condition of the Poor, untitled flyers on nutrition, Jan. 1927 and March 1937, CSS Papers, Box 69, "Nutrition" Folder.

35. Virginia Britton, "The Cost of Food and the Adequacy of Income," *JHE* 29 (May 1937): 294–95. The tubercular woman who won the Red Cross prize for feeding her family on seven dollars a week in New York City won mainly because the judges were impressed that she still managed to buy two quarts of milk a day, at twelve cents a quart, for her husband and two children. *NYT*, Dec. 22, 1932.

36. *NYT*, Jan. 26, 1935.

37. "Industrial Nutrition and the War," *NR* 1 (Jan. 1943): 26. The last person to do this had been Joseph Goldberger, who had used studies employing quite different methods to demonstrate that deficiencies in niacin led to pellagra.

38. James S. McClester, "Influence of the Depression on the Nutrition of the American People," *JAMA* 106 (May 30, 1939): 1865–69; "Industrial Nutrition and the War"; H. Borsook, letter to the editor, *NR* 1 (March 1943): 159–60.

39. C. E. Palmer, "Further Studies on Growth and the Economic Depression," *Public Health Reports* 49 (Dec. 7, 1934): 1453; "Height and Weight of Children of the Depression Poor," ibid. 50 (Aug. 16, 1935): 1106.

40. She herself had admitted that they were mere "bits of evidence," and so they were: reports that malnutrition among some schoolchildren in Manhattan had increased from 16 percent in 1929 to 29 percent in 1932; that it had increased from 30 to 42 percent among children at a community health center in Philadelphia in the same year; and that while only 8 percent of high school students in one Baltimore playground athletic league were underweight in 1931, 17 percent in a league in another Maryland county were in 1932. Eliot, "Some Effects of the Depression," 585–98.

41. Deaths from scurvy, rickets, pellagra, and beri-beri fell from 6,900 in 1930 to 2,656 in 1939. *NYT*, June 15, 1941.

42. *NYT*, Feb. 27, 1935.

43. Charles S. Johnson, *Shadow of the Plantation* (Chicago: University of Chicago Press, 1934), 100–101.

44. Margaret Cussler and Mary L. DeGive, "Foods and Nutrition in Our Rural Southeast," *JHE* 35 (May 1943): 281.

45. Wilbur O. Atwater and Charles Dayton Woods, *Dietary Studies with Reference to the Negro in Alabama*, U.S. Dept. of Agriculture, Office of Experiment Stations Bulletin no. 8 (Washington: USGPO, 1897).

46. Jacqueline Jones, *Labor of Love, Labor of Sorrow: Black Women, Work, and the Family from Slavery to the Present* (New York: Basic Books, 1985), 182–90; David Weber, "Anglo Views of Mexican Immigrants: Popular Perceptions and Neighborhood Realities in Chicago, 1900–1940," (Ph.D. diss., Ohio State Univ., 1982), 186–99.

47. Vicki Ruiz, *Cannery Women, Cannery Lives: Mexican Women, Unionization, and the California Food Processing Industry, 1930–1950* (Albuquerque: Univ. of New Mexico Press, 1987) 51.

48. Hazel K. Stiebeling, *Economic and Social Problems and Conditions of the Southern Appalachians*, U.S. Dept. of Agriculture Misc. Pub. no. 205 (Washington: USGPO, 1935), 145–60.

49. James Patterson, *America's Struggle Against Poverty, 1900–1980* (Cambridge: Harvard Univ. Press, 1981), 41.

50. A sampling of the numerous individual case files of the New York Association for Improving the Condition of the Poor failed to turn up even one instance where hunger or severe malnutrition was a factor prompting the agency to take up the case. Nor was either mentioned as "Types of Problems Referred to Nutrition Bureau" in the manuscript history of the AICP. AICP Individual Case Files, CSS Papers; untitled history of NYAICP, p. 24, ibid., Box 6.

51. James T. Patterson concluded that had it not been for the Civil Works Administration, predecessor to the WPA, which in January 1934 had over four million people on its payrolls at wages averaging fifteen dollars a week (triple the average relief benefit for a family of five) "the country would have faced widespread malnutrition and unrest." By February 1934 New Deal relief agencies were supporting twenty-eight million people, 22 percent of the population. Patterson, *America's Struggle*, 57.

52. Sheila Kischler Bennett and Glen Elder, Jr., "Women's Work in the Family Economy: A Study of Depression Hardship in Women's Lives," *Journal of Family History* (Summer 1979): 153–75; Glen Elder, Jr., and Jeffrey K. Liker, "Hard Times in Women's Lives: Historical Influences Across Forty Years," *American Journal of Sociology* 88, no. 3 (1982): 241–69.

53. E. Wright Bakke, *The Unemployed Worker: A Study of the Task of Making a Living Without a Job* (New Haven: Yale Univ. Press, 1940), 266–68; D. G. Wiehl, "Diets of Low-Income Families in New York City," *Milbank Memorial Fund Quarterly Bulletin* 11 (1933): 299.

54. Bakke, *Unemployed Worker*, 268.

55. Jacqueline Dowd Hall et al., *Like a Family: The Making of a Southern Cotton Mill World* (Chapel Hill: Univ. of North Carolina Press, 1987), 311.

56. Bakke, *Unemployed Worker*, 268.

57. The items they would buy more of mentioned most, in order of frequency, were meat, vegetables, fruits, and dairy products—particularly milk. George Gallup, *The Gallup Poll, 1935–1971* (New York: Random House, 1972), vol. 1, pp. 254–55.

58. *FFR*, Feb. 20, 1939.

59. *FFR*, March 20, 1939; U.S. Senate, Committee on Agriculture, Nutrition,

and Forestry, *The Food Stamp Program: History, Description, Issues, and Options: Committee Print*, 99th Cong., 1st sess., 1985, pp. 4–8; Maurice MacDonald, *Food Stamps and Income Maintenance* (New York: Academic Press, 1977), 2; Jeffrey M. Berry, *Feeding Hungry People: Rulemaking in the Food Stamp Program* (New Brunswick: Rutgers Univ. Press, 1984), 22.

60. Faith M. Williams, "Food Consumption at Different Economic Levels," *Monthly Labor Review* 42 (April 1936): 889–92.

61. Gallup, *Gallup Poll, 1935–1971*, vol. 1, p. 192.

62. Claude Wickard to Franklin D. Roosevelt, May 20, 1920, including draft of FDR to Jon Winant, n.d., U.S. Dept. of Agriculture, War Food Administration, Records, National Archives, RG 16, Entry 218, Box 15.

63. Agricultural Marketing Administration, "Distribution Branch Programs, Aug. 22, 1942," memorandum in U.S. Dept. of Agriculture, Food Distribution Administration, Records, National Archives, RG 136, Entry 138, Box 66.

64. It had been set up by executive order, so scrapping it did not need congressional approval. Senate Agriculture Committee, *Food Stamp Program*, 7–8; Berry, *Feeding Hungry People*, 22; Janet Poppendieck, *Breadlines Knee-Deep in Wheat: Food Assistance in the Great Depression* (New Brunswick: Rutgers Univ. Press, 1986), 241.

65. Norwood A. Kerr, "Drafted into the War on Poverty: USDA Food and Nutrition Programs, 1961–1969," *Agricultural History* 64 (Spring 1990): 156; *The School Lunch Program and Agricultural Surplus Disposal*, U.S. Dept. of Agriculture Misc. Pub. no. 467 (Washington: USGPO, 1941), 27.

66. *SNL*, May 31, 1941.

67. Gove Hambidge, "Malnutrition as a National Problem," *JHE* 31 (June 1939): 360–62.

68. *NYT*, Sept. 15, 1940.

69. *NYT*, March 2, 1941.

70. Middle-income people did mention health more frequently than the poor. For some reason, respondents were not asked about taste. Kurt Lewin, "Forces Behind Food Habits and Methods of Change," in *The Problem of Changing Food Habits: Report of the Committee on Food Habits*, 1941–1943, National Research Council Bulletin no. 108 (Washington: NRC, 1943), 45.

Chapter 5. Oh What a Healthy War

1. *NYT*, Jan. 22, March 8, 1941.
2. *NYT*, Jan. 21, 1941.
3. *NYT*, March 8, 1941.
4. M. L. Wilson, "Nutrition and Defense," *JADA* 17 (Jan. 1947): 13–14.
5. *NYT*, Jan. 21, 1941.
6. *NN*, Sept. 1940.
7. Two hundred thousand of the unfit, however, were judged capable of "limited military service," and another one hundred thousand were declared "capable of rehabilitation to the point where they can do full military service"—something that was not bruited about as much. Lewis Hershey, "Selective Service and Its Relation to Nutrition," U.S. Federal Security Agency (hereafter FSA), *Proceedings of the National Nutrition Conference for Defense, May 26–28, 1941* (Washington: USGPO, 1942), 67.
8. Since vitamin C was thought to prevent tooth decay and three times as many draftees were rejected because of poor teeth in 1940–1941 as in 1917, some research-

ers concluded that the diets of children born twenty years earlier must have been particularly deficient in it. *SNL*, May 3, 1941.

9. *NYT*, Jan. 22, 1941.

10. His official title was "coordinator of all health, medical, welfare, nutrition, recreation, and other related fields of activity affecting the national defence." *NYT*, Jan. 22, 1941.

11. It had been set up in September 1940 specifically to advise the government on defense matters.

12. Frank Gunderson, "Peacetime and Wartime Functions of the Food and Nutrition Board, National Research Council," *NR* 1 (April 1943): 161–62; Lydia J. Roberts, "Beginnings of the Recommended Dietary Allowances," *JADA* 34 (1958): 902–3; Margaret Conner, "A History of Dietary Standards," in Adelia Beeuwkes et al., eds., *Essays in the History of Nutrition and Dietetics* (Chicago: American Dietetic Association, 1967), 109; "Nutrition in the Defense Program," *JHE* 43 (April 1941): 250.

13. The thiamin recommendations, however, went along with Wilder's very high estimates, something which in turn greatly inflated estimates for niacin, the need for which was calculated at ten times that of thiamin. Lydia Roberts, "Beginnings of the Recommended Dietary Allowances," in Beeuwkes et al., *Essays*, 107–10; "Recommended Daily Dietary Allowances," *JHE* 43 (Sept. 1941): 476–77; Margaret Reid, "Food, Liquor, and Tobacco," in J. Friedrich Dewhurst et al., eds., *America's Needs and Resources: A New Survey* (New York: Twentieth Century Fund, 1955), 155–57.

14. Russell M. Wilder, "Misinterpretation and Misuse of the Recommended Dietary Allowances," *Science* 101 (March 23, 1945): 285–87.

15. FSA, *Proceedings, National Nutrition Conference, 1941*.

16. Indeed, in July 1943 a poll indicated that 59 percent of Americans wanted Hoover to "take over the entire food problem in the United States." George Gallup, *The Gallup Poll, 1935–1971* (New York: Random House, 1972), vol. 1, p. 396.

17. A further complication is that it was made part of the Office of Emergency Management of the Executive Office, not the FSA. U.S. Executive Office of the President, Executive Order Establishing the Office of Defense Health and Welfare Services, Sept. 3, 1941.

18. There was a vague agreement that the OPA and the USDA would somehow share decision-making in this field, but in practice the OPA usually called the shots.

19. Helen Mitchell, "U.S.A. Nutrition Program: How Is It Organized," *JHE* 35 (Jan. 1943): 32–33. By war's end the Interdepartmental Nutrition Coordinating Committee had fifty members. National Research Council, Committee on Food Habits (NRC-CFH), "Translation of Scientific Findings into Living Habits," Liaison Session, May 19, 1945, mimeo, pp. 1–2, in Margaret Mead Papers, Manuscripts Division, Library of Congress, Box F12.

20. This was the new name for the Food and Nutrition Committee.

21. Russell Wilder, "Nutrition in the United States: A Program for the Present Emergency and the Future," *Annals of Internal Medicine* 14 (1941): 2190–91.

22. "The American Diet," *NR* 1 (June 1943): 239.

23. H. D. Kruse et al., *Inadequate Diets and Nutritional Deficiencies in the United States: Their Prevalence and Significance*, National Research Council Bulletin no. 109 (Washington: NRC, 1943), 46.

24. Frank G. Boudreau, "Appraisal of Nutritional Status," *AJPH* 31 (Oct. 1941): 1061.

25. William Schmidt, "Newer Medical Methods of Appraisal of Nutritional Status," *AJPH* 30 (Feb. 1941): 165–66.

26. Kruse et al., *Inadequate Diets*, 47.

27. Schmidt, "Newer Medical Methods," 166–68.

28. *SNL*, May 31, 1941.

29. Wilder, "Nutrition in the United States," 2192.

30. Russell Wilder, "Nutritional Problems as Related to National Defense," *Journal of Digestive Diseases* 8 (July 1941): 244–45.

31. Wilder, "Nutrition in the United States," 2192.

32. FSA, *Proceedings, National Nutrition Conference, 1941*, 180–83.

33. Increased supplies of riboflavin and the development of a new strain of high-thiamin yeast helped undermine opposition among smaller millers, who had generally held out against it. By then, 75 to 80 percent of the nation's family flour was enriched. Six states, mainly in the South—where public health officials were particularly positive about it because it involved adding the antipellagra factor nicotinic acid—had made it mandatory. Russell M. Wilder and Robert R. Williams, *Enrichment of Flour and Bread: A History of the Movement*, National Research Council Bulletin no. 110 (Washington: NRC, 1944), 5–7.

34. Wilder, "Nutritional Problems," 245; Wilder, "Nutrition in the United States," 2192–95. Vitamin A was already commonly added to margarine, but on a voluntary basis.

35. Russell Wilder, "The Quality of the Food Supply; Control a Pressing Need," memo attached to Cummings to Wilson, Oct. 10, 1942, U.S. Executive Office of the President, Office of the Coordinator of Health, Welfare, and Related Defense Activities, Nutrition Division, Records, National Archives RG 136, Entry 218, (hereafter Nutrition Division Records), Box 31.

36. "Shotgun Vitamins Rampant," *JAMA* 117 (Oct. 21, 1941): 1447.

37. "Council on Foods and Nutrition," *JAMA* 121 (April 24, 1943): 1369–70.

38. Ten of the thirty-three members of the board were physicians, and six were also members of the AMA's Council on Foods and Nutrition, which had taken a very conservative stand on enrichment. James S. McLester, "Nutrition and the Nation at War," *Bulletin of the New York Academy of Medicine*, August 1942, p. 504.

39. Lydia J. Roberts, "Report of the Committee on Dietary Allowances," Nov. 28, 1944, Mead Papers, Box F24; "Vitamin Pills for Industrial Workers," *NR* 2 (Nov. 1944): 342; "Present Knowledge of the Vitamin B-Complex in Human Nutrition (Part 1)," *NR* 4 (May 1946): 130–32; "Mental Response to Thiamin Supplements," *NR* 5 (Nov. 1946): 343–45; Lydia J. Roberts, "Beginnings of the Recommended Dietary Allowances," in Beeuwkes et al., *Essays*, 110–11.

40. Nathan Schefferman, "Vitamins and Labor Accord," in American Federation of Labor, *Labor's Conference on Food and Nutrition, Oct. 9, 1943: Proceedings* (n.p., n.d. [1943?]), 47.

41. "Industrial Nutrition and the War," *NR* 1 (Jan. 1943): 70–71.

42. Henry Borsook, "Industrial Nutrition and the National Emergency," *AJPH* 32 (May 1942): 526; "The American Diet," *NR* 1 (June 1943): 239.

43. McLester, "Nutrition and the Nation at War," 500; Shefferman, "Vitamins and Labor Accord," 47.

44. National Research Council, Food and Nutrition Board (NRC-FNB), Committee on Milk, "Report on Milk Fortification, Dec. 1, 1944," Mead Papers, Box F24. The vitamin advocates argued—not without merit—that it was easier to per-

suade Americans to take something that looked like medicine than to have them change their diets. Shefferman, "Vitamins and Labor Accord," 48.

45. NRC-FNB, Committee on Nutrition of Industrial Workers, *The Food and Nutrition of Industrial Workers in Wartime* (Washington: USGPO, 1942), 15.

46. "Meat, poultry or fish, dried beans, peas or nuts, occasionally" were another group; eggs constituted a group of their own, even though only three or four of them had to be eaten each week. Margaret Reid, *Food for People* (New York: Wiley, 1943), 466.

47. People were to ensure that each day they drank one pint of milk (more for children) and ate least the following: one egg (or more milk, or cheese, beans, or peanuts); one or more servings of meat, fish, or fowl; one potato; two servings of vegetables (one green, one yellow); two fruits; and some whole grain or enriched cereal products. *NN*, March 1943; *JAMA* 122 (June 26, 1943): 607.

48. Perhaps the committee's most important contribution came in 1944, when it came up with extensive tables of the nutritional composition of common foods. *Tables of Food in Terms of Eleven Nutrients*, U.S. Dept. of Agriculture Misc. Pub. no. 572 (Washington: USGPO, 1944); Conrad Elvehjem, "Seven Decades of Nutrition Research," in American Association for the Advancement of Science, *Centennial* (Washington: AAAS, 1950), 100.

49. M. L. Wilson to Ruth Benedict, Nov. 6, 1940, Mead Papers, Box F6, cited in Rebecca Sprang, "The Cultural Habits of a Food Committee," *Food and Foodways* 2, no. 4 (1988): 371.

50. "Vibio" would be the name for flour with Vitamin B_1, iron and calcium; "Bicapt" for that with Vitamin B_1, calcium, and phosphorus; "Bermaco" for gently milled flour with added wheat germ and bran flakes; and "Vito" for flour that had undergone a long extraction process. *NYT*, March 2, 1941.

51. Carl Guthe, "History of the Committee on Food Habits," in *The Problem of Changing Food Habits: Report of the Committee on Food Habits, 1941–1943*, National Research Council Bulletin no. 108 (Washington: NRC, 1943), 14–15; Margaret Mead, "The Problem of Changing Food Habits," in ibid., 21.

52. NRC-CFH, "Translation of Scientific Findings," 3–5.

53. She seemed to mean there were some foods that locally connoted "prestige" and "modernity" to low-status tenants wanting to emulate higher-status sharecroppers, who in turn aspired to small-town working-class diets. Mead to Wilson, Aug. 10, 1942, Mead Papers, Box F21.

54. Margaret Mead, "Background Material for Agenda, CFH, Apr. 26–27, 1942," Mead Papers, Box F9.

55. Margaret Mead to M. L. Wilson, June 23, 1943, U.S. Dept. of Agriculture, War Food Administration, Nutrition Programs Branch, Records, National Archives, RG 136, Entry 139 (hereafter WFA Nutrition Program Records), Box 144. Wilson would not endorse it as policy, perhaps because it was practically unintelligible. The garbled bureaucratese in the memos of Richard O. Cummings, author of the excellent *The American and His Food* (Univ. of Chicago Press, 1940), who worked in the Nutrition Division, also contrasts starkly with the clarity of his academic prose, suggesting that there was something inherently muddling about either the bureaucratic mind-set or the memorandum form itself. See Richard Cummings to Wilson, Dec. 12, 1942, Nutrition Division Records, Box 32.

56. *SNL*, June 7, 1941; Common Council for American Unity, *What's Cooking in Your Neighbor's Pot* (New York: Common Council, n.d. [1944]).

57. Patricia Woodward to Everett Hughes, July 8, 1953; George Abernathy to Woodward, July 16, 1943. Both in Mead Papers, Box F7.

58. Margaret Cussler to Wilson, Aug. 24, 1942; W. C. Hallenbeck to J. C. Leukhardt, June 26, 1942. Both in Nutrition Division Records, Box 1.

59. Earl L. Koos, "A Study of the Use of the Friendship Pattern in Nutrition Education," in *Problem of Changing Food Habits*, 74–81.

60. In 1946 people in the Office of Education remarked to Wilson that the anthropologists "knew so little about food problems when they began." M. L. Wilson to Margaret Mead, Feb. 5, 1946, Mead Papers, Box 6.

61. NRC-CFH, "Translation of Scientific Findings," p. 6.

62. Margaret Mead, preface to Genoeffa Nizzardini and Natalie F. Joffe, "Italian Food Patterns and Their Relationship to Wartime Problems of Food and Nutrition" (Washington: NRC, 1942), 1.

63. Sprang, "Cultural Habits," 16–17.

64. NRC-CFH, "The Relationship Between Food Habits and Problems of Wartime Emergency Feeding" (Washington: NRC, 1942), 4.

65. It did admit that because of the nation's cultural diversity this would usually be impractical. In those cases, it recommended that evacuees be served simple food with few seasonings and be given a wide choice of condiments to enable them to alter it to their tastes. Ibid., 1–3.

66. One need only look at the incredible number of interrelated factors listed in the committee's *Manual for the Study of Food Habits* to gain an appreciation of the gigantic task which the kind of study Mead had in mind involved. *Manual for the Study of Food Habits*, National Research Council Bulletin no. 111 (Washington: NRC, 1945).

67. Mead, "Problem of Changing Food Habits," 21–22.

68. It took only one day to collect the data for the monograph on Japanese food habits. Sprang, "Cultural Habits," 29.

69. Natalie Joffe and Sula Benet, "Polish Food Patterns" (Washington: CFH, 1943), 12.

70. E.g., Margaret Mead to Liaison Members of the Committee on Food Habits, Dec. 1, 1943, WFA Nutrition Program Records, Box 144.

71. Wilson to Carl Guthe, July 27, 1944, WFA Nutrition Program Records, Box 144.

72. Patricia Woodward to Thelma Dreis, June 14, 1944; Dreis to Mead, August 11, 1944. Both in WFA Nutrition Program Records, Box 144.

73. Jane Howard, *Margaret Mead* (New York: Simon and Schuster, 1984), 233–37.

74. U.S. Congressional Research Service, *The Role of the Federal Government in Human Nutrition Research* (Washington: USGPO, 1976), 116.

75. Minutes of Conference, "Contributions from the Field of Market Research," in *Problem of Changing Food Habits*, 141–44.

76. See the voluminous correspondence with General Foods, General Mills, and other large producers in Nutrition Division Records, Boxes 17, 18, 19, and 20.

77. See correspondence in Nutrition Division Records, Box 19. When the Food and Nutrition Board was asked for its advice it hemmed and hawed until it finally allowed those with added thiamin, niacin, and iron to be called "restored" and be part of the National Nutrition Program. NRC-FNB, "Statement of Principle Concerning Breakfast Food Cereals," May 5, 1942, in ibid., Box 31.

78. *GH*, Nov. 1942, p. 124.

79. Herman Toback to Helen Mitchell, Nov. 17, 1942, Nutrition Division Records, Box 1; John Murphy to Louise Griffith, Nov. 22, 1942, ibid., Box 16.

80. "How Advertisers Set Good Food and Good Health Standards for the American Public," *RM*, Jan. 1943, pp. 34–35.

81. Advertisement in Nutrition Division Records, Box 1.

82. *American Restaurant*, July 1942, p. 4.

83. J. I. Sugarman to Wilson, Apr. 9, 1942; W. H. Sebrell to Sugarman, Aug. 6, 1942. Both in Nutrition Division Records, Box 17.

84. Mabel Stimpson to A. J. Lorenz, July 24, 1942, Nutrition Division Records, Box 16.

85. Reid, *Food for People*, 466–67.

86. *GH*, Nov. 1942, p. 163.

87. *AC*, Sept. 1945.

88. Beatrice Hunter, *Consumer Beware! Your Food and What's Been Done to It* (New York: Simon and Schuster, 1971), 32.

89. *RM*, Jan. 1943, p. 2.

90. *RM*, June 1943, p. 45.

91. Agnes Fay Morgan to Helen Mitchell, July 21, 1942, Nutrition Division Records, Box 16.

92. McNutt, meanwhile, had left to become head of the new Manpower Commission.

93. The conspiracy seemed to extend far beyond communists and their fellow travelers, however, for he also speculated that some federal officials were "financially interested in foreign coconut plantations or in manufacturing plants engaged in the production of oleomargarine." J. S. Abbott, "Reply to Hon. A. H. Andersen's Speech on the Radio Broadcast," Donald Montgomery Collection, Archives of Labor History and Urban Affairs, Walter Reuther Library, Wayne State Univ., Detroit, Mich., Box 3-17.

94. This was done under emergency war powers and lasted until 1946. Elizabeth Etheridge, *The Butterfly Caste* (Westport, Conn.: Greenwood, 1971), 214.

95. *AC*, March 1943, p. 304.

96. NRC-CFH, "Impact of the War on Local Food Habits," transcript of conference, March 27, 1943, p. 23, Mead Papers, Box F12.

97. R. C. Atchley to Henry J. Morgenthau, Sept. 26, 1942, Nutrition Division Records, Box 16.

98. Mark Graubard to M. L. Wilson, July 14, 1942, Nutrition Division Records, Box 31. In the last sixteen months of the war, after most war plant construction had stopped, it was given the power to certify that plans for new industrial plants had proper provision for industrial feeding facilities.

99. *NYT*, Dec. 30, 1941.

100. U.S. Dept. of Agriculture, *Industrial Feeding in Manufacturing Establishments, 1944* (Washington: USGPO, 1944); Faith M. Williams, "The Standard of Living in Wartime," *Annals of the American Academy of Political and Social Science* 229 (Sept. 1943): 172; National Research Council, Committee on Industrial Feeding, *The Nutrition of Industrial Workers* (Washington: NRC, 1945), 5–33.

101. Wilder was not even a member of the interdepartmental Civilian Requirements Policy Committee, but only one of many on the much less powerful Food Advisory Committee. Minutes and Memos, Food Advisory Committee; Roy Hendrickson to the War Food Administrator (Marvin Jones), July 3, 1943; War Production Board, Civilian Requirements Policy Committee, Minutes, July 9, 1943. All in

U.S. Dept. of Agriculture, War Food Administration, Records, National Archives, RG 16, Entry 218, Box 17.

102. Mitchell to Wilson, Dec. 17, 1942, Nutrition Division Records, Box 31.

103. J. J. Berrall to Alberta McFarlane, Dec. 2, 1942, Nutrition Division Records, Box 24.

104. War Food Administration, "Statements of Agencies Actively Participating in the Food and Nutrition Program," Apr. 4, 1944, WFA Nutrition Program Records, Box 140.

105. Ibid.; Williams, "Standard of Living," 172.

106. Florence Wyckoff to Elizabeth Magee, Feb. 14, 1945, National Consumers League Papers, Library of Congress (hereafter NCL Papers), Box C19; Helen Pundt, *AHEA: A History of Excellence* (Washington: AHEA, 1980), 180–81. The bill calling for its transfer to the FSA also called for its costs to be shared by the states. Some liberal consumers' groups opposed this on the grounds that southern states needing it most would not be able to afford it. When the whole lunch program came under attack from conservative Republicans, the other liberals realized that the only way to save it was to continue it as a program designed mainly to get rid of food surpluses and that it must therefore remain in the USDA. See correspondence of Florence Wyckoff, 1943-1944, NCL Papers, Box C19.

Chapter 6. Food Shortages for the People of Plenty

1. *NN*, Jan., June 1942; Richard Lingeman, *Don't You Know There's a War On?* (New York: Putnam, 1970), 244–45.

2. *New York World-Telegram*, Jan. 21, 1942. Honey's nutritional advantage over sugar is in fact miniscule.

3. In fact, per capita consumption of refined sugar only dropped from its late 1930s average of about 97 pounds a year to about 81 pounds in 1942 and 1943. It then rose to almost 90 pounds in 1944 before sinking to 74 pounds in 1945. U.S. Bureau of the Census, *Historical Statistics of the United States, Colonial Times to 1970* (Washington: USGPO, 1975) (hereafter *Hist. Stat.*), vol. 2, p. 331.

4. *NYT*, Feb. 22, 1943; Lingeman, *Don't You Know*, 254–55.

5. Martin Bell, *A Portrait of Progress: A Business History of the Pet Milk Company, 1885–1960* (St. Louis: Pet Milk, 1962), 155.

6. Rationing had been imposed, not because of short supplies, but because of a shortage of shipping, mainly from Brazil. Lingeman, *Don't You Know*, 246.

7. *NYT*, Nov. 1, 1947.

8. Canadian dietary adjustments paralleled the relatively mild ones of the Americans, rather than the much more severe ones of the British. United States, United Kingdom, and Canada, Combined Food Board, *Food Consumption Levels in Canada, the United Kingdom, and the United States*, Report of the Special Joint Committee (Ottawa: King's Printer, 1944), 12–23.

9. Rhoda Metraux, "Qualitative Study of Current Attitudes on Food Problem," n.d. [Sept. 1942?], memo in Margaret Mead Papers, Manuscripts Division, Library of Congress, Box F9.

10. J. Stanley Cassidy to Marvin Jones, Aug. 23, 1943, U.S. Dept. of Agriculture, War Food Administration, Records National Archives RG 16, Entry 218, (hereafter WFA Records), Box 98.

11. *Charleston News and Courier*, Aug. 27, 1943, cited in John Hammond Moore,

"No Room, No Rice, No Grits: Charleston's 'Time of Trouble,' 1942–1944," *South Atlantic Quarterly* 85, no. 1 (1986): 29.

12. Moore, "No Room, No Rice," 28–29.

13. *NYT*, Jan. 8, 1944.

14. *NYT*, May 12, June 1, June 18, 1943.

15. D'Ann Campbell, *Women at War with America: Private Lives in a Patriotic Era* (Cambridge: Harvard Univ. Press, 1984), 182; Studs Terkel, *The Good War* (New York: Ballantine, 1985), 133.

16. Terkel, *Good War*, 117.

17. Major General E. B. Gregory, "Food for the Army," Statement to the Special Senate Committee, April 14, 1943, WFA Records, Box 17.

18. George Gallup, *The Gallup Poll, 1935–1971* (New York: Random House, 1972), vol. 1, p. 396.

19. Jane Howard, *Margaret Mead* (New York: Simon and Schuster, 1984), 233.

20. *NN*, Nov. 1942.

21. *NN*, March 1943.

22. This had the same aim—reducing bread wastage. Robert Goodhart to J. T. McCarthy, Dec. 28, 1942, U.S. Executive Office of the President, Office of the Coordinator of Health, Welfare, and Related Defense Activities, Nutrition Division, Records, RG 136, Entry 218, National Archives (hereafter Nutrition Division Records), Box 1.

23. *NN*, Oct. 1942.

24. *NN*, April 1944.

25. *NYT*, May 27, 1944.

26. *NYT*, June 13, 1943.

27. Margaret Mead to J. C. Leukhardt, January 3, 1944, U.S. Dept. of Agriculture, War Food Administration, Nutrition Programs Branch, Records, National Archives, RG 136, Entry 139 (hereafter WFA Nutrition Program Records), Box 144.

28. *NN*, Nov. 1942.

29. *NN*, Feb. 1943.

30. Lingeman, *Don't You Know*, 258–59.

31. *NYT*, March 23, 1943.

32. Lingeman, *Don't You Know*, 260–61.

33. T. Swann Harding, "Food in 1943," *JHE* 35 (Jan. 1943): 3.

34. *Amarillo Globe News*, Aug. 20, 1943.

35. Margery Horstmann to Jones, Aug. 25, 1943, WFA Records, Box 98. Carl Wilsey to Jones, Aug. 25, 1943, and R. J. Meyer to Jones, Aug. 23, 1943, ibid., Box 97.

36. G. E. Oaks to Jones, Aug. 27, 1943; Horstman to Jones, Aug. 25, 1973; Meyer to Jones, Aug. 23, 1943; Harry Johnson to Jones, Aug. 24, 1943. All in WFA Records, Boxes 97 and 98.

37. Gallup, *Gallup Poll, 1935–1971*, vol. 1, p. 488.

38. Richard J. Hooker, *Food and Drink in America: A History* (Indianapolis: Bobbs-Merrill, 1981), 344.

39. Gallup, *Gallup Poll, 1935–1971*, vol. 1, p. 488.

40. Helen Pundt, *AHEA: A History of Excellence* (Washington: AHEA, 1980), 175.

41. *NYT*, Oct. 3, 1945.

42. Susan Hartmann, *The Home Front and Beyond: American Women in the 1940s* (Boston: Twayne, 1982), 21.

43. American Public Health Association, *Newsletter*, May 25, 1942.

44. Louisa Pryor Skilton, "Victory Lunches for Sturdy Men," *AC*, March 1942, p. 348.

45. General Mills, "Outline of the Career in Advertising of Marjorie Child Husted," Feb. 1950, typescript in Marjorie C. Husted Papers, Schlesinger Library of Women's History, Radcliffe College, Cambridge, Mass.; *NYT*, Dec. 28, 1986.

46. Leila J. Rupp, *Mobilizing Women for War* (Princeton: Princeton Univ. Press, 1970), 152.

47. Lawrence A. Keating, *Men in Aprons* (New York: M. S. Mill, 1944).

48. Gallup, *Gallup Poll, 1935–1971*, vol. 1, p. 396.

49. Robert J. Havighurst and H. Gerton Morgan, *The Social History of a Wartime Community* (New York: Longmans, Green, 1951), 127.

50. Gallup, *Gallup Poll, 1935–1971*, vol. 1, p. 225; *NYT*, Dec. 22, 1940. Another survey one year later produced similar results, with a slight decrease in lower income yes answers. Gallup, *Gallup Poll, 1935–1971*, vol. 1, p. 310.

51. In 1941 the richest third of families spent, on average, well over twice as much on food as the poorest third. By 1944 they were spending about one-third more. Campbell, *Women at War*, 184.

52. Sherman Briscoe, in National Research Council, Committee on Food Habits (hereafter NRC-CFH), "Impact of the War on Local Food Habits," transcript of conference, March 27, 1943, p. 14, Mead Papers, Box F12.

53. Arthur Raper, in ibid., 13–14.

54. "The Fabulous Market for Food," *Fortune*, Oct. 1953, pp. 135–36.

55. *Wartime Food Purchases*, U.S. Dept. of Labor Bulletin no. 838 (Washington: USGPO, 1945), 5, cited in Campbell, *Women at War*, 251.

56. *Hist. Stat.*, vol. 2, p. 328; Margaret Reid, "Food, Liquor, and Tobacco," in J. Friedrich Dewhurst et al., eds., *America's Needs and Resources: A New Survey* (New York: Twentieth Century Fund, 1955), 157.

57. In the four years from 1941 to 1945 per capita consumption of eggs went up 30 percent, to a quite amazing 403 eggs per person per year, while milk consumption went up 20 percent. *Hist. Stat.*, vol. 2, p. 331.

58. Pearl S. Buck, preface to 1st ed., in Buwei Yang Chao, *How to Cook and Eat in Chinese*, 3d ed., (New York: Vintage, 1963), xviii–xix.

59. *JHE* 37 (Nov. 1945): 123.

60. Charles Whitney, "The Most Important Subject in the World Today Is Food," *AC*, Sept. 1943.

61. Charlotte Evans and Rose Lubschez, "A Comparison of Diets of School Children in New York City in 1917 and 1942," *Journal of Pediatrics* 24 (May 1944), 518–23.

62. Jessie Whitacre, "What Rural Texans Eat," *JHE* 37 (March 1945): 149-51

63. While overall fresh fruit consumption remained steady, Americans ate fewer apples and more citrus fruits. By 1945 they were eating three times more fresh citrus fruit than apples, a complete reversal of the situation during the previous war. *Hist. Stat.*, vol. 2, p. 330.

64. U.S. Dept. of Health and Human Services and U.S. Dept. of Agriculture, *Nutrition Monitoring in the United States: A Progress Report from the Joint Nutrition Monitoring Evaluation Committee* (Hyattsville, Md.: USDHHS, 1986), 140.

65. Frank Boudreau, "Food and Nutrition Policy Here and Abroad," *AJPH* 34 (March 1944): 217.

66. *NYT*, Nov. 7, 1945.

67. Andrew Brown, in NRC-CFH, "Impact of the War," 19.

68. Ibid.

69. Lingeman, *Don't You Know*, 249.

70. *Life*, June 10, 1946.

71. That year, over one-quarter of the nation's meat supply was taken up by the armed forces and Lend-Lease, mainly the former. *NN*, Oct. 1942; Harding, "Food in 1943," 5.

72. *NYT*, Jan. 30, 1943.

73. *NYT*, Jan. 23, 1943.

74. Ross Gregory, *America 1941: A Nation at the Crossroads* (New York: Free Press, 1989), 38.

75. Eugene Wright, "Bill of Fare à la Guerre," *NYTM*, Feb. 13, 1944.

76. Beard said that the returning GIs had "been everywhere" and "tasted the real thing." Cited in Chris Chase, *The Great American Waistline* (New York: Coward, McCann, 1981), 32.

77. *NYTM*, June 25, 1944.

78. Margaret Mead to Liaison Members of the Committee on Food Habits, Dec. 1, 1943, WFA Nutrition Program Records, Box 144; Claire Jones, *The Chinese in America* (Minneapolis: Leoner, 1972), 39–40.

79. *The Army Cook*, Technical Manual TM-10-405 (Washington: U.S. Dept. of War, 1942), 147–90.

80. U.S. Dept. of the Navy, *The Cook Book of the United States Navy, 1932* (Washington: USGPO, 1932), 26, 48. In 1941 the navy asked the National Restaurant Association for help in revising its cookbook, particularly in light of the fact that more of its personnel were now stationed ashore than at sea, but the new cookbook did not appear until mid-1944. *AC*, April 1944, p. 5.

81. *NYT*, April 6, 1944.

82. Major General W. H. Middleswort, "Problems of Subsistence That May Confront the Quartermaster Corps of the Future," *FT* 5 (Jan. 1951): 6.

83. U.S. Dept. of War, *Army Cook*, 103–6.

84. Erna Risch, *Quartermaster Support of the Army: A History of the Corps, 1775–1939* (Washington: USGPO, 1952), 505–7.

85. U.S. Secretary of War, Commissary-General of Sustenance, *Manual for Army Cooks* (Washington: USGPO, 1896); U.S. Army, Quartermaster-General, *Manual for Army Cooks, 1916* (Washington: USGPO, 1917.)

86. Risch, *Quartermaster Support*, 507; Jean Drew, "GAL in the GI Kitchen," *Collier's*, July 26, 1952.

87. *JADA* 17 (March 1941): 243.

88. Drew, "GAL in the GI Kitchen," 10. The Americans were not alone in ignoring climatic and other differences. One of the reasons for the ultimate defeat of Rommel's German army in the North African desert in 1942–1943 is said to have been its quartermasters' rigid adherence to the heavy traditional diet of *wurst* and *kraut*.

89. Irene Nehrling, "How America's Soldier Is Fed," *AC*, March 1942, p. 342.

90. U.S. Dept. of War, *Army Cook*, 104. Of course, where enough beef was unavailable, other meats would be substituted for it.

91. Some of this section is based on my training and experience as a supply officer in the Royal Canadian Navy Reserve during the late 1950s and early 1960s.

92. Nehrling, "How America's Soldier Is Fed," p. 342.

93. Ida Bailey Allen, "What the Army Can Teach You," *AC*, Sept. 1943, p. 17.

94. *NN*, Oct. 1942.

95. *NYT*, Oct. 4, 1944.

96. *NN*, June 1945; *NYT*, May 22, 1943.

97. *SNL*, June 22, 1940.

98. "Nutrition Survey of American Troops in the Pacific," *NR* 4 (Sept. 1946): 257.

99. Mary Barber, "Army Feeding Is a Big Job," *JHE* 33 (Dec. 1941): 703.

100. *SNL*, June 22, 1942.

101. Nehrling, "How America's Soldier Is Fed," 366–67.

102. U.S. Army, "Biennial Report of the Chief of Staff," Oct. 10, 1945, in *NYT*, Oct. 10, 1945.

103. "Nutrition Survey," 257; "Life at the Front," *NYTM*, Nov. 19, 1944.

104. Jane Stern and Michael Stern, *Square Meals* (New York: Knopf, 1984), 241.

105. *NYT*, Oct. 24, 1944.

106. *NYT*, Oct. 25, 1944.

107. *NN*, Dec. 1945.

108. *NN*, June 1945.

109. *NYT*, Nov. 25, 1945.

110. One-cent pieces were still selling for from five to seventy-three cents in a Lake Forest, Illinois, grade school in May 1946. *Life*, June 10, 1946.

111. *NN*, Feb. 1946.

12. "Food Crisis: 1946," U.S. Dept. of Agriculture, Famine Emergency Committee, Records, National Archives RG 16, Entry 220 (hereafter FEC Records), Box 5.

113. Barton Bernstein, "The Postwar Famine and Price Control, 1946," *Agricultural History* 38, no. 4 (1964): 237.

114. *NN*, Feb. 1946.

115. "Food Crisis: 1946."

116. It was hoped that this would make American bread heavier and allow more wheat to be shipped abroad.

117. FEC, Minutes of Meeting, March 11, 1946, FEC Records, Box 25.

118. *Richmond News Leader*, March 12, 1946.

119. The United States had committed itself to export 225 million bushels of wheat before June 30, 1946, but 205 million more bushels were consumed as animal feed than the Department of Agriculture had estimated, leaving a tremendous shortfall. Keith Hutchison, "Feast and Famine," *Nation*, Feb. 16, 1946, p. 196.

120. Cattle would eat it when mixed with corn.

121. FEC, Minutes of Conference, March 11, 1946, FEC Records, Box 1.

122. Gallup, *Gallup Poll*, vol. 1, pp. 569, 575.

123. Clinton Anderson to Harry S. Truman, April 5, 1946, FEC Records, Box 26.

124. "The Food Scandal," *Fortune*, May 1946, p. 92.

125. Anderson to Truman, April 5, 1946.

126. Alfred Stedman to Chester Davis, April 15, 1946, FEC Records, Box 29.

127. FEC, Minutes of Meeting, White House, June 24, 1946, FEC Records, Box 1.

128. FEC Minutes, June 24, 1946; Robert Shields to Heads of PMA Offices, May 20, 1946. Both in FEC Records, Box 1. F. Straub to J. S. Russell, May 29, 1946, ibid., Box 29.

129. Harry Truman, "Youth Against Famine," remarks before the National Youth Conference at the White House, July 15, 1946, FEC Records, Box 29; Davis to Clinton P. Anderson, June 27, 1946, ibid., Box 9.

130. Robert J. Donovan, *The Presidency of Harry S. Truman. Conflict and Crisis: 1945–1948* (New York: Norton, 1977), 236–37.

131. Paul Porter, radio talk, Sept. 14, 1946, FEC Records, Entry 221, Box 9.

132. All quotes are from A. J. Liebling, *The Press* (New York: Ballantine, 1961), 90–97.

133. Donovan, *Truman*, 236.

134. Ibid.

135. Ibid., 97–99.

136. *Christian Century*, Oct. 15, 1947.

137. Robert G. Whelen, "Luckman Tackles His Biggest Selling Job," *NYTM*, Nov. 2, 1947.

138. Gallup, *Gallup Poll, 1935–1971*, vol. 1, p. 686.

139. *NYT*, Oct. 8, 1947.

140. *NYT*, Nov. 20, 1947.

141. *Christian Century*, Oct. 27, 1947.

142. *NYT*, Jan. 5, 1948.

Chapter 7. The Golden Age of Food Processing

1. "The Fabulous Market for Food," *Fortune*, Oct. 1953.

2. *Twenty-fourth Annual Nielsen Report to Retail Food Stores* (Chicago: A. C. Nielsen, 1958), 30. In 1941 Americans spent $150 per capita on food produced outside the home. By 1958 food prices were slightly more than twice their 1941 level but food spending had increased much more, to $390 per person. Margaret Burk, "Pounds and Percentages," in U.S. Dept. of Agriculture, *Yearbook of Agriculture, 1959* (Washington: USGPO, 1959), 596.

3. Joseph Conlin, *The Troubles: A Jaundiced Glance Back at the Movement of the 60s* (New York: Franklin Watts, 1982), 72–74.

4. U.S. Bureau of the Census, *Historical Statistics of the United States, Colonial Times to 1970* (Washington: USGPO, 1975) (hereafter *Hist. Stat.*), vol. 2, p. 289.

5. The importance of new shopping patterns is reflected in the fact that the number of new stoves sold, 5.5 million, was only a fraction of that of refrigerators. Before the war, most urban homeowners were satisfied with cheap iceboxes, which could keep meat and dairy products cool for two or three days. Once-a-week shopping trips to the supermarket made them, and ice men, relics. Susan Hartmann, *The Home Front and Beyond: American Women in the 1940s* (Boston: Twayne, 1982), 165–67; Elaine Tyler May, *Homeward Bound: American Families in the Cold War* (New York: Basic, 1988), 165–66.

6. General Mills, "Outline of the Career in Advertising of Marjorie Child Husted," Feb. 1950, typescript in Marjorie C. Husted Papers, Schlesinger Library of Women's History, Radcliffe College, Cambridge, Mass.

7. Loren Baritz, *The Good Life: The Meaning of Success for the American Middle Class* (New York: Knopf, 1989), 186.

8. Susan Hartmann, "Prescriptions for Penelope: Literature on Women's Obligations to Returning World War II Veterans," *Women's Studies* 5 (1978): 231–33.

9. Ibid., 231–32.

10. Marjorie Husted, "The Women in Your Lives," typescript of speech, Oct. 13, 1952, Husted Papers, Folder 2.

11. May, *Homeward Bound*, 166–67.

12. *NYT*, Feb. 26, 1948.

13. Recordings of these and other sitcoms were viewed at the Museum of Broadcasting, New York, N.Y.

14. Ruth Mills Teague, *Cooking for Company* (New York: Random House, 1950), vii.

15. Cited, without attribution, in Roberta Seid, *Never Too Thin: Why Women Are at War with Their Bodies* (New York: Prentice-Hall, 1989), 103.

16. Jennifer Colton, "Why I Quit Working," *GH*, Sept. 1951, p. 53.

17. Good Housekeeping Institute, *The Good Housekeeping Cook Book* (New York: Good Housekeeping, 1955), 163–75.

18. "Meat Loaf Meals," *BHG*, Jan. 1948, p. 59.

19. *FE*, Jan. 1960, p. 44.

20. Marshall Adams to Adelaide Hawley, May 17, 1956, Adelaide Hawley Cummings Papers, Schlesinger Library of Women's History, Radcliffe College, Cambridge, Mass. (hereafter Cummings Papers), Box 3.

21. Mary J. Burnley and Genevieve Callahan, "Sea Food: Chef's Specials," *BHG*, March 1948, p. 65.

22. *FE*, Nov. 1970, p. 23.

23. "The Food Market," *Fortune*, Oct. 1953, p. 271.

24. *FE*, Oct. 1957, pp. 7–8.

25. Baritz, *Good Life*, 193.

26. *FE*, Aug. 1957, p. 7, Oct. 1957, p. 8, Feb. 1960, p. 53.

27. Egmont Arens, "Packaging for the Mass Market," in Paul Sayres, ed., *Food Marketing* (New York: McGraw-Hill, 1950), 231.

28. *FE*, Jan. 1960, p. 40.

29. *NYT*, Sept. 12, 1962.

30. Ibid.

31. Alexander McFarlane, "Of Convenience, Food Innovation, and Calling the Tune: The Revolutionary Imperative," *FT* 23 (April 1969): 43.

32. *House Beautiful*, Jan. 1946, p. 51.

33. *BHG*, April 1946, pp. 79, 120.

34. Eleanor Early, in *American Cookery*, quoted in Jane Stern and Michael Stern, *Square Meals* (New York: Knopf, 1984), 243.

35. Harden F. Taylor, Oral History Interview, pp. 46, 77; Donald K. Tressler, Oral History Interview, pp. 33–38. Both in Archives and Manuscript Division, Cornell Univ. Libraries, Ithaca, N.Y.

36. George L. Mentley, "Frozen Foods: A Marketing Case History," in Sayres, *Food Marketing*, 294, 296.

37. By 1945 per capita consumption was triple that of 1940 but was still only 1.9 pounds per year. *Hist. Stat.*, vol. 2, 331.

38. Mentley, "Frozen Foods," 295; "Frozen Foods: Interim Report," *Fortune*, Aug. 1946, p. 107.

39. *Time*, Dec. 7, 1959.

40. *NYT*, Feb. 26, 1948.

41. Because defrosting large blocks of frozen juice seemed a daunting chore, it had been assumed that powdered orange juice would become the market leader. But the idea of reducing the juice to sludge and then freezing it proved to be a breakthrough. Mentley, "Frozen Foods," 286.

42. *NYT*, Nov. 24, 1953.

43. Still, the contrast with France is interesting, for while Americans regard it as normal to bring the food to the television set, working-class French people do the opposite, pulling the TV set up to the dining table, where it seems to sit like a loquacious guest.

44. *FE*, July 1957, p. 7; *Hist. Stat.*, vol. 2, p. 331.

45. *Time*, Dec. 7, 1959.

46. *NYT*, Feb. 2, 1956.

47. *FE*, Feb. 1954, p. 198; *BW*, Jan. 31, 1959.

48. *FE*, May 1951, pp. 17–25, May 1954, p. 29; *BW*, Jan. 31, 1959.

49. The number of canneries declined from 2,265 to 1,758. *FE*, April 1957, p. 7.

50. *USNWR*, Feb. 15, 1957.

51. *Canner*, Sept. 27, 1952.

52. Mildred Boggs and Clyde Rasmussen, "Modern Food Processing," in U.S. Dept. of Agriculture, *Yearbook of Agriculture, 1959* (Washington: USGPO, 1959), 428.

53. *USNWR*, Dec. 7, 1959.

54. George F. Stewart, "Better Food the Chemical Way," *Science Digest*, Nov. 1960, p. 3.

55. Boggs and Rasmussen, "Modern Food Processing," 420.

56. Ross H. Hall, *Food for Nought: The Decline of Nutrition* (New York: Harper and Row, 1974), 37–38.

57. Martin Glicksman, "Fabricated Foods," *CRC Critical Reviews in Food Technology*, April 1971, p. 24.

58. "What Has Happened to Flavor?" *Fortune*, April 1952, p. 130.

59. By 1975 the feed conversion ratio had dropped from four-to-one to two-to-one and the broiler's life span had been reduced from eighteen weeks to nine. Robert E. Cook et al., "How Chicken on Sunday Became an Anyday Treat," in U.S. Dept. of Agriculture, *Yearbook of Agriculture, 1975* (Washington: USGPO, 1975), 125–32; Robert White-Stevens, "Antibiotics Curb Diseases in Livestock, Boost Growth," in ibid., 85.

60. R. P. Niedermeyer et al., "Move Over, Milky Way—Our Cows Are Stars Too," in ibid., 144.

61. "What Has Happened to Flavor?" 130.

62. *USNWR*, Feb. 15, 1959.

63. *NYT*, Feb. 2, 1956.

64. *NYT*, Oct. 3, 1945.

65. *NYT*, Feb. 2, 1956.

66. Dean McNeal, "What Do the Next Ten Years Hold for You?" *FE*, April 1960, p. 46.

67. Food processors normally spent only between 0.2 and 0.5 percent of sales on R and D, versus the 2 to 7 percent spent in other industries. *FT*, May 1954, p. 116; Hall and Stieglitz Co., *The Convenience Food Manufacturing Industry* (New York: the authors, 1963), 35–38.

68. "These Foods Are News!" *BHG*, April 1959, March 1961, Nov. 1961.

69. *Time*, Dec. 7, 1959.

70. *FE*, April 1957, p. 19, May 1957, p. 20.

71. Adelaide Hawley Cummings (Betty Crocker), "The Future," ms. in Cummings Papers, Box 3, Folder 3; U.S. National Commission on Food Marketing, *Studies in Organization and Competition in Grocery Manufacturing* (Washington: USGPO, 1966), 91–97; *BW*, Jan. 31, 1959.

72. *USNWR*, Feb. 15, 1957; USDA, *Yearbook, 1959*, 421.

73. *NYT*, Dec. 3, 1950.

74. National Research Council, Food Protection Committee, *Use of Chemical Additives in Foods* (Washington: NRC, 1951); *NYT*, Dec. 3, 1950, Aug. 1, Dec. 18, 1951.

75. Benjamin Burton, *The Heinz Handbook of Nutrition* (New York: McGraw-Hill, 1959, 1965), 406–8.

76. Conrad A. Elvehjem, "Seven Decades of Nutrition Research," in American Association for Advancement of Science, *Centennial* (Washington: AAAS, 1950), 100.

77. Stewart, "Better Foods the Chemical Way," 8.

78. *NYT*, Jan. 24, 1957.

79. U.S. Food and Drug Administration, *Food Facts vs. Food Fallacies* (Washington: USGPO, 1959); *FP*, Nov. 1961, p. 10.

80. *NYT*, May 27, 1953; *JHE* 45 (Oct. 1953): 584.

81. The most impressive saving over using unprocessed foods came from the price advantage of frozen over fresh orange juice and instant over fresh coffee, and the relatively large amounts of these used by American consumers. Preliminary Report, U.S. Dept. of Agriculture Marketing Bulletin no. 22, cited in *FP*, Nov. 1962, pp. 77–8.

82. Chains were defined as having more than eleven stores. National Commission on Food Marketing, *Organization and Competition in Food Retailing* (Washington: USGPO, 1966), 34–37.

83. *FE*, Oct. 1957.

84. Visitors were particularly impressed by the packaged fresh meat and produce, and even Tito's wife looked as if she would like to dawdle there, but the marshal made a beeline to an exhibit of his favorite—hunting rifles. *NYT*, Sept. 7, 1957; "U.S. Supermarket in Yugoslavia," *NYTM*, Sept. 22, 1957.

85. *USNWR*, Feb. 15, 1957; *NYT*, Sept. 20, 1957.

86. *NYT*, July 25, 1959.

87. *NYT*, Nov. 14, 1962.

88. *NYT*, Nov. 15, 1957.

89. The study also seemed to support Packard, indicating that in 1960 shoppers spent 50 percent more time in the supermarket but emerged with only one item more than they did in 1949. "Better Fix on Shopper Image," *FE*, Feb. 1960, p. 53.

90. Ibid.

91. Walter Weir, "Advertising Tells the Story," in Sayres, *Food Marketing*, 222.

92. Arens, "Packaging," 228–32.

93. Vance Packard, *The Hidden Persuaders* (New York: David Mackay, 1957), 109.

94. *Time*, Dec. 7, 1959.

95. *NYT*, Nov. 4, 1990.

96. *GH*, Sept. 1946, p. 145; *BHG*, Apr. 1948, p. 78.

97. *Sponsor*, Dec. 27, 1954.

98. See *AC*, 1946, passim, for page after page of advertisements for this kind of material.

99. Kraft Theater, "Long Time till Dawn," 1953, recording in Museum of Broadcasting, New York, N.Y.

100. Kraft Theater, "A Night to Remember," 1955, recording in Museum of Broadcasting, New York, N.Y.

101. A 1948 "double-easy" recipe for sweet dough called for sugar, salt, shortening, milk, egg, and lukewarm water to be mixed together. Then a cake of yeast had to be crumbled into the mixture and stirred until dissolved. Gold Medal flour was then added, along with some nutmeg and mace. Then one of four additions—cut-up cherries, citron, raisins, or chopped nuts—was added and the mixture put in a warm place to rise. After baking, it was to be topped with a thick white icing and maraschino cherries. *BHG*, Jan. 1948, p. 77.

102. *Sponsor*, Dec. 27, 1954.

103. Arens, "Packaging," 231.

104. William H. Whyte, Jr., *The Organization Man* (New York: Doubleday, 1956), 330.

105. Richard Polenberg, *One Nation, Divisible: Class, Race, and Ethnicity in the United States Since 1938* (Baltimore: Penguin, 1980), 139–44.

106. *FE*, Nov. 1957.

107. Mary Ann McAvoy and Beverly Sloane, *From Vassar to Kitchen* (New Haven: n.p., 1965), 101–2. See the extensive collection of community cookbooks in the Schlesinger Library of Women's History, Radcliffe College, Cambridge, Mass.

108. David Riesman, *Abundance for What? and Other Essays* (New York: Doubleday, 1964), 133.

109. *NYT*, Sept. 12, 1962.

110. *Consumer Reports*, Sept. 1963, p. 412.

111. *NYT*, Nov. 14, 1962.

Chapter 8. The Best-fed People the World Has Ever Seen?

1. New York Herald Tribune, *The New York Herald Tribune Institute Cookbook* (New York: Scribner's, 1947), v–vi.

2. George Gallup, *The Gallup Poll, 1935–1971* (New York: Random House, 1972) vol. 1, p. 636.

3. Better Homes and Gardens, *Better Homes and Gardens New Cook Book* (Des Moines: Meredith, 1953), 24–25.

4. John E. Gibson, "What Do You Like to Eat?" *Catholic Digest*, Jan. 1964, pp. 60–62.

5. Francois Rysavy, *White House Chef* (New York: Putnam's, 1957), 95–96.

6. *NYT*, July 31, 1961.

7. Gallup, *Gallup Poll, 1935–1971*, vol. 2, p. 936.

8. Good Housekeeping Consumer Panel, *Food Selection and Preparation* (New York: Good Housekeeping, 1945).

9. Richard Polenberg, *One Nation, Divisible: Class, Race, and Ethnicity in the United States Since 1938* (Baltimore: Penguin, 1980), 146.

10. Helen A. Hunscher, review of *Old World Foods for New World Families*, *JHE* 40 (Feb. 1948): 78; American Home Economics Association, *The World's Favorite Recipes* (New York: Harper and Row, 1951), iii–iv.

11. Cambridge Home Information Center, Annual Report, 1953, Cambridge

Home Information Center Papers, Schlesinger Library of Women's History, Radcliffe College, Cambridge, Mass., Folder 9.

12. Mothers' Club, Middletown, Mass., Congregational Church, *What's Cooking?* (n.p., 1951), 91.

13. Department of Food and Nutrition, School of Home Economics, Kansas State College, *Practical Cookery and the Etiquette and Service of the Table* (New York: John Wiley, 1956).

14. In the 1820s the young eaters who patronized Delmonico's new French restaurant in New York considered themselves rebels against their grandmothers' strictures against sauces ("vile greasy compounds") and "converts from the plain-and-roasted doctrine to the rich new gravy faith." Robert Shaplen, "Delmonico: The Rich New Gravy Faith," *New Yorker*, Nov. 10, 1956, p. 197.

15. *NYT*, March 16, 1961.

16. Clementine Paddleford, *How America Eats* (New York: Scribner's, 1960), v.

17. *Betty Crocker's New Picture Cook Book* (New York: McGraw-Hill, 1961), 276, 265.

18. There was also one for tamales, but it was hardly Mexican. Instead of steaming a *masa*-based mixture, as Mexicans do, corn meal mush was deep fried, like southern hush puppies. *The Army Cook*, Technical Manual TM 10-405 (Washington: U.S. Dept. of War, 1942).

19. Herbert Mitgang, "Pizza à la Mode," *NYTM*, Feb. 12, 1956.

20. Claire Jones, *The Chinese in America* (Minneapolis: Leoner, 1972), 40.

21. *LHJ*, June 1951.

22. *Time*, Dec. 7, 1959.

23. "Revolution in the Kitchen," *Fortune*, Feb. 15, 1957.

24. Helen Woodward, *The Lady Persuaders* (New York: Obolensky, 1960), 118.

25. "Freezer Living Is Luxury for Everyone," *GH*, May 1959, p. 76.

26. Tested Recipe, Inc., *COOKINDEX* (New York: Cookindex, 1958), A9, E21.

27. A 1952 study indicated that 76 percent of women prepared meals from such suggested recipes and 67 percent purchased specific products because they had recipes on the package. *NYT*, Dec. 9, 1952.

28. Joshua Gitelson, "Populox: The Suburban Cuisine of the 1950s," unpublished undergraduate seminar paper, Yale Univ., Spring 1991, pp. 5–6.

29. Marion Rombauer Becker, *Little Acorn: The Story Behind the Joy of Cooking, 1931–1966* (Indianapolis: Bobbs-Merrill, 1969), n.p.

30. Jane Stern and Michael Stern, *Square Meals* (New York: Knopf, 1984.)

31. McKinley Kantor, "I'll Take Midwestern Cooking," *Saturday Evening Post*, June 7, 1952.

32. Gallup, *Gallup Poll, 1935–1971*, vol. 2, p. 986.

33. Ibid., vol. 1, p. 1186.

34. Charles Einstein, "Eating Towns," *Harper's*, May 1960, pp. 28–29.

35. Duncan Hines, *Adventures in Good Eating* (Bowling Green, Ky.: Duncan Hines Institute, 1949), vi.

36. Duncan Hines, *Duncan Hines' Food Odyssey* (New York: Crowell, 1955), 237; "Hines Abroad," *New Yorker*, July 24, 1954.

37. Ruth Noble, *A Guide to Distinctive Dining* (Cambridge, Mass.: Berkshire, 1954, 1955).

38. *Holiday*, July 1952, p. 141, July 1955, p. 74, July 1956, p. 65, July 1957, pp. 78–79.

39. "Holiday Handbook: Dining in Southern California," *Holiday*, Nov. 1962, p. 155.

40. Einstein, "Eating Towns," 28. Over the years from 1939 to 1980, when Gallup polled Americans annually on what cities they thought served the finest food New York came out consistently on top, with San Francisco, Chicago, and New Orleans the usual runners-up. Baltimore, however, never made it anywhere near the top rank. Gallup, *Gallup Poll, 1935–1971*, passim.

41. "Coquilles St. Jacques was made with béchamel sauce. White sauces were always glutinous," he recalled. Author's interview with Craig Claiborne, Feb. 23, 1989.

42. Duncan Hines and Frank J. Taylor, "How to Find a Decent Meal," *Saturday Evening Post*, April 27, 1947.

43. John Steinbeck, *Travels with Charley* (New York: Viking, 1962), 126.

44. *Cooking for Profit*, Jan. 1952, pp. 1–2.

45. *New Yorker*, April 7, 1951, p. 22.

46. Hines and Taylor, "How to Find a Decent Meal," 99.

47. Steinbeck, *Travels with Charley*, 127.

48. *RM*, Jan. 1953, pp. 37–38.

49. Richard S. Gutman and Elliot Kaufman, *American Diner* (New York: Harper and Row, 1979), 30–66.

50. Jim Horan, "Speaking Out," *FFF*, Dec. 1952.

51. *BW*, May 3, 1952.

52. *BW*, April 12, 1952.

53. *RM*, March 1963, p. 35.

54. *RM*, March 1963, pp. 31–33.

55. *FF*, Jan. 1959, p. 70.

56. *FFF*, May 1956, p. 14.

57. Carson Gulley, "Take It from Me," *Cooking for Profit*, Jan. 1952, p. 6.

58. *FFF*, April 1952, p. 33.

59. *FF*, May 1959, p. 91.

60. *FF*, June 1959, p. 85.

61. "600,000 Hamburgers a Year Sold at Mawby's," *FFF*, Nov. 1952, p. 18.

62. "The Varsity Sells 3½ Million Franks and Burgers a Year," *FFF*, July 1952, p. 44.

63. "Fountron Electronic Eye Service," *FFF*, Aug. 1952, pp. 24–25.

64. *BW*, July 23, 1949.

65. Like McDonald's, it also featured automatic milkshake machines. *FFF*, April 1956, p. 106.

66. *FF*, January 1959, p. 33.

67. John F. Love, *McDonald's: Behind the Arches* (New York: Bantam, 1986), 41–43.

68. Ibid.

69. John L. Hess and Karen Hess, *The Taste of America* (New York: Grossman, 1977), 200.

Chapter 9. Cracks in the Façade

1. James Gilbert, *A Cycle of Outrage* (New York: Oxford Univ. Press, 1986), 63–78, 143–61, 178–95.

2. In 1931 the average unemployed college-educated woman spent 15.1 hours on food preparation. In 1965 her middle-class counterpart spent 16 hours. Joann Vanek, "Household Technology and Social Status: Rising Living Standards and Status and Residence Differences in Housework," *Technology and Culture* 19 (July 1978): 361–75.

3. Betty Friedan, *The Feminine Mystique* (New York: Norton, 1963), 15–69.

4. *Twenty-fourth Annual Nielsen Report to Retail Food Stores* (Chicago: A. C. Nielsen, 1958), 31; Landon Y. Jones, *Great Expectations: America and the Baby Boom Generation* (New York: Coward McCann, 1980), 41.

5. Harry Botsford, "Outdoor Hospitality: The Gentleman Plays with Fire," in *Esquire's Handbook for Hosts* (New York: Grosset and Dunlap, 1953), 97–98.

6. Ibid., 97.

7. Marjorie Husted, "The Women in Your Lives," typescript of speech, Oct. 13, 1952, in Marjorie C. Husted Papers, Schlesinger Library of Women's History, Radcliffe College, Cambridge, Mass., Folder 2.

8. Vance Packard, *The Hidden Persuaders* (New York: David Mackay, 1957), 107–9.

9. *NYT*, June 30, 1957.

10. *NYT*, Oct. 22, 1969.

11. Ibid.; Jacqueline Verrett and Jean Carpet, *Eating May Be Hazardous to Your Health* (New York: Simon and Schuster, 1974), 70–71; James Turner, *The Chemical Feast* (New York: Grossman, 1970), 206; Wallace Jensen, "The Story of the Laws Behind the Label," *FDA Consumer*, June 1981, p. 37. This argument did not inhibit them from advertising the vitamins in their products, even though most vitamin research was based on analogous work with rats.

12. *USNWR*, Dec. 7, 1959.

13. *USNWR*, Dec. 28, 1959.

14. Rachel Carson, *Silent Spring* (Boston: Houghton Mifflin, 1962); *FP*, July 1961, pp. 1–2, Sept. 1961, pp. 42–43.

15. Daniel Maynard, Oral History Interview, Archives and Manuscripts Division, Cornell Univ. Libraries, Ithaca, N.Y., pp. 69–81.

16. Margaret Reid, "Food, Liquor, and Tobacco," in J. Friedrich Dewhurst et al., eds., *America's Needs and Resources* (New York: Twentieth Century Fund, 1955), 165–66.

17. Lydia Roberts, *Nutrition Work with Children* (Chicago: Univ. of Chicago Press, 1927), 131–38; ibid., rev. ed. (1935), 201–8; Ethel Martin, *Roberts' Nutrition Work with Children* (Chicago: Univ. of Chicago Press, 1954) chaps. 4 and 5.

18. Theodore Sorensen, *Kennedy* (London: Hodder and Stoughton, 1965), 199.

19. A. J. Liebling, *The Press* (New York: Ballantine, 1961), 103.

20. Michael Harrington, *The Other America* (New York: Macmillan, 1962).

21. E. Neige Todhunter, "The Food We Eat," *JHE* 50 (Sept. 1958): 512–13; Ross H. Hall, *Food for Nought: The Decline in Nutrition* (New York: Harper and Row, 1974), 247–48.

22. *FP*, Dec. 1961, p. 63.

23. *NYT*, Feb. 25, 1962.

24. Hall and Stieglitz Co., *The Convenience Food Manufacturing Industry* (New York: the authors, 1963), 47; *FFR*, Mar. 1962.

25. George Orwell, "It Looks Different from Abroad," *New Republic*, Dec. 2, 1946.

26. *Cosmopolitan*, Nov. 1954, p. 6, April 1954, p. 16.

27. *SNL*, July 4, 1953, April 3, 1954, Jan. 22, 1955.
28. *NYT*, Nov. 15, 1952; *BW*, Dec. 6, 1952.
29. Peter Wyden, *The Overweight Society* (New York: Morrow, 1965), 15.
30. Helen Woodward, *The Lady Persuaders* (New York: Obolensky, 1960), 8, 156.
31. Hall and Stieglitz Co., *Convenience Food*, 47.
32. The idea itself was by no means new. The new products were not far removed from a number of Depression-era powdered potions or even its famed bananas-and-milk diet. Hillel Schwartz, *Never Satisfied: A Cultural History of Diets, Fantasies, and Fat* (New York: Free Press, 1986), 198.
33. Hall and Stieglitz Co., *Convenience Foods*, 47.
34. *NYT*, July 18, 1962.
35. Avis De Voto to Julia Child, June 15, 1957, Avis De Voto Papers, Schlesinger Library of Women's History, Radcliffe College, Cambridge, Mass., Box 2.
36. Elaine Whitelaw, quoted in *NYT*, Oct. 18, 1989.
37. J. Edgar Hoover, who made up in power what he may have lacked in breeding, was a regular. Soulé would send him a bottle of Romanée-Conti to accompany his *filet de boeuf perigourdine*, and when he died he bequeathed Hoover his watch. *NYT*, March 15, 1966.
38. Joseph Wechsberg, "The Ambassador in the Sanctuary," *New Yorker*, March 28, 1953.
39. "The Grand Tour," *Time*, April 11, 1949; "The Return of Tourism," *Fortune*, June 1949, p. 99.
40. Appropriately enough for a clientele who were uncertain as to whether the English or French elite was more worthy of emulation, Voisin's "French" chef was an Englishman who had worked under Escoffier in London in 1914, and his specialty—a purée of game and chestnuts—seemed to blend the tastes of the two. *New Yorker*, Dec. 24, 1955.
41. Author's interview with Julia Child, Dec. 28, 1988.
42. *NYT*, Jan. 28, 1966. He later concluded the snobbishness was the result of deep personal insecurity. Author's interview with Craig Claiborne, Feb. 23, 1989.
43. Lucius Beebe, "The Perfect Restaurateur," *Holiday*, May 1953, pp. 81–84.
44. Waverley Root and Richard de Rochemont, *Eating in America: A History* (New York: Morrow, 1976), 338.
45. Jerome Beatty, "Sherman Packs 'Em In," *American Magazine*, June 1941, pp. 44–45, 117–18.
46. Charles Wertenbaker, "Manhattan's Haughtiest Eatery," *Saturday Evening Post*, Dec. 27, 1952.
47. Beebe, "Perfect Restaurateur," 81–84.
48. *BW*, Aug. 23, 1958.
49. *NYT*, March 16, 1961.
50. *NYT*, Sept. 26, 1972.
51. *NYT*, Jan. 21, 1968.
52. *NYT*, April 7, 1961.
53. *NYT*, Dec. 11, 1964.
54. *NYT*, July 30, 1961.
55. Claiborne interview.
56. Betty Fussell, *Masters of American Cookery* (New York: Times Books, 1983), 40.

57. As early as 1958 and 1959, soon after he took over at the *Times*, he was writing feature articles on Persian, Pakistani, Chilean, and Sicilian cuisines.

58. Calvin Tomkins, "Good Cooking," *New Yorker*, Dec. 23, 1974.

59. Child interview.

60. Claiborne interview.

61. E.g., "Napoleon and Friends," *McCall's*, Jan. 1966, pp. 88–90; E. Graves, "Ballet of Flames for Christmas," *Life*, Dec. 16, 1966; E. Alston, "Napoleon's Chicken Marengo," *Look*, Sept. 9, 1969.

Chapter 10. The Politics of Hunger

1. U.S. Senate, Committee on Labor and Public Welfare, Subcommittee on Employment, Manpower, and Poverty (hereafter Senate Poverty Subcommittee), *Examination of the War on Poverty: Hearings*, 90th Cong., 1st sess., 1967, p. 697.

2. Senate Poverty Subcommittee, *Hunger in America: Chronology and Selected Background Materials: Committee Print*, 90th Cong., 2nd sess., 1968, p. 1; *NYT*, April 10, 11, 12, 1967; Nick Kotz, *Let Them Eat Promises* (Englewood Cliffs, N.J.: Prentice-Hall, 1969), 2–4; Arthur Schlesinger, Jr., *Robert Kennedy and His Times* (Boston: Houghton Mifflin, 1978), 794–95; Robert Sherrill, "It Isn't True That Nobody Starves in America," *NYTM*, June 4, 1967.

3. *NYT*, Feb. 2, 1953.

4. U.S. Senate, Select Committee on Nutrition and Human Needs (hereafter Senate Nutrition Committee), *Nutrition and Health 2: Committee Print*, 94th Cong., 2nd sess., 1976, p. 56; *NYT*, June 6, 1963; Irene Wolgamot, "The World Food Congress," *JHE* 55 (Aug. 1963): 603.

5. *Farm Policy: The Politics of Oil, Surpluses, and Subsidies* (Washington: Congressional Quarterly, 1984), 127; *NYT*, Nov. 18, 1966. In 1971 it was revealed that $693 million of the money was used by the foreign governments to buy American military equipment. Over the next three years, after this was partially rectified, over half of the money went to two countries, South Vietnam and Cambodia. *NYT*, Jan. 5, 1971; Roy Freund, "The Politics of Hunger," *Progressive*, Dec. 1979, pp. 38–39.

6. Supporters of the project were enraged when George Larrick, head of the FDA, called their product "filthy." *Newsweek*, June 17, 1963.

7. *Chemical Week* 103, no. 24 (1969): 61.

8. *NYT*, April 7, Aug. 5, Nov. 21, Nov. 25, 1968.

9. D. G. Snider et al., "The Fish Protein Concentrate Story," *FT* 21, no. 7 (1967): 70–72; R. Dabbah, "Protein from Microorganisms," *FT* 24, no. 12 (1970): 659; M. Milner, ed., *Protein Enriched Cereals for World Needs* (St. Paul: American Ass'n. of Cereal Chemists, 1969.)

10. Senate Nutrition Committee, *Nutrition and Private Industry: Hearings*, 90th Cong., 2nd sess., and 91st Cong., 1st sess., 1969, pp. 4620–21; *NYT*, Aug. 13, 1967.

11. *NYT*, June 6, 1963.

12. *NYT*, Sept. 27, 1968.

13. Martin Glicksman, "Fabricated Foods," *CRC Critical Reviews in Food Technology*, April 1971, pp. 21–22.

14. Sherrill, "It's Not True," 101.

15. Schlesinger, *Robert Kennedy*, 795–96; Kotz, *Let Them Eat Promises*, 176–79.

16. Sherrill, "It's Not True," 100–101; Janet Poppendieck, *Breadlines Knee-Deep*

in Wheat: Food Assistance in the Great Depression (New Brunswick: Rutgers Univ. Press, 1986), 244–45; *WSJ*, March 26, 1968.

17. Joseph Brenner et al., "Report to the Field Foundation, June 1967," in Senate Poverty Subcommittee, *Hunger in America*, 42–46.

18. Ibid.; *NYT*, June 17, July 12, 13, 1967.

19. *NYT*, June 12, 1967; Citizens' Board of Inquiry into Hunger and Malnutrition in the United States, *Hunger, U.S.A.* (Boston: Beacon, 1968), 3–5.

20. *NYT*, July 9, 12, 13, 1967, Aug. 8, 1968.

21. *NYT*, July 2, 1967.

22. *NYT*, May 23, 1968.

23. In the mid-1930s around 50 percent of the poor were farmers of some kind. By 1960, 55 percent of the poor lived in cities, 30 percent in small towns, and only 15 percent on farms. James Patterson, *America's Struggle Against Poverty, 1900–1980* (Cambridge: Harvard Univ. Press, 1981), 80.

24. *Farm Policy*, 123.

25. CBS News, "Hunger in America," May 21, 1968, recording in Museum of Broadcasting, New York, N.Y.

26. The figure, like much of the information in the show, came from the hearings of the Citizens' Board of Inquiry into Hunger and Malnutrition. *NYT*, Oct. 24, 1967.

27. CBS News, "Hunger in America."

28. *NYT*, May 24, 1969.

29. While there was some malnutrition, they said, it was the result of "local custom or ignorance," not of an inability to buy food. *NYT*, June 16, Aug. 18, 1968.

30. Norwood A. Kerr, "Drafted into the War on Poverty: USDA Food and Nutrition Programs, 1961–1969," *Agricultural History* 64, no. 2 (Spring 1990): 154–63; Senate Poverty Subcommittee, *Hunger and Malnutrition in America: Hearings*, 90th Cong., 2nd sess., 1968, pp. 203–25.

31. Jeffrey M. Berry, *Feeding Hungry People: Rulemaking in the Food Stamp Program* (New Brunswick: Rutgers Univ. Press, 1984), 24–33, 41. Another problem was that many of the nation's poorest counties—eighty-nine of them in May 1968—had not signed on for either the food stamps or free surplus foods programs. Many of those in the South signed on after Freeman threatened that the federal government would distribute the food itself, circumventing the local white political establishments.

32. *NYT*, Aug. 7, 1969; Berry, *Feeding Hungry People*, 60–61.

33. Kotz, *Let Them Eat Promises*, 170–79.

34. *NYT*, May 23, 1968.

35. *NYT*, Oct. 17, 1968.

36. Senate Nutrition Committee, *Review of the Results of the White House Conference on Food, Nutrition, and Health: Hearings*, 92nd Cong., 1st sess., 1971, p. 305.

37. *Farm Policy*, 127; Kerr, "Drafted into the War on Poverty," 165–66.

38. This was based on the calculation that the poor spent one-third of their income on food. (The better-off spent 17 percent.) Mary Goodwin, "Can the Poor Afford to Eat?" in Helen Wright and Laura S. Sims, eds., *Community Nutrition: People, Policies, and Programs* (Monterey, Calif.: Wadsworth, 1981), 148–50.

39. Even World Health Organization publications mistakenly entitled the RDA tables "requirements." D. M. Hegsted, "On Dietary Standards," *NR* 36 (Feb. 1978): 33. As a result, surveys that measured the nutrient intake of sample households against the RDAs continued to produce incredible estimates of rampant malnutri-

tion at all economic levels. In 1965, they said, only 50 percent of Americans consumed the recommended allowances of seven important nutrients. This was 10 percent fewer than the proportion in 1955, even though real income had risen 25 percent since then. Senate Nutrition Committee, *The Nixon Administration Program: Hearings*, 90th Cong., 2nd sess., and 91st Cong., 1st sess., 1969, p. 2583. This was blamed on increased consumption of baked goods at the expense of milk, cereals, vegetables, and fruit. Kenneth W. Clarkson, *Food Stamps and Nutrition* (Washington: American Enterprise Institute, 1975), 75.

40. Hegsted, "On Dietary Standards," 35.

41. A. Wretland, "Standards for Nutritional Accuracy of the Diet: European and WHO/FAO Perspectives," *American Journal of Clinical Nutrition* 36 (1982): 366–75. Furthermore, not all poor people actually spent their food dollars in as nutritionally wise a manner as the USDA budget assumed; nor did all those below the poverty line fail to buy nutritionally adequate diets.

42. *NYT*, Dec. 19, 1968.

43. Senate Nutrition Committee, *Nixon Administration*, 2580–84; *NYT*, Feb. 18, 1969.

44. Senate Nutrition Committee, *SCLC and East St. Louis: Hearings*, 90th Cong., 2nd sess., and 91st Cong., 1st sess., 1969, p. 3287.

45. Senate Nutrition Committee, *Nutrition and Human Needs: Hearings*, 90th Cong., 2nd sess., and 91st Cong., 1st sess., 1969, pp. 677–723.

46. *NYT*, April 17, 1969.

47. *Ten-State Nutrition Survey*, vols. 1–5 (Washington: U.S. Dept. of Health, Education and Welfare, 1972); *NYT*, Oct. 14, 1970.

48. Senate Nutrition Committee, *The Role of the Federal Government in Human Nutrition Research: Committee Print*, 94th Cong., 2nd sess., 1976, pp. 16–17.

49. *NYT*, Aug. 27, 1973.

50. Senate Nutrition Committee, *Review of Results*, 76.

51. Robert Pear, "Counting the Hungry—A Contentious Issue," *NYT*, May 25, 1989.

52. Berry, *Feeding Hungry People*, 45.

53. *NYT*, Feb. 16–20, 1969.

54. Byron Shaw, "Let Us Praise Dr. Gatch," *Esquire*, June 1968; Charles and Bonnie Remberg, "America's Hungry Families," *GH*, Oct. 1968. Both rpt. in Senate Poverty Subcommittee, *Hunger in America*, 210–25.

55. *St. Louis Globe-Democrat*, June 21, 1969.

56. *Chicago Sun-Times*, April 13–21, 1969.

57. *NYT*, Feb. 20, 1969.

58. Senate Nutrition Committee, *SCLC and East St. Louis*, 3166–67. See David Garrow, *Bearing the Cross: Martin Luther King and the Southern Christian Leadership Conference* (New York: Morrow, 1986), chap. 11, on the vagueness of the Poor People's Campaign's goals.

59. *Chicago Daily News*, May 15, 1969.

60. Senate Nutrition Committee, *SCLC and East St. Louis*, 3285–87.

61. *Chicago Sun-Times*, May 17, 1969. As Saul Alinsky pointed out, there would have been little fanfare had it been a white group setting up the little breakfasts. Panther enthusiasm seemed to flag when the white media lost interest. Joseph Conlin, *The Troubles* (New York: Franklin Watts, 1982), 188–89.

62. Senate Poverty Subcommittee, *Hunger and Malnutrition in America: Hearings*, 90th Cong., 1st sess., 1967, pp. 8–62.

63. Robert Sherrill, "Why Can't We Just Give Them Food?" *NYTM*, March 22, 1970.

64. Senate Nutrition Committee, *Nutrition and Human Needs*, 1162–74; *NYT*, Feb. 18, 1969.

65. Berry, *Feeding Hungry People*, 59.

66. Senate Nutrition Committee, *Nixon Administration*, 2437–521; *NYT*, May 9, 11, 12, 14, 1969.

67. Clark Mollenhoff to the President, Dec. 1, 1969, in Bruce Oudes, ed., *From the President's Desk: Richard Nixon's Secret Files* (New York: Harper and Row, 1989), 77.

68. The administration backtracked somewhat on the point, saying that stamps would not be eliminated completely but would likely continue to be given to single individuals and married people without children who were not full beneficiaries of the family income maintenance program. Senate Nutrition Committee, *Welfare Reform and Food Stamps: Hearings*, 90th Cong., 1st sess., 1969, pp. 3760–3820; *NYT*, May 1, Aug. 9, 11, 12, 15, 1969; *WP*, Aug. 12, 1969.

69. Jean Mayer, "The Nutritional Status of American Negroes," *NR* 23 (June 1965): 161–65.

70. *NYT*, Dec. 18, 20, 1968.

71. The McGovern committee disputed the $1.3 billion the administration had budgeted for food aid in the coming fiscal year, saying it would take $4 billion a year plus up to $10 billion in Moynihan-style income maintenance to eliminate hunger. *NYT*, Aug. 7, 1969.

72. U.S. Office of the President, *Welfare Reform—A Message from the President of the United States*, House Doc. no. 91-146, in *Congressional Record*, Aug. 11, 1969; *WP*, Aug. 15, 27, Sept. 10, 1969; *NYT*, Aug. 20, 1969; *New York Daily News*, Sept. 11, 1969; Senate Nutrition Committee, *Welfare Reform and Food Stamps*, 3767–71.

73. Daniel Moynihan to Nixon, Oct. 8, 1969, in Oudes, *From the President's Desk*, 58.

74. *NYT*, Sept. 17, 1969.

75. *NYT*, Nov. 8, 1969.

76. *NYT*, Sept. 14, Nov. 7, Dec. 1, 1969.

77. *NYT*, Dec. 2, 1969.

78. *NYT*, Dec. 2, 3, 1969.

79. *NYT*, Dec. 3, 1969.

80. Elizabeth Drake, "The White House Conference—A Personal View," *Journal of the American Dietetic Association* 56, no. 4 (April 1970): 328.

81. *NYT*, Dec. 3, 1969.

82. *NYT*, Oct. 27, 1972.

83. Even within the USDA, food aid programs had finally been liberated from farm bloc pressure and left to stand on their own as measures to help the poor. Poppendieck, *Breadlines*, 248.

84. Consumer Nutrition Center, "Food Consumption and Dietary Levels of Low-income Households, Nov. 1977–March 1978," Nationwide Food Consumption Survey Preliminary Report no. 8 (Washington: U.S. Dept. of Agriculture, 1981).

85. Jeffrey M. Berry, "Consumers and the Hunger Lobby," *Proceedings of the American Academy of Political Science* 34, no. 3 (1982): 72–73.

86. *NYT*, Oct. 27, 1972.

87. Robert Greenstein, "An End to Persistent Poverty and Hunger in Amer-

ica," in Catherine Lerza and Michael Jacobson, eds., *Food for People, Not for Profit* (New York: Ballantine, 1975), 312.

88. *Nation*, July 6, 1974.

89. David Tabacoff, "The Food Stamp Program—Target for Slander," *Nation*, April 3, 1976.

90. Berry, *Feeding Hungry People*, 82–83.

91. *Time*, Nov. 3, 1975; Tabacoff, "Food Stamp Program," 402.

92. Berry, *Feeding Hungry People*, 90–93.

93. David Potter, "Conflict, Consensus, and Comity," in Don Fehrenbacher, ed., *History and Society: Essays of David Potter* (New York: Oxford Univ. Press, 1973), 189.

94. The prediction that this would strike the Soviet Union and China particularly hard stimulated CIA plans to have the United States use its food supplies as a strategic weapon. *NYT*, March 17, 1975.

95. Berry, *Feeding Hungry People*, 83.

96. Phillip Goldblatt, Mary Moore, and Albert Stunkard, "Social Factors in Obesity," *JAMA* 102 (1965): 97–102; A. Stunkard et al., "Influence of Social Class on Obesity and Thinness in Children," *JAMA* 221 (1972): 579.

97. CBS News, "Hunger in America."

98. *NYT*, Oct. 29, 1974.

99. V. M. Gladney, *Food Practices of Black Americans in Los Angeles County* (Los Angeles: County Department of Health Services, 1972); *NYT*, Oct. 14, 1970.

100. John P. Walker, "Internal and External Poverty and Nutritional Determinants of Urban Slum Youth," *Ecology of Food and Nutrition* 2 (1973): 8–10.

101. Sherrill, "Why Can't We Just Give Them Food?"

102. It was the most unpopular of ten programs respondents in a 1981 Gallup poll were asked to evaluate; 61 percent thought that the government spent too much on it. Berry, *Feeding Hungry People*, 98.

103. Senate Nutrition Committee, *Review of Results*, 76–77.

104. It turned its attention back to food aid mainly to head off the Ford administration's attempted depredations.

Chapter 11. Nutritional Terrorism

1. James Whorton, *Before Silent Spring: Pesticides and Public Health in Pre-DDT America* (Princeton: Princeton Univ. Press, 1974), 212–35.

2. U.S. House of Representatives, Select Committee to Investigate the Use of Chemicals in Food Products, *Chemicals in Food Products: Hearings*, 82nd Cong., 1st sess., 1951, pp. 90–91, and 81st Cong., 2nd sess., 1950, passim.

3. *NYT*, Dec. 10, 1951.

4. *NYT*, Oct. 16, 1954.

5. Rachel Carson, *Silent Spring* (Boston: Houghton Mifflin, 1962), 17–51.

6. Ibid., 15, 157.

7. *NYT*, Aug. 10, Sept. 27, 1962; Frank Graham, Jr., *Since Silent Spring* (Boston: Houghton Mifflin, 1970), 69–74.

8. Beatrice Hunter, *Consumer Beware! Your Food and What's Been Done to It* (New York: Simon and Schuster, 1971), 41; *NYT*, July 22, 1962.

9. His argument was the one commonly used by the USDA and agribusiness: "If the use of pesticides in the U.S.A. were to be completely banned, crop losses would probably soar to 50%, and food prices would increase 4 to 5 fold. Who then

would provide for the needs of the low-income groups? Certainly not the privileged environmentalists." N. E. Borlaug, "Mankind and Civilization at Another Crossroad," *Bioscience* 22, no. 1 (1972): 41–44; Vance Bourjaily, "One of the Green Revolution Boys," *Atlantic Monthly*, Feb. 1971, p. 76.

10. *NYT*, Oct. 29, 1968.

11. National Analysts, Inc., *A Study of Health Practices and Opinions* (Springfield, Va: National Technical Information Service, 1972), table 22, reproduced in Howard A. Schneider and J. Timothy Hesla, "The Way It Is," *NR* 31 (Aug. 1973): 236.

12. Robert Rodale, "J. I. Rodale's Greatest Contribution," *OGF*, Sept. 1971, p. 30.

13. The circulation of *Prevention* magazine, which Rodale began in 1950 as a vehicle for organic and health food theories, also soared. By 1966 it was selling several hundred thousand copies a month; in the early 1970s it passed the two million mark. Victor Herbert and Stephen Barrett, *Vitamins and "Health Foods": The Great American Hustle* (Philadelphia: Stickley, 1981), 99; Barbara Griggs, *The Food Factor* (London: Viking, 1986), 177–78; *NYT*, Sept. 7, 1970; Ronald Deutsch, *The New Nuts Among the Berries* (Palo Alto: Bull, 1977), 305–9.

14. *Ramparts*, July 1968.

15. *NYT*, Sept. 7, 1970.

16. J. I. Rodale, "Why I Started Organic Gardening [1967]," *OGF*, Sept. 1971, p. 39; R. Rodale, "J. I. Rodale's Greatest Contribution," 32.

17. See note 11 above.

18. James Connif, "Those Chemicals in Our Food," *McCall's*, Feb. 1965, p. 83.

19. Griggs, *Food Factor*, 178–80.

20. Adelle Davis, *Let's Eat Right to Keep Fit*, rev. ed. (New York: Harcourt, Brace, Jovanovich, 1970). Many of them, it seems, were irrelevant to what she was saying, or she distorted their contents. Herbert and Barrett, *Vitamins*, 94–95.

21. Daniel Yergin, "Supernutritionist," *NYTM*, May 20, 1973; Griggs, *Food Factor*, 181–83; Davis, *Let's Eat Right*, 203.

22. Davis, *Let's Eat Right;* Adelle Davis, *Let's Cook It Right* (New York: Harcourt Brace, 1947), 4–5; *Time*, Dec. 18, 1972.

23. Deutsch, *New Nuts*, 4–5; Yergin, "Supernutritionist."

24. Griggs, *Food Factor*, 186–87.

25. *NYT*, Sept. 7, 1970.

26. *Newsweek*, May 25, 1970; M. C. Goldman, "Southern California—Food Shopper's Paradise," *OGF*, Nov. 1970, pp. 38–39.

27. *NYT*, Sept. 7, 1970; *Time*, Apr. 12, 1971.

28. National Analysts, *Study of Health Practices*, cited in Schneider and Hesla, "The Way It Is," 236.

29. Ibid., 235.

30. National Research Council, Food and Nutrition Board, "Supplementation of Human Diets with Vitamin E," *NR* 31 (Oct. 1973): 327.

31. Yergin, "Supernutritionist."

32. American Medical Association, Council on Foods and Nutrition and Council on Industrial Health, "Indiscriminate Administration of Vitamins to Workers in Industry," *JAMA* 118 (Feb. 21, 1942): 618–21.

33. *NYT*, July 8, 1954.

34. *JAMA* 174 (Nov. 5, 1960): 1332.

35. *FP*, Nov. 1961, pp. 19–20.

36. *NYT*, April 25, 1961.

37. Ibid.

38. *NYT*, Oct. 7, 1961.

39. "FDA Fact Sheet," May 1967, in U.S. Senate, Select Committee on Nutrition and Human Needs (hereafter Senate Nutrition Committee), *Nutrition and Private Industry: Hearings*, 90th Cong., 2nd sess., and 91st Cong., 1st sess., 1969, pp. 3956–61.

40. See note 11 above.

41. Deutsch, *New Nuts*, 291.

42. J. W. Buchan, "America's Health: Fallacies, Beliefs, Practices," *FDA Consumer*, Oct. 1972, p. 5. Twenty percent also believed that cancer and arthritis might be linked to vitamin and mineral deficiencies. Griggs, *Food Factor*, 292.

43. In fact, the principle had already been accepted with the approval of vitamin A enrichment of margarine and iodization of salt. *JAMA* 205 (Sept. 16, 1968): 160–61; Senate Nutrition Committee, *Nutrition and Private Industry*, 4247. The NRC's Food and Nutrition Board approved the wider application of the principle in 1968; see its *General Policies in Regard to Improvement in Nutritional Quality of Foods* issued that year.

44. Deutsch, *New Nuts*, 323; James Harvey Young and Robert S. Stitt, "Nutrition Quackery: Upholding the Right to Criticize," *FT*, Dec. 1981, p. 42.

45. Charles McCarry, *Citizen Nader* (New York: Saturday Review Press, 1972), 139.

46. Ibid., 140–45.

47. *New Republic*, July 15, 1967.

48. Ralph Nader, "Don't Eat That Dog," *NR*, March 18, 1972; Harrison Wellford, *Sowing the Wind* (New York: Grossman, 1972), 1–25.

49. Ibid., 24.

50. Ibid., 25.

51. Senate Nutrition Committee, *Nutrition and Private Industry*, 3888–4081.

52. *WP*, Oct. 21, 1969; Chris Lecos, "The Sweet and Sour History of Saccharin, Cyclamate, and Aspartame," *FDA Consumer*, Sept. 1981, pp. 8–11.

53. *WP*, Nov. 17, 1968; Senate Nutrition Committee, *Nutrition and Private Industry*, 3922; James Turner, *The Chemical Feast* (New York: Grossman, 1970), 209.

54. Technically, this turned out not to be the case, for cyclamate had been "grandfathered" in, but the government could not resist the moral pressure to ban it.

55. *WP*, Oct. 19, 1969; *WSJ*, Oct. 21, 1969.

56. Peter Flanagan to the President, Oct. 20, 1969, in Bruce Oudes, ed., *From the President's Desk: Richard Nixon's Secret Files* (New York: Harper and Row, 1989), 63.

57. Senate Nutrition Committee, *Nutrition and Private Industry*, 4976–5142; *Chicago Sun-Times*, Oct. 25, 1969; *Washington Evening Star*, Oct. 25, 1969.

58. *WSJ*, Oct. 27, 1969.

59. Cited in Jacqueline Verrett and Jean Carpet, *Eating May Be Hazardous to Your Health* (New York: Simon and Schuster, 1974), 98–99.

60. Elizabeth Whelan and Frederick Stare, *Panic in the Pantry* (New York: Atheneum, 1975), 154–5, 158, 163.

61. *NYT*, Jan. 10, 1971.

62. *NYT*, July 24, 28, 30, Aug. 5, 21, Sept. 15, 29, 30, 1971.

63. Deutsch, *New Nuts*, 317–19.

64. From 1956 to 1965 the percentage of grocery manufacturers' advertising spending devoted to newspapers declined from 21.2 percent to only 6.8 percent,

while that spent in magazines fell from 19 to 14.9 percent. Television spending, meanwhile, grew from 57 percent to over 77 percent. Moreover, all of the increase in TV spending came in "spot" commercials, which give advertisers practically no influence over program content. In women's magazines the trend was to a smaller proportion of food ads, as the variety of other products grew. *Summary of Special Report to Cereal Institute, Inc.: Retail Trends of Breakfast Cereals to August 1, 1965* (New York: A. C. Nielsen, 1965). Arlene Leonhard-Sparks, "A Content Analysis of Food Ads Appearing in Women's Consumer Magazines in 1965 and 1977" (Ed.D. diss., Teachers College, Columbia Univ., 1980), 2.

65. Even in 1963, while sponsor control of programming was still tight, Fred W. Friendly's "CBS Reports" was able resist to the defection of Standard Brands and two other sponsors and broadcast a program on Rachel Carson's *Silent Spring* (Graham, *Since Silent Spring*, 75). Cigarette sponsorship was a different story, and publicizing health claims against them was dicier.

66. *NYT*, Feb. 22, 1969.

67. Senate Nutrition Committee, *Food Additives: Hearings*, 92nd Cong., 2nd sess., 1972; *NYT*, May 10, 1969.

68. "Food Regulation: A Consumer Advocate's View," *FDA Consumer*, May 1977, p. 8.

69. It was only thanks to the royalties from the process for irradiating milk with vitamin D and a relative pittance in food industry money that a number of them could do so at the University of Wisconsin.

70. Senate Nutrition Committee, *The Role of the Federal Government in Human Nutrition Research: Committee Print*, 94th Cong., 2nd sess., 1976, p. 15.

71. *WP*, Oct. 29, 1969; Charles Homer, Michele Habibi, and Michael Jacobson, "The Frozen Fruits of Government Research," in Catherine Lerza and Michael Jacobson, eds., *Food for People, Not for Profit* (New York: Ballantine, 1975), 382.

72. Paul Starr, *The Social Transformation of American Medicine* (New York: Basic Books, 1982), 338–47.

73. "The Better Way," *GH*, May 1961, pp. 137–39.

74. Jean Mayer, letter to the editor, *NYT*, June 9, 1968.

75. William Goodrich to John Harvey, April 12, 1965; R. E. Newberry to M. R. Stephens, Nov. 10, 1965. Both in Senate Nutrition Committee, *Nutrition and Private Industry*, 4288–95.

76. Sheila Harty, *Hucksters in the Classroom: A Review of Industry Propaganda in Schools* (Washington: Center for the Study of Responsive Law, 1979), 23.

77. *FP*, April 1964, p. 65.

78. Joseph Sadusk to George Larrick, Aug. 18, 1964, in Senate Nutrition Committee, *Nutrition and Private Industry*, 4289–91.

79. Victor Herbert, *Nutrition Cultism: Facts and Fiction* (Philadelphia: Stickley, 1980), 11–12.

80. Whelan and Stare, *Panic in the Pantry*, 159; "The Hysteria about Food Additives," *Fortune*, March 1972, p. 141.

81. "Nutrition Foundation Announces New Goals and Leadership," *NR* 30 (Jan. 1972): 1.

82. "A Consumer Looks at FDA," *FDA Consumer*, June 1974, p. 6.

Chapter 12. The Politics of Food

1. Judith Van Allen, "Eating It! From Here to 2001," *Ramparts*, May 1972, pp. 26–31.

2. Charles McCarry, *Citizen Nader* (New York: Saturday Review Press, 1972), 155–56.

3. George Gallup, *The Gallup Poll: Public Opinion, 1972–1977* (Wilmington: Scholarly Resources, 1978), vol. 1, p. 1. Nader did come in ahead of Pope Paul VI. Richard Nixon topped the list, followed by Billy Graham and Edward Kennedy.

4. Soon thereafter Harper and Row put out Ross Hume Hall's *Food for Nought: The Decline in Nutrition* (New York: Harper and Row, 1974).

5. The fact that about half of the book was easy-to-follow recipes for tasty high-protein meatless dishes no doubt played a large role in its brisk sales, particularly from 1973 to 1976, when meat prices soared. But Lappé resisted being slotted as a cookbook author. She focused increasingly on the political aspects of the problem, emphasizing what Third World nations themselves had to do to solve their food problems. Frances Moore Lappé, *Diet for a Small Planet* (New York: Ballantine, 1971); idem and Joseph Collins, *Food First: Beyond the Myth of Scarcity* (Boston: Houghton Mifflin, 1977).

6. Frances Moore Lappé, *Diet for a Small Planet*, rev. ed. (New York: Ballantine, 1975), xvii.

7. David Riesman, "The Uncommitted," in idem, *Abundance for What? and Other Essays* (New York: Doubleday, 1964).

8. Nicholas von Hoffman, *We Are the People Our Parents Warned Us Against* (1968; rpt. Chicago: Ivan Dee, 1989), 44.

9. Ruth Goerling, "Nutrition," *Free Spaghetti Dinner* (Santa Cruz, Calif.), Oct. 2–16, 1970.

10. Von Hoffman, *We Are the People*, 101. The original Diggers were seventeenth-century radicals who invaded and cultivated some London commons in the hope of leading the nation to the socialization of all property.

11. The aphorism has been variously ascribed to Feuerbach, Goethe, and other Germans, for it involves a play on German words. The French writer Brillat Savarin's "Tell me what a man eats and I will tell you who he is" was also often quoted in the same context, but it had a rather different connotation.

12. *Time*, Nov. 16, 1970.

13. "Dorothy to Jan," *Freedom News* (Richmond County, Calif.), Jan. 1970.

14. The editor, a meat-eater, added a note pointing out that New Zealanders and Australians ate more meat than Americans and yet were not as aggressive. *Mother Earth News*, July 1970, p. 40.

15. *NYT*, Nov. 15, 1972.

16. *Time*, Nov. 16, 1970.

17. Ohsawa Foundation, *Zen Macrobiotics* (Los Angeles: Ignoramus Press, 1966), n.p.; *New Yorker*, Aug. 25, 1962; James Taggart, *The Big, Fertile, Rumbling, Cast-Iron, Growling, Aching, Unbuttoned Belly Book* (New York: Grossman, 1972), 499.

18. Ohsawa Foundation, *Zen Microbiotics*, n.p.

19. "The Pure, the Impure, and the Paranoid," *Psychology Today*, Oct. 1978, p. 67.

20. The death—of a twenty-four-year-old New Jersey woman—occurred in 1965, early in the cult's trajectory, and seemed to hardly hamper its spread. U.S. Senate, Select Committee on Nutrition and Human Needs (hereafter Senate Nutrition Committee), *Obesity and Fad Diets: Hearings*, 93rd Cong., 1st sess., 1973, pp. 79–100.

21. American Medical Association, Council on Foods and Nutrition, "Zen Macrobiotic Diets," *JAMA* 218 (Oct. 18, 1971): 397.

22. Lappé, *Diet for a Small Planet*, rev. ed., 69.
23. *Free Spaghetti Dinner*, Oct. 31, 1970.
24. Ibid., Oct. 2–16, 1970.
25. Joan Weiner, "New Food Freaks," *Seventeen*, March 1972, p. 134.
26. Sam Keen, "Eating Our Way to Enlightenment," *Psychology Today*, Oct. 1978, p. 66.
27. Warren Belasco, *Appetite for Change* (New York: Pantheon, 1990), 48.
28. *Time*, Nov. 16, 1970.
29. Ita Jones, *The Grubbag: An Underground Cookbook* (New York: Random House, 1971), cited in Warren Belasco, "Ethnic Fast Foods: The Corporate Melting Pot," *Food and Foodways* 2 (1987): 5.
30. *Fox Valley Kaleidoscope*, Oct. 23–Nov. 15, 1970; *Rochester Patriot*, March 24–April 6, 1976.
31. Belasco, *Appetite*, 97.
32. *Eugene* (Ore.) *Augur*, Jan. 6–12, 1969.
33. Charles A. Reich, *The Greening of America* (New York: Random House, 1971), 178, 389.
34. *Mother Earth News*, May 1970, p. 70, Jan. 1971, p. 46, March 1973, p. 60, Aug. 1971, pp. 171–72, Dec. 1971, pp. 34–36, Sept. 1973, pp. 68–72.
35. Belasco, *Appetite*, 58.
36. *Eugene Augur*, Nov. 19–Dec. 3, 1970.
37. *NYT*, March 6, 1972.
38. George Alexander, "Brown Rice as a Way of Life," *NYTM*, March 12, 1972.
39. "But our motivation is internalized," she added (as if her mother's were not.) *Time*, Nov. 16, 1970.
40. *Berkeley Barb*, 1969–1970; *East Village Other*, July 1, 1966. On the other hand, radical women's groups did echo New Left criticisms of the food industries. "It is upsetting," said the first feminist guide to health, "that we often have to pay twice—first to have the important natural ingredients removed from our food, then to have a few of those nutrients restored." Boston Women's Health Book Collective, *Our Bodies, Ourselves: A Book by and for Women*, 2nd ed. (New York: Simon and Schuster, 1976), 108.
41. *Hamilton* (Ont.) *Spectator*, Sept. 10, 1991. Adelle Davis took LSD several times in the 1950s, but although she found it a "deeply religious" experience it had no effect on her ideas about food. Daniel Yergin, "Supernutritionist," *NYTM*, May 20, 1973.
42. Belasco, *Appetite*, 58.
43. Von Hoffman, *We Are the People*, 49.
44. See the "Underground Press" collection in the Manuscripts Division of the Library of Congress.
45. *NYT*, Jan. 7, 1973.
46. It was also charged that the interchange went the other way. Michael Jacobson noted that twenty-two of the fifty-two top FDA officials had worked for regulated industries or their trade associations. Some of their meetings with industry representatives "resemble a reunion more than a conference," said the ex-FDA scientist Jacqueline Verrett. Jacqueline Verrett and Jean Carpet, *Eating May Be Hazardous to Your Health* (New York: Simon and Schuster, 1974), 94–95.
47. Jim Hightower and Susan De Marco, "Hard Tomatoes, Hard Times," *Environment Action Bulletin*, Aug. 1972, rpt. in Catherine Lerza and Michael Jacobson, eds., *Food for People, Not for Profit* (New York: Ballantine, 1975), 389.

48. Senate Nutrition Committee, *Nutrition and Private Industry: Hearings*, 90th Cong., 2nd sess., and 91st Cong., 1st sess., 1969, pp. 4564–4595, 4976–5142.

49. Beatrice Hunter, *Consumer Beware! Your Food and What's Been Done to It* (New York: Simon and Schuster, 1971), 30–50; Verrett and Carpet, *Eating May Be Hazardous*, 85.

50. Hunter, *Consumer Beware!* 42–43.

51. *Time*, Aug. 26, 1974.

52. Senate Nutrition Committee, *Food Additives: Hearings*, 92nd Cong., 2nd sess., Sept. 12, 1972, pp. 821–29.

53. David Reuben, *Everything You Always Wanted to Know About Nutrition* (New York: Simon and Schuster, 1978); "Book Purged of Academy Slur," *Science* 206 (Dec. 7, 1979): 1166.

54. John B. Klis, "Nutrition—Needs and Deeds," *FT*, June 1970, p. 633; Stare in *Life* cited by James Harvey Young and Robert S. Still, "Nutrition Quackery: Upholding the Right to Criticize," *FT*, Dec. 1981, p. 42; Frederick Stare, "How to Live 5 Years Longer," *Mechanics Illustrated*, May 1968, pp. 74–75.

55. James Turner, *The Chemical Feast* (New York: Grossman, 1970), 213–14; Hunter, *Consumer Beware!* 36.

56. U.S. Senate, Committee on Commerce, Consumer Subcommittee (hereafter Senate Consumer Subcommittee), *Nutritional Content and Advertising for Dry Breakfast Cereals*, 92nd. Cong., 2nd sess., 1972; L. Hoffman, *The Great American Nutrition Hassle* (Palo Alto: Mayfield, 1978), 380–81.

57. Senate Nutrition Committee, *Obesity and Fad Diets*, 18.

58. Benjamin Rosenthal, Michael Jacobson, and Marcy Bohm, "Professors on the Take," *Progressive*, Nov. 1976, pp. 42–43.

59. Barbara Griggs, *The Food Factor* (London: Viking, 1986), 295–96.

60. *Nutrition Action*, Dec. 1975, p. 4.

61. Rosenthal, Jacobson, and Bohm, "Professors on the Take," 43.

62. T. G. Harris, "Affluence, the Fifth Horseman of the Apocalypse," *Psychology Today*, April 1970, pp. 42–45.

63. Stare and Mayer, Letters to the Editor, *NYTM*, Sept. 12, 1976.

64. Claude Fischler, "Attitudes toward Sugar and Sweetness in Historical and Sociological Perspective," in J. Dobbing, ed., *Sweetness* (Berlin: Springer-Verlag, 1987), 83–98; idem, "Les images changeant du sucre: Saccharophilie et saccharophobie," *Journal d'Agriculture Traditionelle et de Botanique Appliquée* 35 (1988): 241–60.

65. Hunter, *Consumer Beware!* 32–33.

66. They both asked a cross-section of adults if they agreed that "too much sugar causes diabetes" and "sugar is more fattening than other foods." Fischler, "Attitudes," 91.

67. *NYT*, Dec. 21, 1974.

68. John Yudkin, *Sweet and Dangerous* (New York: Bantam, 1973), 85–92, 118–60.

69. John Yudkin, *Pure, White, and Deadly* (London: David Poynter, 1974).

70. Fischler, "Attitudes," 91.

71. Victor Herbert and Stephen Barrett, *Vitamins and "Health Foods": The Great American Hustle* (Philadelphia: Stickley, 1981), 101.

72. Elizabeth Walker Mechling and Jay Mechling, "Sweet Talk: The Moral Rhetoric Against Sugar," *Central States Speech Journal* 34 (Spring 1938): 19–32.

73. "What You Can Do to Help Your Husband Avoid a Heart Attack," *GH*,

May 1965, p. 181; Stare, "How to Live 5 Years Longer," 76–77; Senate Nutrition Committee, *Sugar in Diet, Diabetes, and Heart Diseases: Hearings*, 93rd Cong., 1st sess., 1973, pp. 145–287; idem, *Hearings on S. 2830*, 93rd Cong., 1st sess., 1973, pp. 450–67. This more moderate position was later adopted by Michael Jacobson and the Center for Science in the Public Interest. Letitia Brewster and Michael Jacobson, *The Changing American Diet* (Washington: Center for Science in the Public Interest, 1978), 47–48.

74. Sucrophiles still outnumbered sucrophobes by a very wide margin—those who had turned to cyclamates did so mainly to lose weight. Only the accusation that sugar caused tooth decay was widely accepted, and this was mainly with regard to children. Fischler, "Attitudes," 91.

75. It was also criticized for including calcium and vitamin D, which are hardly present in the grains out of which cereals are made but are abundant in the milk that is normally added to them. On the other hand, if a nutrient scored over 100, the excess points canceled out deficiencies in others in the final score.

76. *NYT*, July 24, 1970.

77. Senate Consumer Subcommittee, *Nutritional Content and Advertising*, 87; John Keats, *What Ever Happened to Mom's Apple Pie? The American Food Industry and How to Cope with It* (Boston: Houghton Mifflin, 1976), 16.

78. *NYT*, July 24, 1970.

79. Yudkin, *Sweet and Dangerous*, 58, 181–82.

80. Jean Mayer, "The Bitter Truth about Sugar," *NYTM*, June 6, 1976.

81. Michael Jacobson, "Our Diets Have Changed, but Not for the Best," *Smithsonian*, April 1975, p. 99.

82. William Dufty, *Sugar Blues* (New York: Warner, 1975), 1, 14.

83. Over half the ads on children's television were for food, he said, but mainly for cereals, candies, and snacks. None was for fruits or vegetables, and they were devoid of useful nutritional information. Senate Consumer Subcommittee, *Nutritional Content and Advertising*, 25–37; Robert Choate, "The Sugar-Coated Children's Hour," in Lerza and Jacobson, *Food for Profit*, 145.

84. *NYT*, Nov. 9, 1973; Joan Barthel, "Boston Mothers Against Kidvid," *NYTM*, Jan. 5, 1975.

85. Fischler, "Attitudes," 91.

86. E.g., a 1977 doctoral dissertation that cited unnamed "experts" who said sugar "might" play a role in causing diabetes and thirteen pages later said "excessive sugar intake has been shown to have a role in diabetes and possibly in arteriosclerosis." Lois Kurman, "An Analysis of Messages Concerning Food, Eating Behaviors, and Ideal Body Image on Prime-time American Network Television" (Ph.D dissertation, New York Univ., 1977), 46–47; 60.

87. James Beard, *Beard on Bread* (New York: Knopf, 1973), xii.

88. Claude Fischler, "Le Ketchup et la pilule," *Perspective et Santé* 25 (Spring 1983): 114–15.

Chapter 13. Natural Foods and Negative Nutrition

1. *NYT*, Nov. 10, 1969.

2. Nika Hazelton, "Keep It Natural," *National Review*, Aug. 1972, p. 912.

3. Aaron L. Brody, "Flexible Packaging of Foods," *CRC Critical Reviews in Food Technology* 1 (Feb. 1970): 95–96.

4. *NYT*, Nov. 5, 1971; U.S. Senate, Committee on Commerce, Consumer Subcommittee (hereafter Senate Consumer Subcommittee), *Nutritional Content and Advertising for Dry Breakfast Cereals*, 92nd Cong., 2nd sess., 1972, p. 37.

5. *Advertising Age*, Aug. 10, 1970.

6. Albert Zanger, "What Was the Impact of the Choate Study on the Cereal Market?" in William T. Kelley, ed., *The New Consumerism* (Columbus, Ohio: Grid, 1973), 285; *Advertising Age*, Aug. 10, Sept. 25, 1972.

7. Daniel Yergin, "Supernutritionist," *NYTM*, May 20, 1973; "Organic," *New Yorker*, May 22, 1971.

8. *NYT*, Jan. 30, 1974.

9. *Time*, Dec. 18, 1972. Eighty-six percent of food manufacturers developing new products in 1968 were interested in convenience, versus only 16 percent who wanted lo-cal products and 6 percent out for the dietetic market. *FE*, Jan. 1968, p. 68.

10. "The Hysteria About Food Additives," *Fortune*, March 1972, p. 64.

11. Alexander McFarlane, "Convenience, Food Innovation, and Calling the Tune," *FT* 23 (April 1969): 43–45.

12. *FP*, Feb. 1966, p. 10.

13. *FT* 24 (April 1970): 206.

14. "Nutritional Quality and Food Product Development," *NR* 313 (July 1973): 226.

15. "Advertising Nutrition," *New Republic*, March 6, 1971.

16. Rodale tried to combat organic fakery by setting up a program to certify organic foods. He also supported a bill providing for federal certification, sponsored by Representative Edward Koch, because "we need all the help we can get to fight the organic phonies." Robert Rodale, "Congress Gets an Organic Food Bill," *OGF*, July 1972, p. 29.

17. *NYT*, Nov. 17, 1974.

18. Ex-Director of Special Products Arthur Odell, quoted in "The Pure, the Impure, and the Paranoid," *Psychology Today*, Oct. 1978, p. 67.

19. *NYT*, Nov. 14, 1976; *WSJ*, Oct. 27, 1981.

20. *NYT*, Aug. 17, 1976; General Foods advertisement reproduced in Joan Gussow, *The Feeding Web: Issues in Nutritional Ecology* (Palo Alto: Bull, 1978), 201.

21. "Hysteria," 63–64.

22. *NYT*, Dec. 27, 1976.

23. Boston Women's Health Book Collective, *Our Bodies, Ourselves: A Book by and for Women*, 2nd ed. (New York: Simon and Schuster, 1976), 108.

24. L. Hoffman, *The Great Nutrition Hassle* (Palo Alto: Mayfield, 1978), 334.

25. Ibid.

26. Sheila Harty, *Hucksters in the Classroom: A Review of Industry Propaganda in Schools* (Washington: Center for the Study of Responsive Law, 1979). In 1975 they became the first billion-dollar advertisers on TV. *NYT*, March 25, 1975.

27. Yet only 3 percent of the foods advertised were "natural" or "health" foods of the kinds favored by industry critics. Arlene J. Leonhard-Spark, "A Content Analysis of Food Ads Appearing in Women's Consumer Magazines in 1965 and 1977" (Ed.D. diss., Teachers College, Columbia Univ., 1980), 98.

28. *NYT*, Aug. 7, 1975.

29. *NYT*, Aug. 8, 1977.

30. The Dancer-Fitzgerald Sample, which used a more restrictive definition of "new product," reported that the number of new products introduced annually tri-

pled in the next fourteen years. John M. Connor, *Food Processing: An Industry Powerhouse in Transition* (Lexington, Mass.: Heath, 1988), 63.

31. *NYT*, March 15, 1978.

32. Brett Silverstein, *Fed Up: The Forces That Make You Fat, Sick, and Poor* (Boston: South End, 1984), 51.

33. Stephanie Crocco, "What the Industry Can Do About Growing Public Mistrust," *FE*, Aug. 1975, p. 35.

34. Herbert Abelson et al., *Food and Nutrition: Knowledge, Beliefs* (Princeton: Response Analysis Corp., 1974), 39, typescript in Library of Congress; Alice Fusillo, "Food Shoppers' Beliefs: Myths and Realities," *FDA Consumer*, Oct. 1974, p. 17.

35. Fergus Clydesdale, "The Reality of Change in Illusionary Society," *Processed Prepared Foods*, June 1978, p. 66.

36. See my *Revolution at the Table: The Transformation of the American Diet* (New York: Oxford Univ. Press, 1988), chap. 10; *NYT*, June 15, 1975; U.S. Dept. of Health and Human Services and U.S. Dept. of Agriculture, *Nutrition Monitoring in the United States: A Progress Report from the Joint Nutrition Monitoring Evaluation Committee* (Hyattsville, Md.: USDHHS, 1986), 212.

37. Hoffman, *Nutrition Hassle*, 333.

38. Abelson et al., "Food and Nutrition," 30–32.

39. "The Food Fad Boom," *FDA Consumer*, Dec. 1973–Jan. 1974, pp. 5–12.

40. This was particularly irksome to corn oil producers, who were forced to stand by while producers of oils with much lower percentages of polyunsaturated fats stole much of the low-cholesterol market from them. Dorothy M. Rathmann, J. Richard Stockton, and Daniel Merrick, "Dynamic Utilization of Recent Nutritional Findings: Diet and Cardiovascular Disease," *CRC Critical Reviews in Food Technology*, Sept. 1960, pp. 331–378.

41. *FT* 24 (Aug. 1970): 866.

42. *NYT*, Nov. 11, 1973, June 3, 1974, Sept. 7, 1971, Aug. 7, 1975. The order seems to have headed off a rise in advertised health claims. Leonhard-Sparks, "Content Analysis," 101.

43. Senate Consumer Subcommittee, *Nutritional Content and Advertising*, 17.

44. *NYT*, July 7, 1974.

45. Judith Brown, "Graduate Students Examine TV Ads for Food," *Journal of Nutrition Education* 9 (July–Sept. 1977): 121–22.

46. *NYT*, Dec. 19, 20, 22, 29, 1974; *NYT*, Feb. 2, Oct. 11, 22, 1975.

47. Colman McCarthy, "A Regular Family Meal (Aargh!)," *Saturday Review*, Sept. 2, 1972; *NYT*, Sept. 10, 1975.

48. *NYT*, July 16, Sept. 10, Oct. 20, 22, 1975. Butz's attitude toward the new food concerns was perhaps best exemplified by his response to ex–Vice President Hubert Humphrey's 1973 suggestion that if everyone gave up one McDonald's hamburger a week it would release ten billion tons of grain a year for the needy world. He pointed out that this much meat was consumed by America's pets. "Americans are not going to do away with their pets," he said, "and they are not going to eat one less hamburger a week. If anything, they will eat one more hamburger a week. That's the way our standard of living has always improved, and that's the way it will continue to improve, at least in our generation." Maya Pines, "Meatless, Guiltless," *NYTM*, Nov. 24, 1974.

49. An extensive review of the scientific literature for a food technologists' journal concluded that there was a direct relationship between the amount of nitrites in bacon and the amount of nitrosopyrrolidine, a precursor of nitrosamines, "an ex-

tremely dangerous compound [whose] appearance in foods justifiably causes concern." Richard A. Scanlan, "N-Nitrosamines in Food," *CRC Critical Reviews in Food Technology* 5 (Apr. 1975): 357–402.

50. *NYT*, Sept. 16, 1978; "Time for a Cease-Fire in the Food-Safety Wars," *Fortune*, Feb. 1979, p. 94.

51. *NYT*, March 13, 1977; "Time for a Cease-Fire," 94; Chris Lecos, "The Sweet and Sour History of Saccharin, Cyclamate, Aspartame," *FDA Consumer*, Sept. 1981, p. 9.

52. James Harvey Young, "The Agile Role of Food: Some Historical Reflections," in *Nutrition and Drug Interrelations* (New York: Academic Press, 1978), 13–14; Jane Brody, *Jane Brody's Nutrition Book: A Lifetime Guide to Good Eating for Better Health and Weight Control* (New York: Norton, 1981), 12.

53. *Science* 203 (March 23, 1979): 1221–22.

54. *OGF*, Sept. 1972, p. 12.

55. *NYT*, May 24, 1978.

56. The large majority of 1950s dieters (about three-quarters) remained women, who dieted mainly for cosmetic purposes. Roberta Seid, *Never Too Thin: Why Women Are at War with Their Bodies* (New York: Prentice-Hall, 1989), 122–27, 152.

57. *JAMA* 206 (Oct. 28, 1968): 999–1000, 1002–3.

58. U.S. Senate, Committee on Nutrition and Human Needs (hereafter Senate Nutrition Committee), *Diet Related to Killer Diseases, 2. Cardiovascular Disease: Hearings*, 95th Cong., 2nd sess., 1977, p. 18.

59. "Diet and Coronary Heart Disease," *NR* 30 (Oct. 1972): 223.

60. *NYT*, Oct. 25, 1974.

61. Senate Nutrition Committee, *Diet Related to Killer Diseases, re Meat: Hearings*, 95th Cong., 1st sess., 1977, p. 6. Dissenters such as Yudkin kept challenging the evidence, which was by no means overwhelming. Not until 1990 was a clear link between lowering serum cholesterol levels and reducing the incidence of heart attacks established. *NYT*, Oct. 5, 1990.

62. With hopes for legislation to force the government to mount an antisugar campaign in the guise of an antidiabetes campaign thus dimmed, the committee could only recommend that money be given the National Institutes of Health for more research on the subject. Senate Nutrition Committee, *Hearings on S. 2830*, 93rd Cong., 1st sess., 1973, p. 467.

63. Warren Belasco, *Appetite for Change* (New York: Pantheon, 1989), 175.

64. U.S. Dept. of Health and Human Services, Public Health Service, *The Surgeon General's Report on Nutrition and Health: Summary and Recommendations* (Washington: USGPO, 1988), 7.

65. *AC*, May 1942, p. 162.

66. William F. Kannel, "Obesity and Heart Disease," in Senate Nutrition Committee, *Hearings on National Nutrition Policy: Background Reading Documents* (Washington: USGPO, 1974), 79, cited in Seid, *Never Too Thin*, 171.

67. Richard F. Spark, "Fat Americans," *NYTM*, Jan. 6, 1974.

68. Michael Jacobson and Catherine Lerza, "Introduction," in C. Lerza and M. Jacobson, eds., *Food for People, Not for Profit* (New York: Ballantine, 1975), 165–67.

69. Letitia Brewster and Michael Jacobson, *The Changing American Diet* (Washington: Center for Science in the Public Interest, 1978). Their argument that the diet was less healthy than that of 1910 is unconvincing. Even if one grants that there is something unhealthy about increasing consumption of sweets, they emphasize the great rise in refined sugar consumption but ignore the simultaneous fall in

consumption of molasses, sorghum, maple sugar, and other traditional sweeteners, which left overall sweetener consumption little changed since 1925. They condemned higher levels of fat consumption but did not take into account the great drop in consumption of high-cholesterol lard and butter, which were replaced by vegetable shortening, salad oils, and margarine, all of which contain lower levels of low-density lipids. Even with the postwar rise in meat consumption, the total of saturated fatty acids in food supply had risen only 9 percent since 1900. Perhaps most important, however, is that the USDA Food Supply Series upon whose statistics their book was based is hardly a reliable source for the use to which they put it. The USDA itself has said that using it for "interpretation of trends in nutrient intake is problematic because of changes in survey methods over time." USDHHS and USDA, *Nutrition Monitoring: Progress Report*, 77; idem, *Nutrition Monitoring in the United States: An Update Report on Nutrition Monitoring* (Washington: USGPO, 1989), 34–35, 123. Infant mortality rates and stature statistics, which are more useful indicators of nutritional status, indicate that there was a more or less steady improvement in the average American diet at least until the 1980s. For a more detailed discussion of this see my *Revolution at the Table*, 194–95.

70. Ivan Illich, *Medical Nemesis: The Expropriation of Health* (London: Calder and Boyers, 1975).

71. Barbara Griggs, *The Food Factor* (London: Viking, 1986), 276.

72. T. G. Harris, "Affluence, the Fifth Horseman of the Apocalypse," *Psychology Today*, April 1970, p. 43.

73. "It is also true that many of the nation's poor are obese because their diets are overloaded with inexpensive, fattening, but in nutritious foods." *NYT*, Aug. 27, 1973.

74. Senate Nutrition Committee, *Dietary Goals for Americans: Committee Print*, 95th Cong., 1st sess., Feb. 1977, p. 3.

75. Senate Nutrition Committee, *Diet Related to . . . Cardiovascular Disease*, 35.

76. *NYT*, Dec. 26, 1974.

77. *FT* 24 (June 1970): 27.

78. National Analysts, Inc., *A Study of Health Practices and Opinions* (Springfield, Va.: National Technical Information Service, 1972), cited in Howard A. Schneider and J. Timothy Hesla, "The Way It Is," *NR* 31 (Aug. 1973): 235.

79. Quoted in Don F. Hadwiger, "Nutrition, Food Safety, and Farm Policy," *Proceedings of the American Academy of Political Science 34, no. 3* (1982): 81.

80. *GH*, March 1972, pp. 175–77.

81. Senate Nutrition Committee, *Dietary Goals*, 1–13.

82. Jean Hewitt, *The New York Times Natural Foods Cook Book* (New York: Quadrangle, 1971), 41, 87.

83. U.S. Senate, Committee on Agriculture, Nutrition, and Forestry, *1978 Food and Agriculture Outlook: Committee Print*, 95th Cong., 1st sess., 1977, p. 6.

84. The argument was that prepubescent and young males and premenstrual women hardly ever developed atherosclerosis and reducing cholesterol levels in males over fifty seemed not to make a difference. *NYT*, Feb. 1, 1978; Senate Nutrition Committee, *Diet Related to Killer Diseases, 6: Hearings*, 95th Cong., 1st sess., 1977, pp. 7–11.

85. The main concern seemed to be a drop in iron in the national diet that would accompany falling beef consumption. Senate Nutrition Committee, *Dietary Goals for the United States: Committee Print* (rev.) 95th Cong., 1st sess., Dec. 1977, p. viii.

86. It added that individuals in high-risk categories for hypertension might benefit from lowering sodium intake, that diabetics should lose weight, and that doctors might recommend that patients with family histories of early heart disease lose weight and lower cholesterol intake, but it was careful to limit these recommendations specifically to those at risk and avoid dictating to the population at large. American Medical Association, Council on Scientific Affairs, "AMA Concepts of Nutrition and Health," *JAMA* 242 (1979): 2335.

87. National Research Council, Food and Nutrition Board (hereafter NRC-FNB), *Toward Healthful Diets* (Springfield, Va.: National Technical Information Service, 1980).

88. U.S. Dept. of Health, Education, and Welfare, Office of the Assistant Secretary for Health, Education, and Welfare, Office of the Assistant Secretary for Health and Surgeon General, *Healthy People: The Surgeon General's Report on Health Promotion and Disease Prevention* (Washington: USGPO, 1979), 153. The summary of scientific background papers was more cautious, calling many of these things "prudent for most Americans." Vicki Kalmar, "An Integrative Summary," in Surgeon General, *Healthy People: Background Papers, 1979* (Washington: USGPO, 1979), 48.

89. NRC-FNB, *Toward Healthful Diets*, 2–16; Mark Hegsted, "Comment on Toward a Healthful Diet," in Victor Herbert, *Nutrition Cultism: Facts and Fictions* (Philadelphia: Stickley, 1980), 224–25.

90. National Academy of Sciences–National Research Council, *Recommended Dietary Allowances*, 9th ed. (Washington: NAS, 1980); USDHHS and USDA, *Nutrition Monitoring: Progress Report*, 7.

91. American Society for Clinical Nutrition Symposium, "The Evidence Relating Six Dietary Factors to the Nation's Health, Consensus Statements," *American Journal of Clinical Nutrition* 32 (Dec. 1979): supplement; Hadwiger, "Nutrition, Food Safety," 79–88.

92. Nick Mottern, "Dietary Goals," *Food Monitor*, Apr. 1978, pp. 8–10.

93. *NYT*, Jan. 24, 1978.

94. It also eased the warning against eating too many eggs for premenopausal women, young children, and the elderly but added an admonition that only as many calories as were expended should be consumed. Senate Nutrition Committee, *Dietary Goals* (rev.), 4.

95. *NYT*, Jan. 23, 25, 1978; Mottern, "Dietary Goals," 9.

96. *BW*, Dec. 1, 1973.

97. Ray Walsh, "Quaker Oats Company," in *International Directory of Company Histories* (Chicago: St. James Press, 1989), vol. 2, p. 559.

98. Connor, *Food Processing*, 42–44; Walsh, "General Mills," "Quaker Oats Company," in *International Directory*, vol. 2, pp. 502, 559 (the diversification pattern also emerges in the other company histories in the "Food Products" section of this work, many of which were written by Ray Walsh); *BW*, Dec. 1, 1973.

99. Bruce W. Marrion, *The Organization and Performance of the U.S. Food System* (Boston: Heath, 1985), 227–29. The number of food processing companies declined by 60 percent from 1947 to 1982. Connor, *Food Processing*, 36. Among the major sectors controlled by four firms by 1974 were breakfast cereals (90 percent of sales), canned goods (80 percent), bread and prepared flour (75 percent), dairy products (70 percent), baking (65 percent), fluid milk (65 percent), sugar (65 percent), and processed meats (56 percent). One corporation (Campbell's, of course) sold 90 per-

cent of the nation's soup. Daniel Zwerdling, "The Food Monopolies," *Progressive*, Jan. 1975, p. 15.

100. Senate Nutrition Committee, *Nutrition and Private Industry: Hearings*, 91st Cong., 2nd sess., and 92nd Cong., 1st sess., 1969, p. 3890.

101. Michael E. Porter, "Consumer Behavior, Retailer Power and Market Performance in Consumer Goods Industries," *Review of Economics and Statistics* 56 (1974) 419–36; Phillip Nelson, "The Economic Consequences of Advertising," *Journal of Business*, April 1975, p. 237, cited in Marrion, *Organization and Performance*, 502.

102. Senate Nutrition Committee, *Dietary Goals*, 59–63.

103. Lynne Masover, unpublished thesis material, Northwestern Univ. Medical School, Chicago, Ill., cited in Senate Nutrition Committee, *Dietary Goals*, 59–66.

104. Sam Keen, "Eating Our Way to Enlightenment," *Psychology Today*, Oct. 1978, p. 62; *FP*, Sept. 1962, p. 15.

105. "The American Diet Shifts and Pays Off," *Psychology Today*, April 1979, p. 104.

106. "Nutrition Beliefs: More Fashion than Fact," *FDA Consumer*, June 1976, pp. 15–17; Belasco, *Appetite*, 194.

107. S. Q. Haider and M. Wheeler, "Nutritive Intake of Black and Hispanic Mothers in a Brooklyn Ghetto," *JADA* 75 (1979): 670–74; K. G. Dewey, M. A. Strode, and Y. R. Fitch, "Dietary Change Among Migrant and Non-Migrant Mexican-American Families in Northern California," *Ecology, Food, and Nutrition* (1984): 11–24; Mary Moore, Albert Stunkard, and Leo Srole, "Obesity, Social Class, and Mental Illness," *JAMA* 181 (1962): 962–66; Albert Stunkard, *The Pain of Obesity* (Palo Alto: Bull, 1976); USDHHS and USDA, *Nutrition Monitoring: Progress Report*, 59–62, 301–6. This USDHHS/USDA survey, done in 1977–1978, also indicated that little had changed with regard to American avoirdupois since two previous surveys in the early 1960s and early 1970s.

108. Almost one-third of poor females between 25 and 35 were labeled overweight and over one-half of those between 35 and 55 were classified as such. Black women formed a disproportionate part of that category: about 60 percent of all middle-aged black women were declared overweight (301–6).

109. M. L. Day, M. Lenter, and S. Jaquez, "Food Acceptance Patterns of Spanish-Speaking New Mexicans," *Journal of Nutrition Education* 10 (1978): 121–23.

110. Claude Fischler, *L'Homnivore* (Paris: Odile Jacob, 1990), 194.

111. Keen, "Eating Our Way," 62.

Chapter 14. Darling, Where Did You Put the Cardamom?

1. "How America Entertains," *LHJ*, Oct. 1966, pp. 99–112.

2. Francois Rysavy, *A Treasury of White House Cooking* (New York: Putnam's, 1972), 54.

3. Stephen E. Ambrose, *Nixon: The Triumph of a Politician, 1962–1972* (New York: Simon and Schuster, 1989), 18, 62.

4. Henry Haller, *The White House Family Cook Book* (New York: Random House, 1987), 78–94.

5. He acknowledged that they could not be banished from the "Western White House" in San Clemente, California. By the next year, he had relented in regard to Washington, although he reaffirmed that he still preferred Bordeaux. H. R. Haldeman to Lucy Winchester, Aug. 19, 1969; the President to Herbert Butterfield, March

16, 1970; the President to Haldeman, March 16, 1970. All in Bruce Oudes, ed., *From The President's Desk: Richard Nixon's Secret Files* (New York: Harper and Row, 1989), 40, 108, 109.

6. *NYT*, March 3, 1970.

7. Julia Child to Simone Beck, Nov. 12, 1973, Avis DeVoto Papers, Schlesinger Library of Women's History, Radcliffe College, Cambridge, Mass., Box 2.

8. *NYT*, Sept. 26, 1972.

9. Author's interview with Craig Claiborne, Feb. 23, 1989. In France it was listed as a major specialty of a number of Michelin three-rosette restaurants in the mid-1930s and from 1951 to 1965, but after 1965 it disappeared from the listing. Claude Fischler, *L'Homnivore* (Paris: Odile Jacob, 1990), 247–48.

10. Claiborne interview.

11. "How America Entertains," 103.

12. *NYT*, Jan. 1, 1970.

13. Elizabeth Rozin and Paul Rozin, "Culinary Themes and Variations," *Natural History* 90 (Feb. 1981): 6–7.

14. Southeast Asian cooking achieved a certain degree of chic in France only after its empire there was gone.

15. *NYT*, Aug. 24, 1976.

16. The "northern" Chinese restaurants themselves were opened by better-off, better-educated post–World War II refugees from China, mainly from the north, who were disdainful of their lowly Cantonese predecessors. Chi Kien Lee, "The Chinese Restaurant Industry in the United States: Its History, Development, and Future" (M.P.S. monograph, School of Hotel Administration, Cornell Univ., 1975).

17. Elizabeth Lambert Ortiz's popular *The Art of Mexican Cooking* (New York: Bantam, 1968) came out of New York City, as did Diane Kennedy's influential *Cuisines of Mexico* (New York: Harper and Row, 1972), which was vigorously promoted by the *New York Times*'s Craig Claiborne.

18. I use the term "ethnic food" in its commonly used sense—to refer to food originating with a foreign group. Anthropologists prefer the term "nationality food" for this, reserving "ethnic" for the food of one's own group, but in the present context that would be confusing. Susan Kalcik, "Ethnic Foodways in America: Symbol and the Performance of Identity," in Linda Brown and Kay Mussell, eds., *Ethnic and Regional Foodways in America: The Performance of Ethnic Identity* (Knoxville: Univ. of Tennesee Press, 1984), 44.

19. *NYT*, Aug. 24, 1976.

20. Leon Ullenswang, "Food Consumption Patterns in the Seventies," *Vital Speeches of the Day* 36 (Feb. 1, 1970): 240.

21. Christopher Lasch, *The Culture of Narcissism* (New York: Norton, 1978); Steven Mintz and Susan Kellogg, *Domestic Revolutions: A Social History of Family Life* (New York: Free Press, 1988), 205–7.

22. Carol Reuss, "*Better Homes and Gardens:* Consistent Key to Long Life," *Journalism Quarterly* 51 (Summer 1974): 296.

23. Alex Comfort, *The Joy of Sex: A Cordon Bleu Guide to Lovemaking* (New York: Crown, 1972); idem, *The Joy of Sex: A Gourmet's Guide to Lovemaking*, (New York: Crown, 1983.)

24. In 1960 it was running an average of just two food articles per issue, down from the three or four per issue of the 1930s and 1940s, despite being full of food advertisements. By 1966 it was running seven food-related articles per issue. *LHJ*, 1935–1966, passim.

25. Claiborne interview.

26. The paper edition, published in October 1965, went through six printings before the year was even out. Craig Claiborne, *An Herb and Spice Cook Book* (New York: Bantam, 1965).

27. *NYT*, Oct. 14, 1970.

28. *Time*, Dec. 19, 1977.

29. *NYT*, Jan. 1, 1971.

30. *NYT*, Feb. 27, 1972. In 1973 and 1974 the *New York Times* ran regular feature articles on ethnic food stores—Portuguese, Spanish, kosher, Greek, Chinese, Near Eastern, Indian, and so on, in New York, New Jersey, and California.

31. *Time*, Dec. 19, 1977.

32. The Rozins note that "it is those cultures with a bland starch staple with small amounts of meat which have the most marked cuisines in terms of flavor" ("Culinary Themes," 8).

33. Richard Beardsley, "Chinese Food: Inexpensive, Nutritious, Delicious . . . and Fun!" *Mother Earth News*, Sept. 1971, pp. 79–90; *Lifestyle*, Nov. 1973.

34. *Time*, Dec. 19, 1977.

35. "On and Off the Avenue," *New Yorker*, Dec. 17, 1979.

36. *Time*, Feb. 21, 1977.

37. Fischler has a number of discussions of how these processes work with particular reference to France (*L'Homnivore*, esp. chaps. 2 and 9).

38. Ibid., 229–31.

39. Alice Waters, *The Chez Panisse Menu Cookbook* (New York: Random House, 1982), x; Elizabeth David, *French Country Cooking*, rev. ed. (London: Penguin, 1966).

40. Jane Stern and Michael Stern, "American Gourmet," paper presented at Radcliffe College, Cambridge, Mass., November 8, 1990.

41. *Time*, Dec. 19, 1977.

42. Kalcik, "Ethnic Foodways," 39.

43. *Time*, Dec. 19, 1977.

44. Haller, *White House Cook Book*, 96.

45. Paul Prudhomme, *Chef Paul Prudhomme's Louisiana Kitchen* (New York: Morrow, 1984), preface.

46. *Time*, Dec. 19, 1977.

47. Ibid.

48. Robert Reich, "Secession of the Successful," *NYTM*, Jan. 20, 1991.

49. Stern and Stern, "American Gourmet."

50. In this context, the one classic downscale recipe, for spaghetti and meatballs, seems to be included to add a charming little antisnobbish note. Alice Waters, *Chez Panisse Pasta, Pizza, and Calzone* (New York: Random House, 1984), 1, 142–43.

51. Ullenswang, "Food Consumption in the Seventies," 214.

52. Community Garden Club, *Cohasset Entertains* (Cohasset, Mass.: n.p., 1978.)

53. Junior Branch, Franklin Country Public Hospital, *Kiss Me Quick, I'm Cooking* (Greenfield, Mass.: n.p., 1978).

54. *Time*, Dec. 19, 1977.

55. *BW*, March 8, 1976.

56. *PF*, May 1983, p. 129, April 1983, p. 124.

57. *PF*, 1972–1983, passim.

58. *PF*, Nov. 1983.

59. However, they later became popular in France too.

60. Bob Messenger, "Changing Demographics Light Latest Trends," *PF*, Nov. 1986, p. 41.

61. Fischler, *L'Homnivore*, 159.

62. *BW*, March 8, 1978; Maxine Margolis, *Mothers and Such: Views of American Women and Why They Changed* (Berkeley: Univ. of California Press, 1984), 225.

63. Ibid., 228.

64. *NYT*, Dec. 16, 1990.

65. Barbara Ehrenreich, *For Her Own Good: 150 Years of the Experts' Advice to Women* (New York: Anchor, 1979), 288–89.

66. By the mid-1970s, one study showed, although the amount of time households spent on cooking had not changed during the past ten years, married men were helping more in the kitchen. John P. Robinson, "Household Technology and Household Work," in Sarah F. Burk, ed., *Women and Household Labor* (Berkeley Hills: n.p., 1980), 53–67, cited in Margolis, *Mothers*, 179.

67. *Time*, Dec. 19, 1977.

68. Even what is probably the nation's oldest continuously reprinted community cookbook, Milwaukee's *The Way to a Man's Heart: The Settlement Cook Book*, whose title had remained unchanged since 1901, felt the pressure to change. In 1976 its title was reversed to *The Settlement Cook Book: The Way to A Man's Heart* (New York: Simon and Schuster, 1976), with the subtitle in tiny lettering.

69. Margolis, *Mothers*, 231. This is not to say no differences remained. Women, for example, ate twice as many frozen chicken dinners as men, double the number of frozen croissants, and three times the number of frozen bagels. Men ate twice as many Salisbury steak dinners as women. *PF*, Sept. 1985, p. 18.

Chapter 15. Fast Foods and Quick Bucks

1. Their instructions were often too complicated, their food too mediocre, and their prices no better than their fast food competitors'. Phillip Langdon, *Orange Roofs and Golden Arches: The Architecture of American Chain Restaurants* (New York: Knopf, 1986), 82.

2. *TGM*, Aug. 31, 1991.

3. J. Anthony Lukas, "As American as a McDonald's Hamburger on the Fourth of July," *NYTM*, July 4, 1971.

4. John F. Love, *McDonald's: Behind the Golden Arches* (New York: Bantam, 1986), 143.

5. Robert L. Emerson, *Fast Food: The Endless Shakeout* (New York: Liebhar-Friedman, 1979), 2.

6. Paul Hirshorn and Steven Izenhour, *White Towers* (Cambridge: MIT Press, 1979), 23. Kroc also sensed that there was a gap in the franchised drive-in market and positioned his chain between Dairy Queen, which had been very successful at the low end of the market, and full-service franchises like Big Boy, which demanded very large investments by franchisees. Love, *McDonald's*, 52–53.

7. *Time*, Sept. 17, 1973. When White Tower did belatedly move onto suburban strips, it had trouble shedding its inner-city image. It ultimately became a Burger King franchisee. Hirshorn and Izenour, *White Tower*, 23–4.

8. Love, *McDonald's*, 142; Lukas, "As American," 25.

9. Love, *McDonald's*, 142.

10. Lukas, "As American," 24; *Time*, Sept. 17, 1973; Raymond Sweet, "What Ronald Macdonald Won't Reveal," *Rochester Patriot*, Aug. 11–31, 1976. Its original

mascot, a round cartoon "Speedie," was abandoned in 1962, in part because of its uncomfortable resemblance to "Speedy Alka-Seltzer." Langdon, *Orange Roofs*, 147.

11. John E. Pearce, *The Colonel: The Captivating Biography of the Dynamic Founder of a Fast-food Empire* (Garden City, N.Y.: Doubleday, 1982), 140.

12. It also diluted the fun image by simultaneously going for the teenage market with its large "two-handed" burgers. *FF*, March 1972, pp. 84–85.

13. Vicenzo (Jimmy) Bruno, of Bruno's Little Italy restaurant in Little Rock, Arkansas, claimed that his father began making America's first "commercially sold pizza" in New York City in 1903, but Lombardi's at 53½ Spring Street in New York had a more widely accepted claim for the same year. *Food for Profit*, June 1953, p. 11; Erik Amfithreatrof, *Children of Columbus* (Boston: Little, Brown, 1973), 239.

14. Joe Vergara, *Love and Pasta: A Recollection* (New York: Scholastic Resources, 1969), 15.

15. Doris Friedensohn, "A Kitchen of One's Own," ms. in possession of the author.

16. Herbert Mitgang, "Pizza à la Mode," *NYTM*, Feb. 12, 1956, p. 64.

17. Amfithreatrof, *Children of Columbus*, 239.

18. Lawrence A. Lovell-Troy, "Kinship Structure and Economic Organization Among Ethnic Groups: Greek Immigrants and the Pizza Business" (Ph.D diss., Univ. of Connecticut, 1979), 38–40.

19. Warren Belasco, "Ethnic Fast Foods: The Corporate Melting Pot," *Food and Foodways* 2 (1987): 18.

20. Pearce, *Colonel*, 162–63.

21. Lukas, "As American," 5; Ted Hickley and Richard Johnson, "Ray Kroc, Embodiment of Mid-Twentieth Century America," *Journal of the West* 25 (Jan. 1986): 99.

22. Pearce, *Colonel*, 170.

23. *Pizza and Pasta*, Aug. 1976, p. 22.

24. *RB*, March 1, 1980, p. 76.

25. As early as 1978 the top three companies—McDonald's, KFC, and Burger King—themselves owned from one-quarter to one-third of their franchises. *RH*, June 1978, p. 92.

26. Emerson, *Fast Food*, 3. In 1985 the twenty-five largest chains spent close to 2 billion dollars on television advertising alone. *RH*, Aug. 1988, p. 108.

27. Mom-and-pop grocery stores continued to close at a fearsome rate. In 1986 there were only ninety-two thousand of them, 50 percent fewer than in 1972. U.S. Bureau of the Census, *Statistical Abstract of the United States: 1987* (Washington: USGPO, 1988), 756.

28. Langdon, *Orange Roofs*, 108–9; *NYT*, May 9, 1971.

29. *RB*, Sept. 1, 1976, p. 77, Sept. 1, 1978, p. 166.

30. *NYT*, April 11, 1977.

31. Malcolm Knapp, "11th Annual Restaurant Growth Index," *RB*, Sept. 1, 1978, p. 165.

32. Ibid.

33. Emerson, *Fast Food*, 25–37.

34. The only explanation for the strange final array of spices must be that the author, under the misapprehension that Chinese star-anise was the same as European fennel and anise seed, was trying to make something like five-spice powder. *Fast Service*, March 1978, p. 36.

35. Langdon, *Orange Roofs*, 148.

36. *NYT*, Dec. 17, 1990.

37. *RH*, June 1978, p. 82; *NYT*, Dec. 17, 1990.

38. *RB*, Mar. 1, 1980, p. 124.

39. Its founder's last name was Bell. Langdon, *Orange Roofs*, 178.

40. *RB*, March 1, 1980, p. 124.

41. *NYT*, Dec. 17, 1990; Belasco, "Ethnic Fast Foods," 1–30.

42. This was reflected in statistics showing that the main change was not so much that an increasing proportion of the food dollar was spent away from home as that those purchases were redistributed toward simpler, shorter eating experiences. Corinne Le Bovitt, "The Changing Pattern of Eating Out," *National Food Situation* 144 (May 1973): 30–34.

43. *BW*, Oct. 27, 1975.

44. John L. Hess, "Restaurant Food: Frozen, Cooked, Then Refrozen and Recooked," in Joan Gussow, ed., *The Feeding Web: Issues in Nutritional Ecology* (Palo Alto: Bull, 1978), 158.

45. Joan Bakos and Malcolm Knapp, "What's Going on Here?" *RB*, Sept. 1976, pp. 77–78.

46. Belasco, "Ethnic Fast Foods," 22.

47. *BW*, Oct. 27, 1975.

48. Langdon, *Orange Roofs*, 189.

49. *NYT*, April 11, 1977.

50. Newspaper Advertising Bureau, *Eat and Run: A National Survey of Fast Food Patronage* (New York: 1978), cited in Emerson, *Fast Food*, 33–34; U.S. Dept. of Health and Human Services and U.S. Dept. of Agriculture, *Nutrition Monitoring: An Update Report on Nutrition Monitoring* (Hyattsville, Md.: USDHHS, 1989), 89–90.

51. *NYT*, May 3, 1977.

52. *FDA Consumer*, May 1983, p. 13.

53. U.S. Dept. of Health and U.S. Dept. of Agriculture, *Nutrition Monitoring: A Progress Report from the Joint Nutrition Monitoring Evaluation Committee* (Hyattsville, Md.: USDHHS, 1986), 221–22.

54. *NYT*, April 16, 1976.

Chapter 16. Paradoxes of Plenty

1. Nicholaus Mills, "The Culture of Triumph and the Spirit of the Times," in idem, ed., *Culture in an Age of Money: The Legacy of the 1980s in America* (Chicago: Ivan Dee, 1991), 11; Debora Silverman, "China, Bloomie's, and the Met," in ibid., 194.

2. *NYT*, June 7, 1987.

3. "Greed is all right . . . Greed is healthy," said the later-jailed Wall Street arbitrager Ivan Boesky in his commencement speech to the University of California-Berkeley Business School. Mills, "Culture of Triumph," 21.

4. Claude Fischler, *L'Homnivore* (Paris: Odile Jacob, 1990), 219.

5. George Gallup, *The Gallup Poll, 1935–1971* (New York: Random House, 1972), vol. 2, p. 1355.

6. David Garner et al., "Psychoeducational Principles in the Treatment of Bulemia and Anorexia Nervosa," in David Garner and Paul E. Garfinkel, eds., *Handbook of Psychotherapy for Anorexia and Bulemia* (New York: Guildford, 1985), 514.

7. J. Rodin et al., "Women and Weight: A Normative Discontent," in T. B.

Sondregger, ed., *Psychology and Gender: Nebraska Symposium on Motivation* (Lincoln: Univ. of Nebraska Press, 1985), 267–307.

8. Susan C. Wolley and O. Wayne Wolley, "Intensive Outpatient and Residential Treatment for Bulemia," in Garner and Garfinkel, *Handbook of Psychotherapy*, 392.

9. David M. Garner et al., "Cultural Expectations of Thinness in Women," *Psychological Reports* 47 (1980): 483–91.

10. Lois Kurman, "An Analysis of Messages Concerning Food, Eating Behaviors, and Ideal Body Image on Prime-time American Network Television" (Ph.D diss., New York Univ., 1977), 1907–8; S. B. Beck et al., "Variables Related to Women's Somatic Preferences of the Male and Female Body," *Journal of Personality and Social Psychology* 34 (1976): 1200–1210; T. Horvath, "Correlates of Physical Beauty in Men and Women," *Social Behavior and Personality* 7 (1979): 145–51; idem, "Physical Attractiveness: The Influence of Selected Torso Parameters," *Archives of Sexual Behavior* 10 (1981): 21–24. All cited in Garner et al., "Psychoeducational Principles," 515.

11. Paul Rozin and April Fallon, "Body Image, Attitudes to Weight, and Misperceptions of Figure Preferences of the Opposite Sex: A Comparison of Men and Women in Two Generations," *Journal of Abnormal Psychology* 97 (1988): 342–45.

12. Fischler, *L'Homnivore*, 348.

13. Garner et al., "Psychoeducational Principles," 518.

14. Roberta Seid, *Never Too Thin: Why Women Are at War with Their Bodies* (New York: Prentice-Hall, 1989), 148.

15. April Fallon and Paul Rozin, "Sex Differences in Perceptions of Desirable Body Shape," *Journal of Abnormal Psychology* 24 (1985): 102–5; Rozin and Fallon, "Body Image," 342–45.

16. Seid, *Never Too Thin*, 148.

17. Fischler, *L'Homnivore*, 326–29.

18. Robert Lipsyte, "What Price Fitness?" *NYTM*, Feb. 16, 1986.

19. Richard F. Spark, "Fat Americans," *NYTM*, Jan. 6, 1974. See Hillel Schwartz, *Never Satisfied: A Cultural History of Diets, Fantasies, and Fat* (New York: Free Press, 1986), 282–302, for a discussion of the rise of this idea.

20. Summaries of many of the questions that can be raised may be found in Seid, *Never Too Thin*, 279–303, and Fischler, *L'Homnivore*, 298–318.

21. National Institutes of Health, Consensus Development Panel, *Statement on Health Implications of Obesity* (Bethesda: NIH, 1985), cited in U.S. Dept. of Health and Human Services and U.S. Dept. of Agriculture, *Nutrition Monitoring in the United States: A Progress Report from the Joint Nutrition Monitoring Evaluation Committee* (Hyattsville, Md.: USDHHS, 1986), 5.

22. U.S. Dept. of Health and Human Services, Public Health Service, *The Surgeon General's Report on Nutrition and Health* (Washington: USDHHS, 1988), 1–3. Italics mine.

23. Seid, *Never Too Thin*, 15, 18.

24. Rozin and Fallon, "Body Image," 343–44.

25. Fifty percent of women, on the other hand, were satisfied or wanted to put on weight; the other half wanted to lose, and 33 percent confessed to having tried. Gallup, *Gallup Poll, 1935–1971*, vol. 3, p. 1355.

26. *PF*, Nov. 1986, p. 39.

27. Rozin and Fallon, "Body Image," 344–45.

28. Fischler, *L'Homnivore*, 297.

29. Twenty percent of the new food products introduced in early 1984 had "less" of something, mainly calories. *PF*, July 1984, p. 112.

30. *NYT*, June 25, 1967, May 30, 1975.

31. John M. Connor, *Food Processing: An Industrial Powerhouse in Transition* (Lexington, Mass.: Heath, 1987), 106.

32. "Foods with 'Less,' " *Prepared Foods New Products Directory: 1985*, p. 51. By 1984 Lean Cuisine was selling at a $400 million annual rate. *PF*, Nov. 1984, p. 76.

33. *PF*, Nov. 1984, p. 71.

34. Cited in Warren Belasco, "Light Economics: Less Food, More Profit," *Radical History Review* 28–30 (1984): 272.

35. *PF*, May 1983, p. 72, April 1983, p. 39.

36. *NYT*, Jan. 7, 1974.

37. *PF*, Nov. 1985, p. 48.

38. Schwartz, *Never Satisfied*, 240.

39. *Time*, Dec. 18, 1972; Herman Tarnower, *The Complete Scarsdale Medical Diet* (New York: Rawson, Wade, 1978).

40. Lower consumption of foods such as milk, butter, and beef deemed high in cholesterol seems to have been counterbalanced by rising consumption of foods such as margarine and vegetable oils, which are equally calorific, as well as snack foods, many of which are high in carbohydrates and fats. U.S. Dept. of Health and Human Services and U.S. Dept. of Agriculture, *Nutrition Monitoring: An Update Report on Nutrition Monitoring* (Hyattsville, Md.: USDHHS, 1989), 123.

41. *NYT*, Feb. 23, 1989.

42. Seid, *Never Too Thin*, 16.

43. This was not the case with black men. USDHHS and USDA, *Nutrition Monitoring: Update*, 109.

44. *NYT*, Dec. 16, 1990.

45. The effect on the elderly was particularly striking, especially when two people could collect benefits. For example, the number of childless elderly families living below the poverty line declined from 965,000 to 574,000 from 1970 to 1980. U.S. Bureau of the Census, *Statistical Abstract of the United States, 1989* (Washington: USGPO, 1989), 456.

46. It did warn that many poor people still did not know of the programs, and others were prevented from benefiting from them because of red tape (*NYT*, May 1, 1979). In a national food survey the previous year, only 3 percent of respondents said that sometimes or often there was not enough food to eat in their households. USDHHS and USDA, *Nutrition Monitoring: Progress Report*, 22.

47. Jean Mayer and J. Larry Brown, "More Prosperity, More Hunger," *NYT*, Feb. 25, 1989.

48. The President's Task Force on Hunger reported in 1984 that hunger was a concern among these three groups. USDHHS and USDA, *Nutrition Monitoring: Progress Report*, 22.

49. *WP*, March 27, 1990.

50. J. Fred Giertz and Dennis H. Sullivan, "Food Assistance Programs in the Reagan Administration," *Publius: The Journal of Federalism* 16 (Winter 1986): 133–47.

51. Fischler, *L'Homnivore*, 262.

52. Ellen Brown, *Cooking with the New American Chefs* (New York: Harper and Row, 1985), 37, 74–75.

53. *PF*, July 1984, p. 104.

54. *PF*, Aug. 1982, p. 80; *TGM*, June 2, 1986.

55. *PF*, Oct. 1985, pp. 33–34.
56. *PF*, July 1985, p. 48.
57. *TGM: Business World*, Nov. 1989, p. 99.
58. *PF*, Nov. 1984, p. 80.
59. *RH*, Aug. 1988, pp. 104–13.
60. *TGM: Business World*, Nov. 1990, p. 103.
61. *TGM*, July 5, 1991.
62. *Time*, April 8, 1991.
63. *NYTM*, April 22, 1990.
64. *NYT*, Sept. 20, 1989.
65. *NYT*, Dec. 6, 1989.
66. USDHHS and USDA, *Nutrition Monitoring: Update*, 89–90.
67. *TGM*, Oct. 2, 1990; *NYT*, March 19, 1991.
68. E.g., Joan Jacobs Brumberg, *Fasting Girls* (New York: Basic Books, 1988), and Seid, *Never Too Thin*.
69. Dale Atrens, *Don't Diet* (New York: Morrow), 1989; Jane R. Hirschmann and Carol Munter, *Overcoming Overeating* (New York: Addison-Wesley, 1989). When Janet Polivy and C. Peter Herman published a similar work, *Breaking the Diet Habit* (New York: Basic Books, 1983), earlier in the decade, Polivy found that "most doctors who didn't ignore us thought we were crazy." *NYT*, Jan. 3, 1990.
70. *NYT*, Aug. 5, 1991.
71. Eric Ravussin et al., "Reduced Rate of Energy Expenditure as a Risk Factor for Body Weight Loss," *New England Journal of Medicine* 318 (Feb. 25, 1988): 467–72; *NYT*, May 24, 1990; *WSJ*, May 25, 1990.
72. Fischler, *L'Homnivore*, 308–10. See Seid, *Never Too Thin*, chaps. 2 and 13 for these arguments.
73. *TGM*, Feb. 24, 1992; *Vogue*, June 1991.
74. *NYT*, Feb. 26, 1991.
75. Fischler, *L'Homnivore*, 354–56.
76. Pillsbury, for example, discovered that "mature adults" were more concerned than any other group about salt, fat, cholesterol, fiber, and calcium in their food. *PF*, April 1990, p. 57.
77. Bob Messenger, "Changing Demographics Light Latest Trends," *PF*, Nov. 1986, p. 37; *PF*, Nov. 1984, p. 82.
78. *NYT*, Feb. 23, 1991.
79. The percentage of Americans who identified cholesterol as a cause of heart disease shot up from 26 percent in 1982 to over 40 percent in 1986. In the latter year, fully 24 percent claimed to be on cholesterol-lowering diets. USDHHS and USDA, *Nutrition Monitoring: Update*, 121–22.
80. *NYT*, March 19, 1991.
81. Anne Casale, *The Long Life Cookbook* (New York: Ballantine, 1987); Joseph Piscatella, *The Don't Eat Your Heart Out Cookbook* (New York: Workman, 1983); Anne Lindsay, *The American Cancer Society Cookbook* (New York: Hearst, 1988); Art Ulene and Mary Ward, *The Count Out Cholesterol Cookbook* (New York: Knopf, 1989); Harriet Roth, *Harriet Roth's Cholesterol-Control Cookbook* (New York: New American Library, 1989); Dean Ornish, *Dr. Dean Ornish's Program for Reversing Heart Disease* (New York: Random House, 1990).
82. Also, the AMA had begun allowing cookbooks to sport "American Medical Association Campaign Against Cholesterol" labels. Casale, *Long Life Cookbook;* Ulene and Ward, *Count Out Cholesterol*.

83. *WSJ*, Jan. 25, Feb. 2, 1990.

84. *NYT*, Aug. 16, 1990.

85. Shai Linn et al., "High Density Lipoprotein Cholesterol Levels Among U.S. Adults by Selected Demographic and Socioeconomic Variables," *American Journal of Epidemiology* 129 (1989): 281–93.

86. *NYT*, July 31, 1991.

87. Madeline Dresser, "Let Them Eat Grain," *Boston Globe Sunday Magazine*, July 15, 1991; *NYT*, July 21, 1991.

88. "A Pyramid Topples at the USDA," *Consumer Reports*, Oct. 1991, p. 663.

89. Fischler sees the current wave of anorexia and bulemia as related to this (*L'Homnivore*, 368–71).

90. Fewer than half of those responding to the October 1989 Gallup poll said they enjoyed eating "a great deal." "Hunger is a chore to be finished rather than an event to be savored," said the social psychologist Rober Cialdini. *NYT*, Dec. 6, 1989. Yet the food industry still saw the success or failure of McDonald's new low-fat burgers hinging on what consumers thought of their taste. *NYT*, March 19, 1991.

91. *NYT*, June 25, 1991. Those familiar with demographic statistics would recognize, however, that this would still represent a significantly longer life span for many thousands of Americans. However, the question remained whether the large majority of Americans who would not benefit should be called upon to drastically alter their diets.

92. *NYT*, June 27, 1991.

Index